T0262679

Agricultural Science

Agricultural Science

Edited by **Thelma Bosso**

New York

Published by Callisto Reference,
106 Park Avenue, Suite 200,
New York, NY 10016, USA
www.callistoreference.com

Agricultural Science
Edited by Thelma Bosso

International Standard Book Number: 978-1-63239-058-5 (Hardback)

Printed in the United States of America.

Contents

Preface

Over the recent decade, advancements and applications have progressed exponentially. This has led to the increased interest in this field and projects are being conducted to enhance knowledge. The main objective of this book is to present some of the critical challenges and provide insights into possible solutions. This book will answer the varied questions that arise in the field and also provide an increased scope for furthering studies.

Agricultural science involves the integrative study of various aspects of agriculture including crop production, economic impact, soil fertility and sustainability. This book discusses important areas in agricultural science, such as crop enhancement, production, response to water, nutrients, and temperature, crop protection, agriculture and human health, and animal nutrition. The topics discussed by the authors involve manipulation of the variables and genetic resources of inheritance of quantitative genes, crop rotation, soil water and nitrogen, and impact of temperature on flowering. Other topics covered include safeguarding crops against insect pests and diseases, associating agriculture topography to recreation by humans, and small ruminant nutrition. This book will be a worthy addition to existing knowledge database and will be particularly useful for university students and professionals in the field of agriculture.

I hope that this book, with its visionary approach, will be a valuable addition and will promote interest among readers. Each of the authors has provided their extraordinary competence in their specific fields by providing different perspectives as they come from diverse nations and regions. I thank them for their contributions.

Editor

Section 1

Crop Improvement

1

Impact of Epistasis in Inheritance of Quantitative Traits in Crops

Bnejdi Fethi and El Gazzeh Mohamed
Laboratoire de Génétique et Biométrie Faculté des Sciences de Tunis, Université Tunis, El Manar, Tunisia

1. Introduction

Epistasis is the interaction between alleles of different genes, i.e. non-allelic interaction, as opposed to dominance, which is interaction between allele of the same gene, called inter-allelic or intra-genic interaction (Kearsey and Pooni, 1996). Statistical epistasis describes the deviation that occurs when the combined additive effect of two or more genes does not explain an observed phenotype (Falconer and Mackay, 1996).

The heritability of a trait, an essential concept in genetics quantitative, "certainly one of the central points in plant breeding research is the proportion of variation among individuals in a population that" is due to variation in the additive genetic (i.e., breeding) values of individuals:

h^2 = VA/VP = Variance of breeding values/ phenotypic variance (Lynch and Walsh, 1998). This definition is now termed "heritability in the narrow-sense" (Nyquist, 1991). Estimation of this parameter was prerequisite for the amelioration of quantitative traits. As well as choosing the selective procedure, that will maximize genetic gain with one or more selection cycles. Various methods were developed in the past, Warner (1952), Sib-Analysis, Parent-offspring regressions etc. Theses methods considered that additive-dominant model is fitted, assuming epistasis to be negligible or non existent. Because of the complexity of theoretical genetics studies on epistasis, there is a lack of information about the contribution of the epistatic components of genotypic variance when predicting gains from selection. The estimation of epistatic components of genotypic variance is unusual in genetic studies because the limitation of the methodology, as in the case of the triple test cross, the high number of generations to be produced and assessed (Viana, 2000), and mainly because only one type of progeny, Half-Sib, Full-Sib or inbred families, is commonly included in the experiments (Viana, 2005). If there is no epistasis, generally it is satisfactory to assess the selection efficiency and to predict gain based on the broad-sense heritability. Therefore, the bias in the estimate of the additive variance when assuming the additive-dominant model is considerable. The preponderance of epistasis effect in the inheritance of quantitative trait in crops was recently reported by many geneticists (Pensuk et al., 2004; Bnejdi and El Gazzah, 2008; Bnejdi et al. 2009; Bnejdi and El-Gazzah, 2010a; Shashikumar et al. 2010). Epistasis can have an important influence on a number of evolutionary phenomena, including the genetic divergence between species.

The aims of our study were to determine the importance of epistasis effects in heredity of quantitative traits and their consequences in the bias of four methods of estimation of narrow-sense heritability.

2. Origin of data and genetic model

Nine quantitative traits with 88 cases of combination cross-site, cross-isolate or cross-treatment of six generations (P_1, P_2, F_1, F_2, BC_1 and BC_2) for three crops (*Triticum Durum, Capsicum annum* and *Avena sp*) were collected from different works realised in our laboratory. Crops, traits and origin of data are reported in Table 1. For each trait parents of crosses were extreme. Transformations (such as Kleckowski transforms (Lynch and Walsh, 1998)) were applied to normalize the distribution of data or to make means independent of variances for several traits.

Durum Wheat   **(*Triticum durum*)**
Two crosses/two sites   Number of head per plant , Spiklets per spike and Number of grains per spike (Bnejdi and El Gazzeh 2010b)
Four crosses/ one site   Resistance to yellowberry (Bnejdi and El Gazzah, 2008)
Four crosses/ one site   Resistance to yellowberry (Bnejdi et al., 2010a)
Four crosses/ Two sites   Grain protein content (Bnejdi and El Gazzeh, 2010a)
Two crosses/ Five salt treatments   Resistance to salt at germination stage (Bnejdi et al., 2011a)
Two crosses/ fifteen isolates   Resistance to *Septoria tritici* (Bnejdi et al., 2011b)
Pepper   **(*Capsicum annuum* L.)**
Two crosses/ Two isolates   Resistance to *Phytophthora nicotianae* (Bnejdi et al., 2009)
Two crosses/ Six isolates   Resistance to *Phytophthora nicotianae* (Bnejdi et al., 2010b)
Oates   **(Avena sp.)**
Two crosses/ Two isolates   Resistance to *P. coronate* Cda. f. sp. *avenae* Eriks (Bnejdi et al., 2010c)

Table 1. Traits assessed in each crop and date of publication

2.1 Best genetic model

Weighted least squares regression analyses were used to solve for mid-parent [M] pooled additive [A], pooled dominance [D] and pooled digenic epistatic ([AA], [DD] and [AD]) genetic effects, following the models and assumptions described in Mather and Jinks (1982). A simple additive-dominance genetic model containing only M, A and D effects was first tested using the joint scaling test described in Rowe and Alexander (1980). Adequacy of the genetic model was assessed using a chi-square goodness-of-fit statistic derived from deviations from this model. If statistically significant at $P < 0.05$, genetic models containing digenic epistatic effects were then tested until the chi-square statistic was non-significant.

3. Phenotypic resemblance between relatives

We now will use the covariance (and the related measures of correlations and regression slopes) to quantify the phenotypic resemblance between relatives. Quantitative genetics as a field traces back to Fisher's 1918 paper showing how to use the phenotypic covariance to estimate genetic variances, whereby the phenotypic covariance between relatives is expressed in terms of genetic variances, as we detail below.

3.1 Parent-offspring regressions

There are three types of parent-offspring regressions: two **single parent - offspring regressions** (plotting offspring mean versus either the trait value in their male parent *Pf* or their female parent *Pm*), and the **mid-parent-offspring regression** (the offspring mean regressed on the mean of their parents, the mid-parent $MP = (Pf+Pm)/2$).

The slope of the (single) parent-offspring regression is estimated by

$$b_{o/p} = \frac{Cov(O,P)}{Var(P)}, \text{ where } Cov(O,P) = \frac{1}{n-1}\left(\sum_{i=1}^{n} O_i P_i - n\overline{O}.\overline{P}\right)$$

Where Oi is the mean trait value in the offspring of parent i (Pi) and we examine n pairs of parent-offspring. One could compute separate regressions using males (Pm) and females (Pf), although the later potentially includes maternal effect contributions and hence single-parent regressions usually restricted to fathers.

$$b_{o/p} = \frac{Cov(O,P)}{Var(P)}$$

$$Cov(O,P) = \frac{\sigma^2 A}{2} + \left(\frac{\sigma^2 AA}{4} + \frac{\sigma^2 AAA}{8} + \frac{\sigma^2 AAAA}{16} + ...\right)$$

$$b_{o/p}^{*} = \frac{Cov(O,P)}{Var(P)} = \frac{\sigma^2 A}{2\sigma_P^2} + \frac{1}{\sigma_P^2}\left(\frac{\sigma^2 AA}{4} + \frac{\sigma^2 AAA}{8} + \frac{\sigma^2 AAAA}{16} + ...\right)$$

$$b_{o/p}^{*} = \frac{Cov(O,P)}{Var(P)} = \frac{h^2}{2} + \frac{1}{\sigma_P^2}\left(\frac{\sigma^2 AA}{4} + \frac{\sigma^2 AAA}{8} + \frac{\sigma^2 AAAA}{16} + ...\right)$$

Assuming an absence of epistasis we have

$$Cov(O,P) = \tfrac{1}{2}\sigma^2 A \text{, giving} \quad b_{o/p} = \frac{\tfrac{1}{2}\sigma^2 A}{\sigma_P^2} = \frac{h^2}{2}$$

$$h^2 = 2b_{o/p}$$

3.2 Full-sib analysis

The covariance full-sib analysis is equal to:

$$Cov(FS) = \frac{1}{2}\sigma_A^2 + \frac{1}{4}\sigma_D^2 + \frac{1}{4}\sigma_{AA}^2 + \frac{1}{8}\sigma_{AD}^2 + \frac{1}{16}\sigma_{DD}^2 + \frac{1}{8}\sigma_{AAA}^2 \ldots\ldots)$$

$$\frac{Cov(FS)}{\sigma_P^2} = \frac{h^2}{2} + \frac{1}{\sigma_P^2}(\frac{1}{4}\sigma_D^2 + \frac{1}{4}\sigma_{AA}^2 + \frac{1}{8}\sigma_{AD}^2 + \frac{1}{16}\sigma_{DD}^2 + \frac{1}{8}\sigma_{AAA}^2 \ldots\ldots)$$

So, when epistasis was considered negligible

$$Cov(FS) = \frac{1}{2}\sigma_A^2$$

$$h^2 = \frac{2Cov(FS)}{\sigma_P^2}$$

3.3 Half-sib analysis

Based on half-sib analysis, narrow-sense heritability was calculated as:

$$Cov(HS) = \frac{1}{4}\sigma_A^2 + \frac{1}{16}\sigma_{AA}^2 + \frac{1}{64}\sigma_{AAA}^2 + \ldots\ldots$$

$$\frac{Cov(HS)}{\sigma_P^2} = \frac{h^2}{4} + \frac{1}{\sigma_P^2}(\frac{1}{16}\sigma_{AA}^2 + \frac{1}{64}\sigma_{AAA}^2 + \ldots\ldots)$$

So, epistasis was considered negligible and the narrow-sense heritability was determined as:

$$Cov(HS) = \frac{1}{4}\sigma_A^2$$

$$h^2 = \frac{4Cov(HS)}{\sigma_P^2}$$

3.4 Method of Warner (1952)

Based on additive dominance model Warner in 1952 revealed that narrow-sense heritability could be estimated as:

$2\sigma^2_{F_2} - (\sigma^2_{BC_1} + \sigma^2_{BC_2})$ Where $\sigma^2_{F_2}$, $\sigma^2_{BC_1}$ and $\sigma^2_{BC_2}$ represented respectively the variance of generation F_2, BC_1 and BC_2

In absence of epistasis

$$2\sigma^2_{F_2} - (\sigma^2_{BC_1} + \sigma^2_{BC_2}) = \frac{1}{2}a^2_A = \sigma^2_A$$

$$\frac{2\sigma^2_{F_2} - (\sigma^2_{BC_1} + \sigma^2_{BC_2})}{\sigma^2_{F_2}} = \frac{\sigma^2_A}{\sigma^2_{F_2}} = h^2$$

Therefore in presence of epistasis

$$2\sigma^2_{F_2} - (\sigma^2_{BC_1} + \sigma^2_{BC_2}) = 2(a + \frac{1}{2}ad)^2 + (d + \frac{1}{2}dd)^2 + \frac{1}{2}aa^2 + \frac{1}{2}ad + ...$$

4. Results and discussion

Separate generation means analysis revealed that the additive-dominance model was found adequate only for 18 cases. Therefore, the digenic epistatic model was found appropriate for 70 cases (Table 2). Additive and dominance effect were significant for all cases of combination. With regard to epistatic effects, the additive x additive effect was significant for 77 cases and the additive x dominance for 42 cases and dominance x dominance effects for 56 cases. Recent studies suggest that epistatic effects are present for inheritance of quantitative traits in many species. Examples are wheat (resistance to leaf rust, Ezzahiri and Roelfs 1989), wheat (resistance to yellowberry, Bnejdi and El Gazzah 2008), common bean (resistance to anthracnose, Marcial and Pastor 1994), barley (resistance to Fusarium head blight, Flavio et al. 2003), chickpea (resistance to *Botrytis cinerea*, Rewal and Grewal 1989), and pepper (resistance to *Phytophthora capsici*, Bartual et al. 1994).

To conclude for this part, the additive dominance model was rarely fitted and digenic epistatic model was frequently appropriate. Therefore epistasis is common in inheritance of quantitative traits and any model or methods assumed that epistasis was negligible were biased.

The comparison of the four methods is reported in Table 3. In absence of dominance and epistatic effect, the methods were not biased. Therefore, in presence of epistasis narrow-sense heritability based on the four methods was underestimated. Based in Full-Sib Analysis and Warner (1952) methods, bias was caused by dominance, interaction between homozygote loci, interaction between heterozygote loci and interaction between homozygote and heterozygote loci. Therefore based in Half-Sib Analysis and Parent-offspring regressions, bias was caused only with the presence of interaction between homozygote loci or fixable effect.

The result of generations means analysis indicate that digenic epistasis model were frequently appropriate. So the additive model in which many methods of genetic quantitative were based was rarely adequate. Based on the result, the methods of Half-Sib Analysis and Parent-offspring regressions were underestimated with additive x additive effect (Table 3). Because additive x additive effect can be fixed by selection, estimation of narrow-sense heritability with theses methods was recommended and efficiency in crops breeding. Linkage disequilibrium and absence of epistasis are compulsorily assumed in almost all the methodologies developed to analyze quantitative traits. The consequence, clearly, is biased estimates of genetic parameters and predicted gains, as linkage and genetic interaction are the rule and not the exception Viana (2004). The prediction of gains from selection allows the choice of selection strategies. Therefore the gain from selection was estimated from narrow-sense heritability. Breeding strategies applied for plant breeding aimed to increase the favourable gene frequency. The efficiency of any methodology of selection was associated with the best estimated of the additive genetic effect value.

Best fit- model	**Number of cases**
M + A + D	18
M + A + D + AA	2
M + A + D+ AA + DD	26
M + A + D + AA + AD	13
M + A + D + DD + AD	3
M + A + D + AA + DD + AD	18
M + A + D + AA + DD + AD + C	8

M, mean; A, additive; D, dominance; AA, additive × additive; AD, additive × dominance; DD, dominance × dominance; C, cytoplasm effect.

Table 2. Best-fit models of nine traits with 88 cases of combinations Cross-site, cross-treatment and or cross-isolate for three crops.

In presence of epistasis effect, Parent-offspring regressions and Half-Sib Analysis were the best methods. In fact, these methods were biased only with interaction between homozygote loci represented by "additive x additive" effect. However, both the methods of Warner (1952) and Full-Sib Analysis were biased with dominance, additive x dominance, dominance x dominance and additive x additive effects. The interaction between the homozygote loci can be fixed by selection. But the fixation of interaction between heterozygote loci prerequisite maintain of heterozygote. Depending upon the methods, the bias in the estimation of narrow-sense heritability in presence of epistasis was more pronounced.

The presence of epistasis complicated the procedure of amelioration of quantitative traits and revealed the limitation of most quantitative studies based on the assumption of negligible epistasis. However, the exploitation of epistasis in the breeding program such as the superiority of heterozygous genotypes over their corresponding parental genotypes was of great importance.

Method	in absence of epistasis	in presence of epistasis
Parent-offspring regressions		
	$b_{o/p}=\frac{Cov(O,P)}{Var(P)}$ $Cov(O,P)=\frac{1}{2}\sigma^2 A$ $b_{o/p}=\frac{\frac{1}{2}\sigma^2 A}{\sigma_P^2}=\frac{h^2}{2}$ $h^2=2b_{o/p}$	$b_{o/p}=\frac{Cov(O,P)}{Var(P)}$ $Cov(O,P)=\frac{\sigma^2 A}{2}+(\frac{\sigma^2 AA}{4}+\frac{\sigma^2 AAA}{8}+\frac{\sigma^2 AAAA}{16}+.........)$ $b_{o/p}^{*}=\frac{Cov(O,P)}{Var(P)}=\frac{\sigma^2 A}{2\sigma_P^2}+\frac{1}{\sigma_P^2}(\frac{\sigma^2 AA}{4}+\frac{\sigma^2 AAA}{8}+\frac{\sigma^2 AAAA}{16}+.........)$ $b_{o/p}^{*}=\frac{Cov(O,P)}{Var(P)}=\frac{h^2}{2}+\frac{1}{\sigma_P^2}(\frac{\sigma^2 AA}{4}+\frac{\sigma^2 AAA}{8}+\frac{\sigma^2 AAAA}{16}+.........)$
Full-Sib Analysais		
	$Cov(FS)=\frac{1}{2}\sigma_A^2$ $h^2=\frac{2Cov(FS)}{\sigma_P^2}$	$Cov(FS)=\frac{1}{2}\sigma_A^2+\frac{1}{4}\sigma_D^2+\frac{1}{4}\sigma_{AA}^2+\frac{1}{8}\sigma_{AD}^2+\frac{1}{16}\sigma_{DD}^2+\frac{1}{8}\sigma_{AAA}^2......)$ $\frac{Cov(FS)}{\sigma_P^2}=\frac{h^2}{2}+\frac{1}{\sigma_P^2}(\frac{1}{4}\sigma_D^2+\frac{1}{4}\sigma_{AA}^2+\frac{1}{8}\sigma_{AD}^2+\frac{1}{16}\sigma_{DD}^2+\frac{1}{8}\sigma_{AAA}^2.....)$
Half-Sib Analysais		
	$Cov(HS)=\frac{1}{4}\sigma_A^2$ $h^2=\frac{4Cov(HS)}{\sigma_P^2}$	$Cov(HS)=\frac{1}{4}\sigma_A^2+\frac{1}{16}\sigma_{AA}^2+\frac{1}{64}\sigma_{AAA}^2+......$ $\frac{Cov(HS)}{\sigma_P^2}=\frac{h^2}{4}+\frac{1}{\sigma_P^2}(\frac{1}{16}\sigma_{AA}^2+\frac{1}{64}\sigma_{AAA}^2+......)$
Warner (1952)		
	$2\sigma_{F_2}^2-(\sigma_{BC_1}^2+\sigma_{BC_2}^2)=\frac{1}{2}a_A^2=\sigma_A^2$ $\frac{2\sigma_{F_2}^2-(\sigma_{BC_1}^2+\sigma_{BC_2}^2)}{\sigma_{F_2}^2}=\frac{\sigma_A^2}{\sigma_{F_2}^2}=h^2$	$2\sigma_{F_2}^2-(\sigma_{BC_1}^2+\sigma_{BC_2}^2)=2(a+\frac{1}{2}ad)^2+(d+\frac{1}{2}dd)^2+\frac{1}{2}aa^2+\frac{1}{2}ad+........$

O, offspring; P, parent; A, additive; D, dominance; AA, additive × additive;
AD, additive × dominance; DD, dominance × dominance; AAA, additive × additive × additive;

Table 3. Bias of four methods of estimation of narrow-sense heritability in presence of epistasis

5. References

Bnejdi F, Saadoun M, Allagui MB, El Gazzah M (2009). Epistasis and heritability of resistance to *Phytophthora nicotianae* in pepper (*Capsicum annuum* L). *Euphytica,* 167: 39-42.

Bnejdi F, El Gazzah M (2008). Inheritance of resistance to yellowberry in durum wheat. *Euphytica,* 163: 225–230.

Bnejdi F, El Gazzah M (2010a). Epistasis and genotype-by-environment interaction of grain protein concentration in durum wheat. *Genet Mol Biol,* 33(1): 125-130.

Bnejdi F, El Gazzah M (2010b). Epistasis and genotype-by-environment interaction of grain yield related traits in durum wheat. *J Plant Breed Crop Sci,* 2 (2): 24-29.

Bnejdi F, Rassa N, Saadoun M, Naouari M, El Gazzah M (2011a). Genetic adaptability to salinity level at germination stage of durum wheat. *Afri J Biot,* 10 (21): 4400-4004

Bnejdi F, Hammami I, Allagui MB, El Gazzah M (2010a). Epistasis and maternal effect in resistance to *Puccinia coronata* Cda. f. sp. *avenae* Eriks in oats (*Avena* sp.) *Agri Sci China,* 9(10): 101-105.

Bnejdi F, Saadoun M, Allagui MB, Colin H, El Gazzah M (2010b). Relationship between epistasis and aggressiveness in resistance of pepper (*Capsicum annuum* L.) to *Phytophthora nicotianae*. *Genet Mol Biol*, 33(2): 279-284

Bnejdi F, Saadoun M, Allagui MB, El Gazzah M (2009). Epistasis and heritability of resistance to *Phytophthora nicotianae* in pepper (*Capsicum annuum* L). *Euphytica*, 167: 39–42.

Bnejdi F, Saadoun M, El Gazzah M (2010c). Cytoplasmic effect on grain resistance to yellowberry in durum wheat. *Czech J Genet Plant Breed*, 46 (4): 145–148

Bnejdi F, Saadoun M, El Gazzah M (2011b). Genetic adaptability of the inheritance of the resistance to different levels of aggressiveness of *Septoria tritici* isolates in durum wheat. *Crop Prot*, 30: 1280–1284

Ezzahiri B, Roelfs AP (1989). Inheritance and expression of adult plant resistance to leaf rust in era wheat. *Plant Dis*, 73:549–551.

Falconer DS, Mackay TFC (1996). Introduction to quantitative genetics. 4th edition, Longman, UK, 464 pp.

Fisher RA (1918). The correlation between relatives on the supposition of mendelian inheritance. *Trans Roy Soc Edin*, 52: 399–433.

Flavio C, Donald CR, Ruth DM, Edward S et al (2003). Inheritance of resistance to fusarium head blight in four populations of barley. *Crop Sci*, 43:1960–1966

Kearsey MJ, Pooni HS (1996). The genetical analysis of quantitative traits. 1st edition. Chapman and Hall, London, 381 pp.

Lynch M, Walsh B (1998). Genetics and Analysis of Quantitative Traits, Sinauer Associates, Inc, Sunderland, 980 pp.

Marcial A, Pastor C (1994). Inheritance of anthracnose resistance in common bean accession G 2333. *Plant Dis*, 78:959–962

Mather K, Jinks JL (1982). Biometrical Genetics. London: Chapman and Hall Ltd, London, 396 pp.

Nyquist WE (1991). Estimation of heritability and prediction of selection response in plant populations. *Crit Rev Plant Sci*, 10: 235–322.

Pensuk V, Jogloy S, Wongkaew S, Patanothai A (2004). Generation means analysis of resistance to peanut bud necrosis caused by peanut bud necrosis tospovirus in peanut. *Plant Breed*, 123: 90-92.

Rewal N, Grewal JS (1989). Inheritance of resistance to Botrytis cinerea Pers. in *Cicer arietinum L. Euphytica*, 44:61–63.

Rowe KE, Alexander WL (1980). Computations for estimating the genetic parameter in joint-scaling tests. *Crop Sci*, 20: 109-110.

Shashikumar KT, Pitchaimuthu M, Rawal RD (2010). Generation mean analysis of resistance to downy mildew in adult muskmelon plants. *Euphytica*, 173:121–127

Viana JMS (2000). Components of variation of polygenic systems with digenic epistasis. *Genet Mol Biol*, 23: 883-892.

Viana JMS (2004). Relative importance of the epistatic components of genotypic variance in non-inbred populations. *Crop Breed App Biot*, 4: 18-27.

Viana JMS (2005). Dominance, epistasis, heritabilities and expected genetic gains. *Genet Mol Biol*, 28: 67-74.

Warner JN (1952). A method for estimating heritability. *Agron J*, 44: 427-430.

2

Genetic Diversity Analysis of *Heliconia psittacorum* Cultivars and Interspecific Hybrids Using Nuclear and Chloroplast DNA Regions

Walma Nogueira Ramos Guimarães[1,*], Gabriela de Morais Guerra Ferraz[1], Luiza Suely Semen Martins[2], Luciane Vilela Resende[3], Helio Almeida Burity[4] and Vivian Loges[1]

[1]*Department of Agronomy, Federal Rural University of Pernambuco, Recife, Pernambuco,*
[2]*Department of Biology, Biochemical Genetics Laboratory/Genome, Federal Rural University of Pernambuco,*
[3]*Department of Agronomy, Federal University of Lavras, Minas Gerais,*
[4]*Agronomic Research Institute of Pernambuco, Recife, Pernambuco, Brazil*

1. Introduction

Heliconia cultivation has intensified in Brazil as a cut flower, especially in the Northeast region. This ornamental rhizomatous herbaceous plant from the *Heliconia* genus, belongs to the Musaceae family, now constitutes the Heliconiaceae family in the Zingiberales order. The various species of *Heliconia* are subdivided into five subgenera: *Heliconia, Taeniostrobus, Stenochlamys, Heliconiopsis* and *Griggisia*; and 28 sections (Kress *et al.*, 1993). In *Heliconia* genus, the number of species ranges from 120 to 257 and there are also a great number of cultivars and 23 natural hybrids (Berry and Kress, 1991; Castro *et al.*, 2007), these plants can be found either in shaded places, such as forests or at full Sun areas, such as forest edges and roadsides (Castro and Graziano, 1997). They are native from Tropical America (Berry and Kress, 1991), found at different altitudes, from sea level up to 2.000 meters in Central and South America, and up to 500 meters in the South Pacific Islands (Criley and Broschat, 1992).

Heliconia hybrids comprise many of the major cultivars as cut flowers, like *H. psittacorum* x *H. spathocircinata* cv. Golden Torch, cv. Golden Torch Adrian, cv. Alan Carle and *H. caribaea* x *H. bihai* cv. Carib Flame, cv. Jacquinii, cv. Richmond Red (Berry and Kress, 1991). Many heliconia species are identified through their morphological differences, such as the size and color of its flowers and bracts. These characteristics can be influenced either by geographic isolation or by environmental factors, such as light and nutrients (Kumar *et al.*, 1998). *H. psittacorum* clones, even when closely grown, can vary in blooming, size and color of bracts, as well as post harvest durability (Donselman and Broschat, 1986).

*Corresponding author

The natural variation among heliconia individuals or populations has led to taxonomic identification doubts among farmers and researchers. Thus, genebanks have played an important role in genetic diversity conservation, providing raw material for crop breeding, including landraces and their wild relatives. DNA markers, which allow the access to variability at DNA level, emerge as an efficient alternative for plant species characterization by quantifying diversity and determining its genetic structure (Bruns *et al.*, 1991).

The choice on which molecular marker technique shall be used depends on its reproducibility and simplicity. Kumar *et al.* (1998) distinguished three cultivars of the hybrid *H. psittacorum* x *H. spathocircinata* cvs. Golden Torch, Red Torch and Alan Carle which showed only slight differences in RAPD markers profile from *H.* x *nickeriensis* Maas and de Rooij (*H. psittacorum* x *H. marginata*), they also observed similarities in RAPD profiles and morphology. The authors concluded that two triploid *H. psittacorum*: cv. Iris and Petra, are supposed to be the same genotype.

Genetic diversity studies grew up in interest during the last years (Jatoi *et al.*, 2008; Kladmook *et al.*, 2010). As a result, nucleotide sequences of ribosomal genes (rDNA) and chloroplast genes (cpDNA) have been exploited to investigate several individuals of the Zingiberales order (Kress, 1990, 1995; Kress *et al.*, 2001) once they are not capable of lateral transfers and are not subject to the same functional limitations, they allow greater confidence in the results (Camara, 2008). The unit of ribosomal eukaryotic organisms consists of three genic and three non-genic regions. On one hand, the genic regions (18S, 5,8S and 26S) are conserved and evolve slowly. On the other hand, non-genic regions, known as ITS - Internal Transcribed Spacer (ITS-1 and ITS-2), evolve rapidly, showing high polymorphism and, therefore, allowing its use at higher hierarchical levels. The variability found in these regions could be the result of mutations in these areas, since they suffer less selection pressure and may be well used to study genetic diversity in plants (Bruns *et al.*, 1991).

This molecular marker is important from the genetic variability assessment point of view, because the rDNA mutltigenic family once subjected to a rapid evolution in concert event, allows greater precision in the reconstruction process of the relationship between species based on sequencing, since this phenomenon increases the intragenomic uniformity (Baldwin *et al.*, 1995). These authors also affirm that due to the biparental inheritance of the nuclear genome it is possible to study the origin of hybrids and their parents. Moreover, chloroplast genes (cpDNA), such as the leucyn and fenilalanyn of RNA transporter (*trnL-trnF*), the treonyn and leucyn of RNA transporter (*trnT-trnL*) and the protein small 4 (*rps4*), have been used successfully to solve genetic diversity doubts in taxonomic lower levels. Johansen (2005), for example, studying the genetic diversity in Zingiberales order, using cpDNA, has positioned all Heliconiaceae and Musaceae within a same clade.

The aim of this study was to evaluate genetic diversity involving *Heliconia psittacorum* cultivars and interspecific hybrids of the Federal Rural University of Pernambuco *Heliconia* Germplasm Collection (UFRPE-HCG), using nuclear and chloroplast DNA regions.

2. Materials and methods

2.1 Plant material and genomic DNA extraction

The *Heliconia* Germplasm Collection (UFRPE-HCG) is located in Camaragibe-PE at 8°1'19'' South, 34°59'33'' West and 100 m above the sea level, in a 0.3 ha experimental area. The average annual temperature is 25.1°C and monthly rainfall of 176 mm, with maximum of 377 mm and minimum of 37 mm (ITEP, 2008). This study evaluated 11 *Heliconia psittacorum* cultivars and interspecific hybrids (Table 1) obtained by exchange with research institutions and farmers from the states of Pernambuco (PE), Alagoas (AL) and Sao Paulo (SP) in Brazil. The analyzed genotypes presented short size, musoid habit and erect inflorescence disposed at a single plan (Berry and Kress, 1991).

Genotypes[a]	Location	Subgenus and Section[b]	Description[c]
Hybrids			
H. x *nickeriensis Maas and de Rooij* (*H. marginata* x *H. psittacorum*)	Paulista - PE	*Heliconia Pendulae*	BC- yellow-orange; OV- dark yellow distally, light yellow proximally, PD- yellow, SE- dark yellow
H. psittacorum L.f x *H. spathorcircinata* Aristeguieta cv. Golden Torch Adrian	Paulista - PE	*uncertain*	BC- yellow-Red; OV- yellow, PD- yellow with indistinct blackish green area distally, SE- dark yellow
H. psittacorum L.f x *H. spathorcircinata* cv. Golden Torch	Paulista - PE	*uncertain*	BC- yellow; OV- yellow, PD- yellow with indistinct blackish green area distally, SE- dark yellow
H. psittacorum L.f. x *H. spathocircinata* cv. Red Opal	Paulista - PE	*uncertain*	BC- orange; OV- yellow, PD- yellow, SE- dark yellow with indistinct blackish green area distally
Triploid[d]			
H. psittacorum L.f. cv. Suriname Sassy	Paulista - PE	*Stenochlamys Stenochlamys*	BC- pink-green; OV- orange distally, yellow proximally, PD- orange to cream, SE- orange with indistinct blackish green area distally
H. psittacorum L.f. cv. Sassy	Paulista - PE	*Stenochlamys Stenochlamys*	BC- pink-green; OV- orange distally, yellow proximally, PD- yellow green, SE- orange with indistinct blackish green area distally
Heliconia sp. (suposed to be *H. psittacorum* cv. Sassy)	Maceió - AL	*Stenochlamys Stenochlamys*	BC- pink-lilac; OV- green distally and yellow green proximally, PD- yellow green, SE- orange with indistinct blackish green area distally
Suposed triploid			
H. psittacorum L.f .cv. Strawberries and Cream	Paulista - PE	*Stenochlamys Stenochlamys*	BC- pink-yellow; OV- yellow to cream, PD- cream, SE- pale yellow with green spot on distally corner
H. psittacorum L.f. cv. Lady Di	Ubatuba - SP	*Stenochlamys Stenochlamys*	BC- red; OV- yellow, PD- light yellow to cream, SE- light yellow with distally dark green band and white tip
H. psittacorum L.f. cv. St. Vincent Red	Ubatuba - SP	*Stenochlamys Stenochlamys*	BC- red-orange; OV-orange distally, orange to cream proximally, PD- orange, SE- orange with indistinct blackish green area distally
H. psittacorum L.f. cv. Red Gold	Paulista - PE	*Stenochlamys Stenochlamys*	BC- red-orange; OV- yellow, PD- yellow, SE- dark yellow with indistinct blackish green area distally

[a]Identification based on Berry and Kress (1991) and Castro *et al.* (2007); [b]Based on Kress *et al.* (1993); [c]BC: bract color; OV: ovary; PD: pedicel; SE: sepals. [d]Ploidy (Costa *et al.*, 2008).

Table 1. Genotypes, location, classification and description for 11 *Heliconia psittacorum* cultivars and interspecific hybrids of the UFRPE Heliconia Germplasm Collection used in this study

Molecular markers analyses occurred in the Plant Biotechnology Laboratory - UFRPE. The optimization of the DNA extraction protocol was performed using fresh young leaves samples of heliconia, harvested in the earliest stage of development and treated under three conditions: harvested, packed in a polystyrene box containing liquid nitrogen and taken to the Laboratory for immediate DNA extraction; harvested and frozen at -20°C for 1 day before extraction; harvest and preserved in silica gel for 5 days before extraction.

In the DNA extraction, Doyle and Doyle (1990) protocol were used with modifications, which was prepared at a 2x CTAB (hexadecyltrimethylammonium bromide) buffer solution. It was added 700 microliter extraction buffer to 200 mg of macerated leaves in test tubes and taken to bath at 65°C. The tubes, after cooled at room temperature, were centrifuged and the supernatant transferred to new tubes. Supernatant was added to 700 microliter (μL) CIA (Chloroform-Isoamyl Alcohol) and then centrifuged was performed.

The supernatant was added to 700 microliter (μL) CIA (Chloroform-Isoamyl Alcohol) and then centrifuged. After this process, supernatant was added to 500 μL of cold isopropanol and stored for 24 hours in a freezer at -20°C. Subsequently, it was washed twice with 70% ethanol and with 95% ethanol. The precipitate was dried at room temperature for 20 minutes and then resuspended with 300 μL TE containing RNAse, incubated at 37°C for 30 minutes, then 5M NaCl and 300 μL of cooled isopropanol were added, in which the DNA was precipitated. Solution was incubated at 4°C throughout the night and the pellet resuspended in 300 μL TE. The DNA was quantified in 0.8% agarose gel.

Primers	Sequence	Number of Basis
ITS 1	5′- TCCGTAGGTGAACCTGCGG -3′	19
ITS 2	5′-GCTGCGTTCTTCATCGATGC-3′	20
ITS 3	5′-GCATCGATGAAGAACGCAGC -3′	20
ITS 4	5′-TCCTCCGCTTATTGATATGC-3′	20
ITS 5	5´-GGAAGTAAAAGTCGTAACAAGG -3´	22
EF11	5-GTGGGGCATTTACCCCGCC-3'	19
EF22	5´-AGGAACCCTTACCGAGCTC-3´	19
trnL	5´-GGTTCAAGTCCCTCTATCCC -3´	20
trnF	5´-ATTTGAACTGGTGACACGAG-3´	20
trnS	5´-TACCGAGGGTTCGAATC -3´	17
rps5′	5´-ATGTCCCGTTATCGAGGACCT -3´	21
rps3′	5′ –ATATTCTACAACTAACAACTC – 3′	21

Table 2. Sequence of primers used in amplification reactions in genotypes of the UFRPE Heliconia Germplasm Collection used in this study

A set of 12 primers (Table 2) and 10 combinations of this primers were selected and tested for the ITS analysis: ITS1-ITS4; ITS5-ITS4; ITS1-ITS2; ITS5-ITS2; ITS3-ITS4 based on White *et al.* (1990); and EF11-EF22; and for chloroplast genes analysis: *rps3'-rps5'* (Sanchez-Baracaldo, 2004); *trnL-trnF* (Sang *et al.*, 1997); *trnS-trnF* and *trnS-trnL.*

2.2 PCR amplification

The DNA amplification using PCR was performed to a final volume of 25 μL containing 1 μL template DNA, 0.3 μL Taq-DNA polymerase (Invitrogen), 2.51 μL Tris-HCl (pH 8, 0), and 0.75 μL $MgCl_2$, 2 μL of each dNTPs, 1 μL primer, 1 μL oligonucleotide 1 and 2; and 15.45 μL milli-Q water to complete the reaction.

Amplifications were performed in a thermocycler MJ Reseach, Inc., PTC100 under the following conditions: step 1 - following a denaturation step of 95°C for 3 minutes; step 2 - 94°C for 1 minute; step 3 - 58°C for 1 minute for annealing temperature; step 4 - 72°C for 1 min (repeat steps 2/3/4 for 29 cycles) followed by a final extension at 72°C for 10 minutes and 10°C for 24h. The PCR product visualization was performed in 1.5% agarose gel stained with SYBER Gold (Invitrogen), visualized under ultraviolet light and recorded on a digital Vilber Lourmat photographer.

2.3 Statistical analysis

Through the interpretation of gels, molecular data were tabulated as presence (1) or absence (0) of DNA fragments by primers for each genotype. Genetic similarities among genotypes were determined based on the Jaccard (1908) coefficients. A dendrogram was then constructed using the unweighted pair-group method of the arithmetic average (UPGMA) based on the similarity matrix. The cluster analyses were conducted using the computer program Gene (Cruz, 2006).

3. Results

The best condition for heliconia DNA extraction was using leaves in the earlyest stage of development, harvested, packed in a polystyrene box containing liquid nitrogen and taken to the Laboratory for immediate DNA extraction.

3.1 Primers selection

Primer combination ITS4-ITS3 resulted in most of the polymorphic band region, while for the primer combination ITS4-ITS5 it was observed the least polymorphism. The primers used amplified from 1 to 6 band regions, with clear polymorphism between the genotypes. The amplifications of the nuclear region that includes the spacers ITS1-ITS2 and EF11-EF22 (Fig. 1) generated fragments of approximately 396 to 506 pb, which agrees with Baldwin *et al.* (1995), by claiming that ITS markers have numerous small sized copies, reaching up to 700 pb.

Chloroplast regions amplifications that used the primers tRNA of leucine and phenylalanine (*trnL-trnF*) generated fragments of approximately 1636 pb (Fig. 2). For the spacers regions *rps3'-rps5'* as well as for the regions *trnS-trnL* and *trnS-trnF*, it was observed monomorphic and polymorphic band patterns for the evaluated cultivars and hybrids.

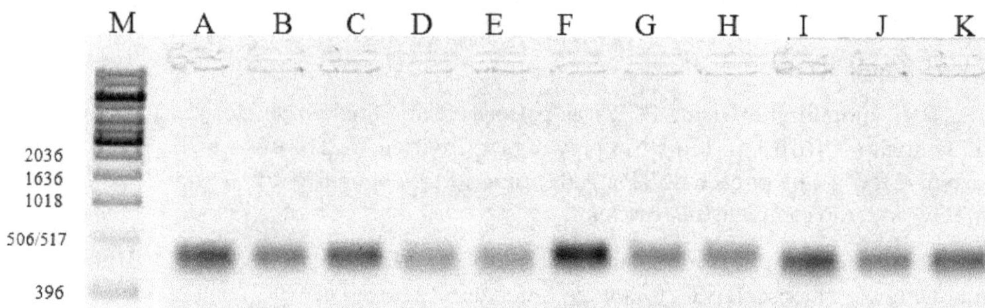

Fig. 1. Nuclear region amplifications that includes the spacers EF11-EF22. Cultivars and interspecific hybrids: A- *H. psittacorum* cv. Sassy; B- *H. psittacorum* cv. Red Gold; C- *H. psittacorum* x *H. spathocircinata* cv. Golden Torch Adrian; D- *H. psittacorum* cv. Suriname Sassy; E- *H. psittacorum* x *H. spathocircinata* cv. Red Opal; F- *H.* x nickeriensis; G- *H. psittacorum* x *H. spathocircinata* cv. Golden Torch; H- *Heliconia* sp.; I- *H. psittacorum* cv. Lady Di; J- *H. psittacorum* cv. Strawberries e Cream; K- *H. psittacorum* cv. St. Vincent Red. (M = 1 kb DNA ladder).

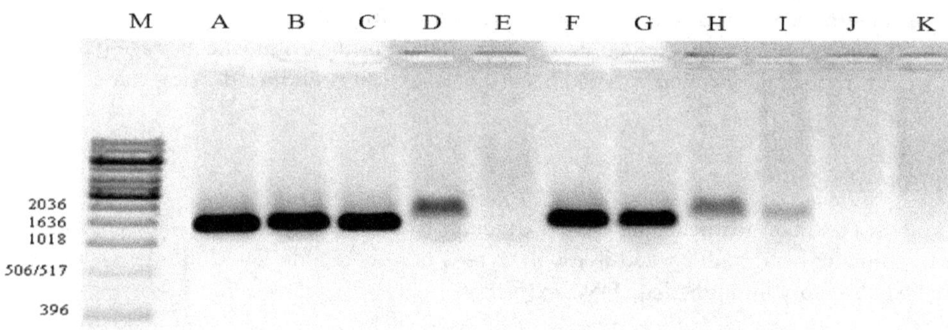

Fig. 2. Chloroplast regions with *primers trnL-trnF* amplifications. Cultivars and interspecific hybrids: A- *H. psittacorum* cv. Sassy; B- *H. psittacorum* cv. Red Gold; C- *H. psittacorum* x *H. spathocircinata* cv. Golden Torch Adrian; D- *H. psittacorum* cv. Suriname Sassy; E- *H. psittacorum* x *H. spathocircinata* cv. Red Opal; F- *H.* x nickeriensis; G- *H. psittacorum* x *H. spathocircinata* cv. Golden Torch; H- *Heliconia* sp.; I- *H. psittacorum* cv. Lady Di; J- *H. psittacorum* cv. Strawberries e Cream; K- *H. psittacorum* cv. St. Vincent Red. (M = 1 kb DNA ladder).

3.2 Internal transcribed spacers

From the data generated by ITS markers and the analysis of the dendrogram (Fig. 3), it was observed the formation of two main groups (GI and GII) well sustained. The GI group, is constituted by *Heliconia* sp., while, the other, more representative, GII, is subdivided into two other subgroups, SG A and SG B.

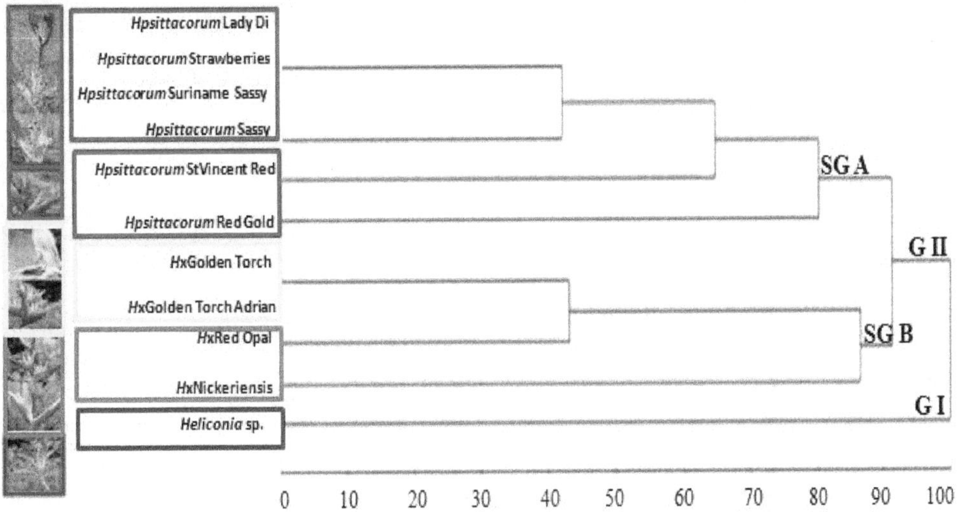

Fig. 3. Cluster analysis on 11 genotypes of the UFRPE Heliconia Germplasm Collection, used in this study, through ITS. G I= Group I; G II= Group II; SG A= Subgroup A; SG B= Subgroup B.

The first group GI, consisting of *Heliconia* sp., that according to farmers, is identified as *H. psittacorum* cv. Sassy and was more divergent from the other genotypes. The hypothesis that it is a new cultivar is supported by the fact that it was the only one that came from the state of Alagoas. It presents floral features intermediate between the triploid cultivars (Costa *et al.*, 2008) of the subgroup SG A, bracts of pink and lilac, which presents individual characteristics such as, ovarian (OV) green distally and yellow green proximally, pedicel (PD) yellow green and sepals (SE) orange with indistinct blackish green area distally.

In the second group GII, subgroup SG A, formed by triploid cultivars of *H. psittacorum*: cvs. Suriname Sassy and Sassy, that present bracts with pink and lilac, was also included genotypes from the state of São Paulo, *H. psittacorum* cv. St Vincent Red and *H. psittacorum* cv. Lady Di.

The subgroup SG B was formed by the hybrid *H. psittacorum* x *H. spathocircinata* cvs. Golden Torch, Golden Torch Adrian and Red Opal, with bracts yellow and red. In this subgroup on an external position, it was observed *H.* x *nickeriensis*, that is supposed to be an hybrid between *H. psittacorum* x *H. marginata*.

The hybrids showed low levels of similarity, around 12% of these comparisons reached levels above 50% (Table 3), probably because they are the result from supposed crosses between genetically distant parents or even the influences of epigenetic factors.

Genotypes	*Hpsittacorum* Sassy	*Hpsittacorum Red Gold*	*Hx GoldenTorch Adrian*	*Hpsittacorum* Suriname Sassy	*Hx* Red Opal	*Hx* Nickeriensis	*Hx GoldenTorch*	*Heliconia* sp.	*Hpsittacorum* Lady Di	*Hpsittacorum* Strawberries	*Hpsittacorum* StVincent Red
Hpsittacorum Sassy	1.00										
Hpsittacorum Red Gold	0.38	1.00									
Hx GoldenTorch Adrian	0.58	0.36	1.00								
Hpsittacorum Suriname Sassy	0.45	0.33	0.44	1.00							
Hx Red Opal	0.18	0.28	0.25	0.40	1.00						
Hx Nickeriensis	0.41	0.44	0.55	0.37	0.33	1.00					
Hx GoldenTorch	0.18	0.28	0.25	0.40	1.00	0.33	1.00				
Heliconia sp.	0.45	0.50	0.44	0.42	0.16	0.37	0.16	1.00			
Hpsittacorum Lady Di	0.09	0.14	0.12	0.13	0.10	0.16	0.12	0.21	1.00		
Hpsittacorum Strawberries	0.09	0.14	0.12	0.13	0.10	0.16	0.12	0.21	0.10	1.00	
Hpsittacorum StVincent Red	0.09	0.13	0.10	0.13	0.10	0.16	0.12	0.20	0.10	0.10	1.00

Table 3. Genetic similarity between 11 cultivars of *H. psittacorum* and interspecific hybrids of the UFRPE Heliconia Germplasm Collection used in this study

4. Discussion

It was not possible to obtain DNA with acceptable quality from *Heliconia* using the conventional methodology, as mentioned by Kumar *et al.* (1998), in an earlier study with molecular markers in heliconia.

4.1 Primers selection

Band patterns variation may be related to high occurrence rate of base substitution and the great possibility of indels accumulation (events of inserts and/or deletions of nucleotides), moreover, these sequences are difficult to identify (Albert *et al.*, 2002). The study with a great number of genotypes aims to explain the inheritance of the chloroplast, which may vary according to the subgenus and be useful for genetic diversity studies of the group. These *primers* (*trnL-F*) have been successfully used in genetic diversity analysis of Orchidaceae (Kocyan *et al.*, 2004) and Bromeliaceae groups (Sousa *et al.*, 2007).

4.2 Internal transcribed spacers

In the absence of more precise evidence, it was decided to keep the genotype, here called *Heliconia* sp., as a specie not yet identified. It is assumed as a new cultivar of *H. psittacorum* cv. Sassy that occurred due to different geographic conditions. In fact, this finding requires further studies. Other molecular markers can be used to solve this issue, as did Kumar *et al.* (1998), that using RAPD, found that two triploid cultivars, Iris and Petra were the same genotype. Sheela *et al.* (2006) by using RAPD, found that cvs. St Vincent Red and Lady Di, were also grouped in the same subgroup. Thus, assuming that these genotypes formed a subgroup brother of triploid cultivars *H. psittacorum* cv. Sassy and cv. Suriname Sassy, presenting 2n = 36 (Costa *et al.*, 2008), leads to the assumption that cvs. St Vincent Red and Lady Di are supposed to be triploid, corroborating with the similar banding pattern among these four genotypes in primer combination ITS3-ITS4.

The group that gathered the hybrids *H. psittacorum* x *H. spathocircinata* cvs. Golden Torch, Golden Torch Adrian and Red Opal was expected, once the nrDNA has biparental inheritance, and it is a nuclear molecular marker. *H.* x *nickeriensis* belongs to the *Heliconia* subgenus and *Pendulae* section (Kress *et al.*, 1993), this subdivision is based on the consistency of vegetative structure, and staminodes and style shape, especially in the pending heliconia. *H. marginata*, alleged parent, has pending inflorescence, and yet, differ from other hybrids that are crosses between *H. psittacorum* x *H. spathocircinata* and belongs to the *Stenochlamys* subgenus and *Stenochlamys* section (Kress *et al.*, 1993). Using RAPD markers to study genetic variability and relationship between 124 genotypes of the genus *Heliconia*, Marouelli *et al.* (2010), managed to gather interspecific hybrids of *H. psittacorum* in the same clade.

The hybrids showed small similarity that can be explained by the coevolution hypothesis, which considers the great genetic diversity of the genotype in the center of origin, once in northeast Brazil is frequently encountered native populations of *H. psittacorum*. Moreover, there is a wide variety of *H. psittacorum* hybrids described in literature, especially *H. spathorcircinata*, confirming the potential of this specie to form hybrids (Berry and Kress, 1991).

The influence of epigenetic factors in the phenotype of an organism and therefore in obtaining hybrids of *Heliconia* should be an issue to be raised. Characteristics of the transmissibility of an individual to other generations are not only linked to genes, the cell should be considered with its cytoplasm, mitochondria and genetic material carried in its structure, as well as the organism as a whole, and the complexity of the environment (Pearson, 2006). Another factor to be considered is the cytosine methylation of the genetic material, also responsible for gene silencing, causing changes in the phenotype, and according to most recent works can be passed to subsequent generations, thus causing greate genetic diversity among individuals of the same specie.

Routinely, new *Heliconia* species have been described and others have been included as synonyms on each revision of the genus or subgenus; but, there is still controversy among authors. This situation suggests the need for a careful review of this group, since the visual botanical identification, may lead to imprecise denomination for the species that are being cultivated.

Although some diversity studies about the Heliconiaceae family have been undertaken in recent years, its classification remains opened, therefore, new genetic markers for the group are required to elucidate these classification issues. The results revealed that there was no repetition of genetic material among the cultivars and interspecific hybrids of *H. psittacorum* evaluated, indicating the necessity to use other regions that could provide potentially informative characters. In conclusion, the genetic diversity nuclear and chloroplast DNA regions observed to study in *Heliconia psittacorum* cultivars and interspecific hybrids, are information promising to be taken in account as a first step towards genetic improvement.

5. Acknowledgements

The authors thank the National Council of Scientific and Technological Development (CNPq) and the Coordination for the Improvement of Higher Education (CAPES) for the scholarship of the first author, the BNB for the financial support, the Bem-Te-Vi Farm, the RECIFLORA association, researcher scientist Dr. Carlos E. F. de Castro Campinas Agronomic Institute (IAC) and trainees of the UFRPE Floriculture Laboratory.

6. References

Albert, B., A.Jonhson, J.Lewis, M.Raff, K.Roberts and P.Walter (2002) Molecular Biology of The Cell, 4nd edn. Garland Publishing, New York, p.1661.

Baldwin, B.G., M.J.Sanderson, J.M.Porter, M.F.Wojcichowiski, C.S.Campbell and M.J. Donoghue (1995) The ITS Region of Nuclear Ribosomal DNA: A Valuable Source of Evidence on Angiosperm Phylogeny. Ann. Missouri Bot. Gard. 82: 247–277.

Berry, F. and W.J.Kress (1991) Heliconia: An Identification Guide. Smithsonian Institution Press, Washington and London, p.334.

Bruns, T.D., T.J.White and J.W.Taylor (1991) Fungal molecular systematics. Ann. Rev. Ecol. Sys. 22: 525–564.

Câmara, P.E.A.S. (2008) Developmental, phyilogenetic, taxonomic study on the moss genus Taxitelium Mitt. (Pylaisiadelphaceae). PhD Tesis, *University of Missouri*, St. Louis.

Castro, C.E.F., C.Gonçalves and A.May (2007) Atualização da nomenclatura de espécies do gênero *Heliconia* (Heliconiaceae). R. Bras. Hortic. Ornam. Campinas, Brazil. 13: 38–62.

Castro, C.E.F. and T.T.Graziano (1997) Espécies do Gênero *Heliconia* (Heliconiaceae). R. Bras. Hortic. Ornam. Campinas, Brazil. 3: 15-28.

Costa, A.S., B.S.F.Leite, V.Loges, E.C.S. Bernardes and A.C. Brasileiro-Vidal (2008) Padrão de distribuição de bandas CMA[3] e localização de sítios de DNAr 5S e 45S na análise de acessos de *Heliconia* (Heliconiaceae). *In*: Proceeding of 54d Congress on Genetic, Salvador, Brazil.

Criley, R.A. and T.K.Broschat (1992) Heliconia: botany and horticulturae of new floral crop. Hortic. Review, New York. 14: 1-55.

Cruz, C.D. (2006) Programa Genes: análise multivariada e simulação. Viçosa, UFV, p.175.

Donselman, H. and T.K.Broschat (1986) Production and postharvest culture of *Heliconia psittacorum* flowers in south Florida. FL. Lauderdale, USA, pp.272-273.

Doyle, J.J. and J.L.Doyle (1990) Isolation of plant DNA from fresh tissue. Focus. 12: 13-15.

ITEP (2008) http://www.itep.br/lamepe.ASP. (*in portuguese*).

Jaccard, P. (1908) Nouvelles recherché sur la distribution florale. Bull. Soc. Vaud. Sci. Nat. 44: 223-270.

Jatoi, S.A., A.Kikuchi, M.Mimura, S.S.Yi and K.N.Watanabe (2008) Relationship of *Zingiber* species, and genetic variability assessment in ginger (*Zingiber officinale*) accessions from ex-situ genebank, on-farm and rural markets. Breed. Sci. 58: 261-270.

Johansen, L.B. (2005) Phylogeny of Orchidantha (Lowiaceae) and the Zingiberales based on six DNA regions. Syst. Botany. 30:106-117.

Kladmook M., S.Chidchenchey and V.Keeratinijakal (2010) Assessment of genetic diversity in cassumunar ginger (*Zingiber cassumunar* Roxb.) in Thailand using AFLP markers. Breed. Sci. 60: 412-418.

Kocyan, Y.L., P.K.Qiu, E.Endress and A.Conti1 (2004) A phylogenetic analysis of Apostasioideae (Orchidaceae) based on ITS, trnL-F and matK sequences. Plant Syst. Evol. 247: 203-213.

Kress, W.J. (1995) Phylogeny of the Zingiberanae: morphology and molecules. *In*: Rudall, P., P.J.Cribb, D.F.Cutler and C.J.Humphries, (eds.). Monocotyledons: systematics and evolution, Royal Botanic Gardens, Kew, UK, pp.443-460.

Kress, W.J. (1990) The phylogeny and classification of the Zingiberales. Ann. Missouri Bot. Gard. 77: 698-721.

Kress, W.J., J.Betancur, C.S.Roesel and B.E.Echeverry (1993) Listapreliminar de las Heliconias de Colombia y cinco espécies nuevas. Caldasia. 17:183-197.

Kress, W.J., L.M.Prince, W.J.Hahn and E.A.Zimmer (2001) Unraveling the evolutionary radiation of the families of the Zingiberales using morphological and molecular evidence. Syst. Biology. 51: 926-944.

Kumar, P.P., J.C.K.Yau and C.J.Goh (1998) Genetic analysis of *Heliconia* species and cultivars with randomly amplified polymorphic DNA (RAPD) makers. J. Ameri. Soc. Hort. Sci. 123: 91-97.

Marouelli, L.P., P.W.Inglis, M.A.Ferreira and G.S.C.Buso (2010) Genetic relationships among *Heliconia* (Heliconiaceae) species based on RAPD markers. Genet. Mol. Res. 9: 1377-1387.

Pearson, H. (2006) What is a gene? Nature, London. 441: 399-401.

Sanchez-Baracaldo, P. (2004) Phylogenetics and biogeography of the neotropical fern genera *Jamesonia* and *Eriosorus* (pteridaceae). Am. J. Bot. 91: 274-284.

Sang, T., D.J. Crawford and T.F. Stuessy (1997) Chloroplast DNA phylogeny, reticulate evolution, and biogeography of Paeonia (Paeoniaceae). Am. J. Bot. 84: 1120-1136.

Sheela, V.L., P.R.Geetha Lekshmi, C.S.Jayachandran Nair and K.Rajmohan (2006) Molecular characterization of *Heliconia* by RAPD assay. J. Tropi. Agric. 44: 37-41.

Sousa, L.O.F., T.Wendt, G.K.Brown, D.E.Tuthill and T.M.Evans (2007) Monophyly and phylogenetic relationships in *Lymania* (Bromeliaceae: Bromelioideae) based on morphology and chloroplast dna sequences. Sys. Botany. 32: 264-270.

White, T.J., T.Bruns, S.Lee and J.Taylor (1990) Amplification and direct sequencing of fungal ribosomal RNA genes for phylogenetics. *In*: Innis, M., D.Gelfand, J.Sninsky and T.White (eds.). PCR Protocols: A Guide to methods and applications. Academic Press, San Diego, pp. 315-322.

Section 2

Crop Production

3

Texture, Color and Frequential Proxy-Detection Image Processing for Crop Characterization in a Context of Precision Agriculture

Cointault Frédéric et al.*
AgroSup Dijon,
France

1. Introduction

The concept of precision agriculture consists to spatially manage crop management practices according to in-field variability. This concept is principally dedicated to variable-rate application of inputs such as nitrogen, seeds and phytosanitary products, allowing for a better yield management and reduction on the use of pesticides, herbicides ... In this general context, the development of ICT techniques has allowed relevant progresses for Leaf Area Index (LAI) (Richardson et al., 2009), crop density (Saeys et al., 2009), stress (Zygielbaum et al., 2009) ... Most of the tools used for Precision Farming utilizes optical and/or imaging sensors and dedicated treatments, in real time or not, and eventually combined to 3D plant growth modeling or disease development (Fournier et al., 2003 ; Robert et al., 2008). To evaluate yields or to better define the appropriated periods for the spraying or fertilizer input, to detect crop, weeds, diseases ..., the remote sensing imaging devices are often used to complete or replace embedded sensors onboard the agricultural machinery (Aparicio et al., 2000). Even if these tools provide sufficient accurate information, they get some drawbacks compared to "proxy-detection" optical sensors: resolution, easy-to-use tools, accessibility, cost, temporality, precision of the measurement ... The use of specific image acquisition systems coupled to reliable image processing should allow for a reduction of working time, a lower work hardness and a reduction of the bias of the measurement according to the operator, or a better spatial sampling due to the rapidity of the image acquisition (instead of the use of remote sensing). The early evaluation of yield could allow farmers, for example, to adjust cultivation practices (e.g., last nitrogen (N) input), to organize harvest and storage logistics. The optimization of late N application could lead to significant improvements for the environment, one of the most important concerns that precision agriculture aims to address.

*Journaux Ludovic[1], Rabatel Gilles[2], Germain Christian[3], Ooms David[4], Destain Marie-France[4], Gorretta Nathalie[2], Grenier Gilbert[3], Lavialle Olivier[3] and Marin Ambroise[1]
[1]*AgroSup Dijon, France*
[2]*Irstea Montpellier, France*
[3]*IMS – Bordeaux University - Bordeaux Sciences Agro, France*
[4]*ULg (Gembloux Agro-BioTech), Belgium*

We propose in this chapter to explore the proxy-detection domain by focusing first on the development of robust image acquisition systems, and secondly on the use of image processing for different applications tied on one hand to wheat crop characterization, such as *the detection and counting of wheat ears per m² (in a context of yield prediction) and the weed detection,* and on the other hand to the evolution of seed development/germination performance of chicory achenes. Results of the different processing are presented in the last part just before a conclusion.

2. Image acquisition system

Image acquisition is the most important step for robust image processing. Indeed, the use of natural images involves some difficulties tied to outdoor conditions (lighting variations) and object complexity (the leaf area presents several planes, high contrast and lighting variations, scale variations, wheat growth stages ....). Photographic slides were taken at different growing stages of the wheatears.

Different solutions exist for image acquisition according to the vertical distance between the camera and the scene (1m, 2m or more), to the illumination used (natural or controlled light), to the kind of images (color, grey level, hyperspectral ...)

Figure 1 shows one solution based on specific boxes with opaque protection and controlled illumination by power-leds, for wheat ear counting, allowing to obtain color or grey level images.

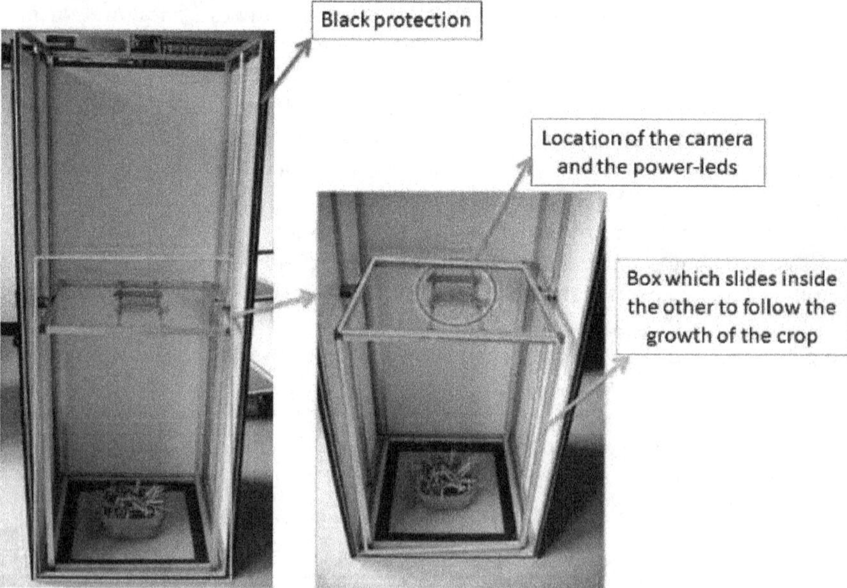

Fig. 1. Example of image acquisition system (left) and its specific illumination by Led (right)

Figure 2 shows some examples of color images of wheat crop at three different growth stages with controlled illumination.

Fig. 2. Example of image acquisition system (left) and its specific illumination by Led (right)

Proxy-Hyperspectral images can also be used and are obtained with specific sensors. As an illustration, images of wheat crop have been acquired in march 2011 on the domain of Melgueil (INRA, 34-France) by means of dedicated apparatus developed by the Cemagref. This device is constituted by a motorized rod installed on a tractor which allows a longitudinal displacement of a push-broom hyperspectral camera located 1 meter above the vegetation (figure 3).

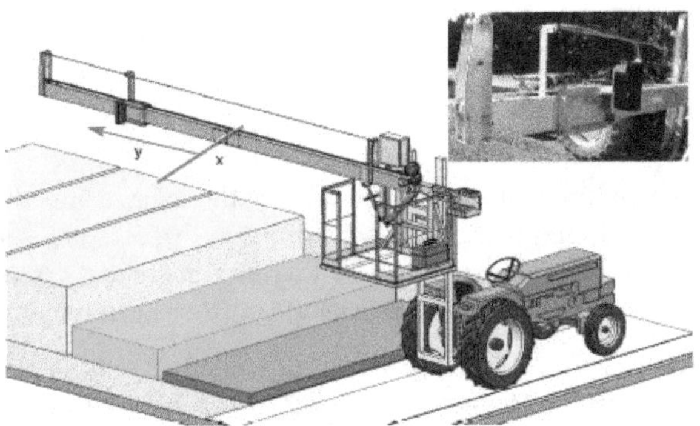

Fig. 3. Hyperspectral acquisition device

Other imaging system is based on chlorophyll fluorescence. A xenon light source (Hamamatsu Lightingcure L8222, model LC5, 150 W) is used to produce a white light which passes through an interference filter (03FIB002, Melles Griot, Carlsbad, USA) with a central wavelength 410 nm and width at half-maximum 80 nm to excite the chlorophyll-a. The light is conducted to the seeds by an optical fiber. The blue light issued from the filter is absorbed by the seeds chlorophylls, which resulted in fluorescence emission. A high-pass filter (665 nm, 03FCG107, Melles Griot 03FCG107) ensured the selection of the fluorescent signal from the blue light reflected by the object.

The images are acquired by a CCD monochrome camera (Hamamatsu C5405-70), with a resolution of 640 x 480 pixels and 256 gray levels. The system is enclosed within a black box to avoid interference from ambient light and is air-conditioned (Tectro TS27, PVG Int. B.V., Oss, The Netherlands) to maintain a temperature of 20°C. The lamp is optically isolated and white light did not escape inside the box. The blue filter transmitted an unexpected, small amount of infrared light between 770 and 900 nm (detected with spectrometer AVS-SD2000, Avantes, Eerbeek, The Netherlands).

3. Proxy-detection image processing for wheat ear counting

Proxy-detection imagery for yield prediction needs the determination of the three components of the yield: number of ears / m^2, number of grains per ear and the weight per thousand seeds. Wheat ear shapes are very variable, especially due to the different inclinations of wheatears. However, wheatears show a very rough surface whereas leaf surfaces are smoother (figure 4).

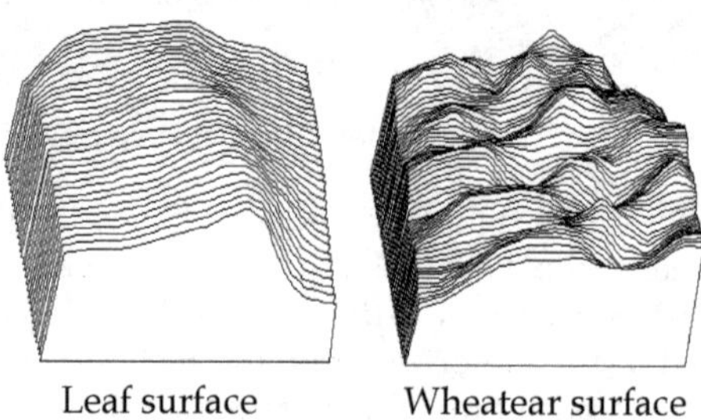

Fig. 4. 3D surface of leaf and ear

Considering this specific roughness (figure 4), texture analysis can be used in order to discriminate wheatears, leaves and ground.

3.1 Texture feature extraction

In order to extract the texture features, a variety of methods have already been proposed in the literature and tested in practice. In this context, three families of texture feature extraction exist: statistical, spectral and structural methods. A successful texture classification or segmentation requires an efficient feature extraction methodology.

Two methods are discussed in this chapter. The first one is based on statistical analysis by cooccurrence matrices, and the second one is based on Fourier filtering.

3.1.1 Cooccurrence matrices

The principle of these matrices is to count how often pairs of grey levels occur in a digital image of texture. In fact, we take into account pairs of pixels that are located along a given direction defined by the angle θ and separated by a fixed distance d.

Let $L_I = \{1,2,...,N_I\}$ and $L_J = \{1,2,...,N_J\}$ be respectively the horizontal and vertical spatial domains and $G = \{0,1,...,\Gamma - 1\}$ the grey levels.

The digital image $f : L_I \times L_J \rightarrow G$ assigns a grey level to each pixel.

Furthermore, we consider four directions (θ = 0°, 45°, 90° and 135°) along which a pair of pixels can lie. Thereafter, for the distance d, we are able to obtain the unnormalized frequencies $\mathrm{P}(\gamma,\gamma',\theta,\mathrm{d})$.

For instance, $\mathrm{P}(\gamma,\gamma',0°,d)$ is defined as the cardinality of the set of pixel pairs having the following properties:

$$\begin{cases} f(i,j)=\gamma \text{ and } f(i',j')=\gamma' \\ |i-i'|=d\ , j=j' \end{cases} \tag{1}$$

For the other angles, the corresponding frequencies are defined similarly. In this document, we consider the distance **d** = 1 (this means we consider neighboring pixels). Moreover, texture of the wheatears, and, more generally, of the pictures, are not oriented along a common direction. Thus, for each pair of grey levels we sum over the four frequencies (corresponding to the four angles) and we note:

$$P_{\gamma\,\gamma'} = \sum_{\theta} P(\gamma,\gamma',\theta°,1) \tag{2}$$

$p_{\gamma\,\gamma'} = \dfrac{P_{\gamma\,\gamma'}}{\sum\limits_{\gamma,\gamma'} P_{\gamma\,\gamma'}}$ is the element of the normalized cooccurrence matrix.

Generally cooccurrence matrices are not used directly because of their size. The principle is to compute a set of measures from these matrices, such as those proposed by Haralick (1973).

Figure 5 shows three samples of ear ,soil and leaf textures: wheatearstexture reveals many transitions between very different grey tones. For leaves and soil the transitions are more progressive.

ears leaves soil

Fig. 5. Examples of the three textural patterns

Thus, we compute four Haralick's features which characterize the particular disposition of the cooccurrence matrix elements for the 3 texture classes:

energy (angular second moment) :

$$\sum_{\gamma,\gamma'} p_{\gamma\,\gamma'}^2 \tag{3}$$

inverse different moment :

$$\sum_{\gamma,\gamma'} \frac{1}{1+(\gamma-\gamma')^2} p_{\gamma\,\gamma'} \tag{4}$$

contrast :

$$\sum_{\gamma,\gamma'} (\gamma-\gamma')^2 p_{\gamma\,\gamma'} \tag{5}$$

entropy :

$$-\sum_{\gamma,\gamma'} p_{\gamma\,\gamma'} \times \log\left(p_{\gamma\,\gamma'}\right) \tag{6}$$

The four coefficients were calculated for windows of 17x17 pixels surrounding each pixel. This window size is smaller than the ears' size and allows for reducing noise effects. The computational cost, the cooccurrence matrices were calculated over posterised pictures (32 grey levels), which avoids obtaining hollow matrices, considering the number of pixel in each window.

Table 1 shows the values of the four features for the 3 samples of the figure 5.

Texture	contrast	energy	entropy	inverse different moment
ears	8.1	0.04	4.9	0.3
leaves	2.5	0.22	3.6	0.6
soil	0.3	0.94	2.0	0.8

Table 1. Values of the four Haralick's parameters associated with the three textural patterns

The features computed from the cooccurrence matrices are used in a learning step in order to construct a discriminant function, called here textural function, which identifies efficiently the pixel belonging to the « ears » class.

Considering a trial set of windows, the use of a stepwise selection leads us to selects a subset of parameters (from the four ones describe above) to produce a good discrimination model.

At the first step, we compute the variance ratio F for each feature:

$$F = \frac{\text{between} - \text{group variance}}{\text{within} - \text{group variance}} = \frac{\sum_i n_i \times \left(\bar{x}_{i.} - \bar{x}_{..}\right)^2 / (g-1)}{\sum_i \left(\bar{x}_{ik} - \bar{x}_{i.}\right)^2 / (n-g)} \tag{7}$$

$g = 2$ is the number of groups (one for ears, one for leaves and ground), n_i is the number of windows for the group i ($\sum n_i = n$), $\bar{x}_{i.}$ is the mean of the group i, $\bar{x}_{..}$ is the mean of the trial set, x_{ik} is the value of the window k belonging to the group i.

The value of the variance ratio reflects the parameter's contribution to the discrimination if it is included in the textural function. The parameter that has the largest ratio is selected at first. Afterwards, at each step, the order of insertion is computed using the partial F-ratio as a measure of the importance of parameters not yet in the equation. As soon as the partial F-ratio related to the most recently entered parameter becomes non-significant the process is stopped.

In a second stage, a discriminant analysis with the selected parameters is computed. Because there are only two groups to discriminate, only one discriminant factor F_D(textural function) is obtained.

3.1.2 Results from the cooccurrence matrices

In order to illustrate the efficiency of this textural function, the values of F_D obtained on the image may be rescaled in the form of a gray level scale. For example, figure 6 shows an original image and the corresponding result after the first stage.

Fig. 6. Example of the textural functionF_D (right) for a given wheatear image (left)

3.1.3 Results and discussion

In order to show its relevance, the Textural function is used in the energy function of a Maximum A Posteriori segmentation algorithm (Martinez de Guerenu et al. 1996). Figure 7 shows separately the Wheatear and Non-wheatear regions resulting from this segmentation stage. Wheatear region pictures show false positive wheatear detecton (leaves or stems classified as wheatears). Non-wheatear region pictures show false negative detection (wheatears that have been forgotten).

Both kind of erroneous classification are circled on the pictures.

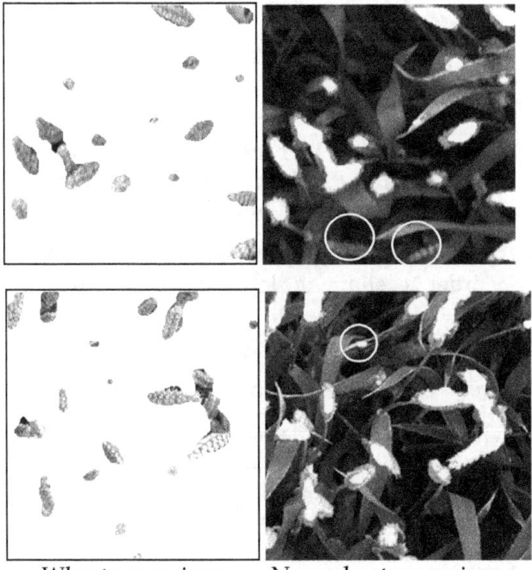

Fig. 7. Wheatear and Non-wheatear region results for MAP segmentation

For the whole set of sample images, 1 to 3 segmentation errors have been found for each image. As the images show an average of 20 wheatears, the error rate can be estimated around 10%. Most of the errors encountered are identified as badly lightened wheatears. This kind of error could be easily overcome by using artificial light instead of natural light conditions, or using color images instead of grey level images, as it will be shown in the next section.

3.2 Fourier filtering

To improve the previous works, one of the first solution has been to represent each image in a colour-texture hybrid space, and to use mathematical morphology tools (Cointault et al., 2008a) in order to extract and count the number of wheat ears.

Although this previous method gives satisfying results (around 10% of well detection, compared to manual counting), the hybrid space construction method is a supervised one and is limited by the objectivity of the operator. Moreover the statistical methods are dependent on the choice of the direction for the process, need an important computing time and are lightning-dependent.

The objective has been thus to propose new detection algorithms more rapids, robusts, and invariants according to image acquisition conditions, based on Fourier filtering and two dimensional discrete fast Fourier transform (FFT) (Cooley &Tukey, 1965). This approach includes three important steps: high-pass filtering, thresholding and cleaning of the image based on mathematical morphology operations (Serra, 1982).

3.2.1 The three important steps

For the high pass Fourier filtering, a two dimensional FFT is performed on the target image (eq. 8).

$$F(k_x,k_y)=\frac{1}{\sqrt{N_xN_y}}\sum_{n_x=0}^{N_x-1}\sum_{n_y=0}^{N_y-1}f(n_x,n_y)e^{\omega\frac{k_xn_x}{N_x}}e^{\omega\frac{k_yn_y}{N_y}} \tag{8}$$

Based on the centered Fourier image, a high pass filter is applied in order to eliminate low frequencies in the FFT image (figure 8). The cut off frequency is empirically sized by a 10 pixels width disk mask as it is shown in the figure 8.

The thresholding of the resulting image is based on anInverse FFT and a predetermined threshold is applied in order to eliminate low pixel values which do not correspond to wheat objects (ground, leaf,...) (figure Fig. 9).

The cleaning step aims at eliminate remaining "non wheat" pixel groups, which are small and scattered. It lies on mathematical morphology operation and is performed with three sub steps:

- First, a dilatation, which aims at making bigger and closer pixel groups in the image.
- Then a blurring convolution with a Gaussian smoothing operator, followed by a thresholding, which eliminate too small groups of pixels. These small groups are considered as miss. This step makes smaller the pixel groups that correspond to wheat and then justify a third step.
- Finally, another dilatation is performed which aims at regenerate size of pixel groups corresponding to wheat ears (figure 10).

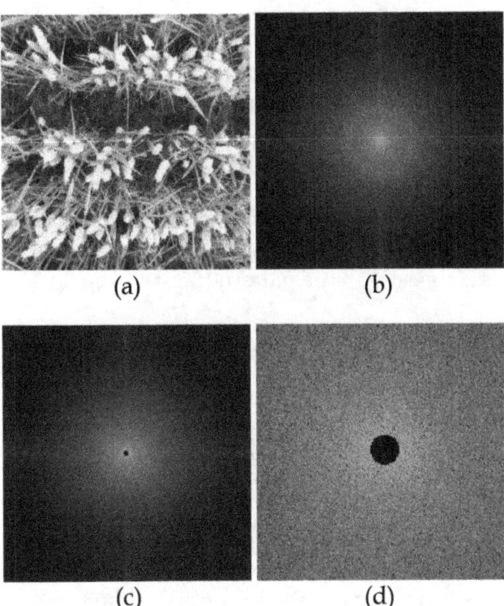

Fig. 8. Example of wheat image (a), (b) its associated FFT projection, (c) cut off disk, (d) zoom of cut off disk.

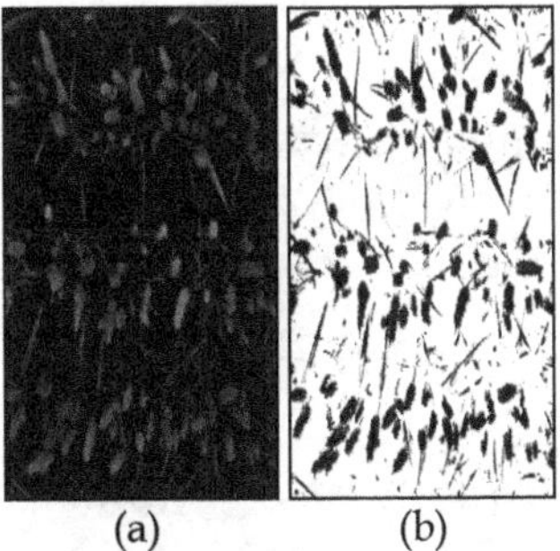

Fig. 9. Image after inverse FFT(a) and threshold image (b).

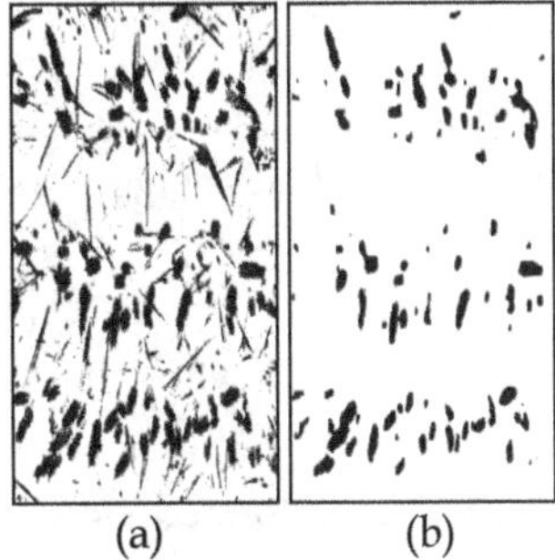

Fig. 10. Threshold image (a) and cleaned image of wheat detection (b).

3.2.2 Wheat ear counting estimation

The first image processing by thresholding and high pass Fourier filtering gives a binary image, composed of several pixel groups. Each group represents one or more wheat ears to be counted. In order to estimate this number we analyze the shape of each group considering two possible configurations:

- Pixel group presents a convex or nearly convex shape pattern: it is considered that only one ear is present in this kind of group (figure 11a).
- Pixel group presents a concave shape pattern (figure 11b): it is consider that several ears are presents.

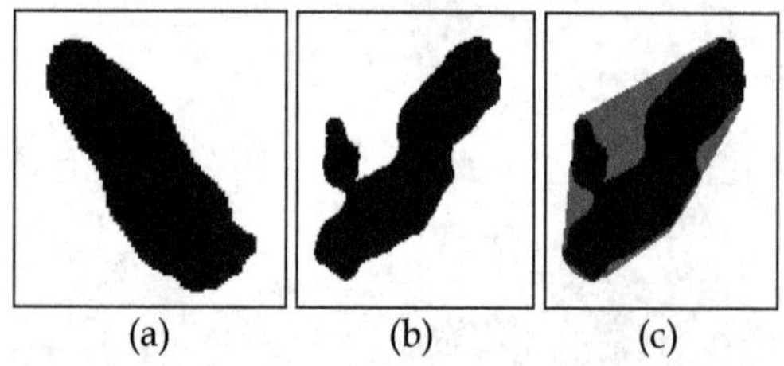

(a) (b) (c)

Fig. 11. Pixel groups with nearly convex (a) and concave shapes (b), convex hull of a concave shaped pixel group (c).

In order to quantify the number of wheat ears, we estimate a shape index based on two features extracted from each pixel group:

$$Q = \frac{S_c - S}{S} - 1 \quad (9)$$

With S the surface and S_C the convex hull surface of the pixel group (Figure 6c).

Coming from this index, a number$X$ of ears is attributed for the group, considering this interval:

$$(X-1)/10 < Q < X/10 \quad (10)$$

Another approach, based on skeleton analysis can be found in (Germain et al, 1995).

3.2.3 Results and discussion

A fast visual observation (figure 12) shows that only a small amount of groups corresponds to non-wheat things in image. More precisely, the biggest wheat ears are well detected and well separated from surrounded leaves.

Ears that lie the nearest of the ground, that are partially hidden or that are a little bit over exposed in the image are not well detected. Small amount of very big leaves also remain after cleaning step.

In order to test our image processing we performed algorithm on 40 images and compare the results with the mean of manual counting done by several experts. For the example, five images sample have been randomly chosen (table 2).

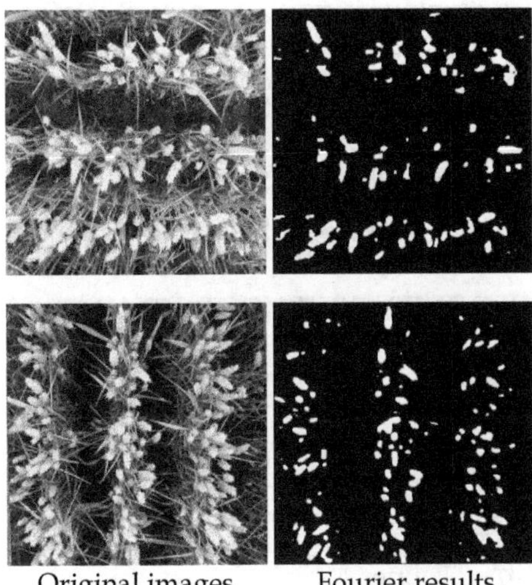

Fig. 12. Results of wheat ear detection using high pass Fourier Filtering.

Image	Counting		Difference (%)
	Manual	image processing	
1	139	142	2,11
2	36	34	-5,88
3	90,5	94	3,72
4	116,5	122	4,51
5	136,5	142	3,87

Table 2. Specifications of five experimental drying runs used for validation.

Within the image set, a high variability of ear's number can be observed. With most of the images, high pass Fourier filtering method returns slightly higher counts. Based on the whole image sample, absolute difference between manual counting and image processing counting is contained under the value of 6%. The mean error obtained is 4,02% for this set of images.

The development of a high pass Fourier filtering approach method aims at creating an easily usable and adaptable method while obtaining a best wheat ear detection. Therefore, it is essential to compare this approach with previously used methods such as hybrid space method (Cointault et al., 2008b). For a visual comparison, high pass Fourier filtering has been applied onto images that have been previously treated with hybrid approach (figure 13).

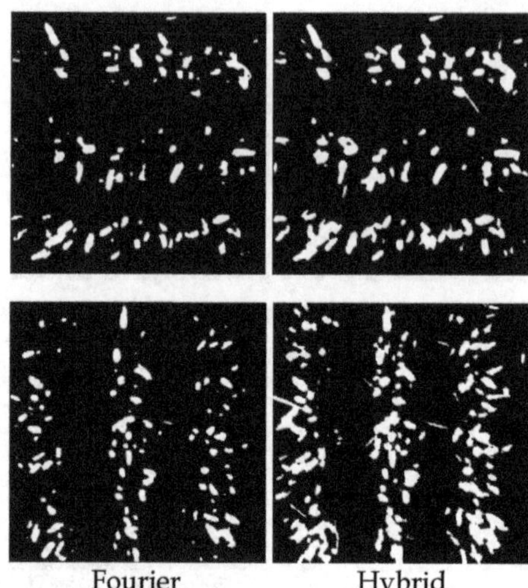

Fig. 13. Visual comparison between Fourier and Hybrid approach for two different images.

Visual comparisons with hybrid space results show that high pass Fourier filtering approach eliminate more non wheat objects within the images. Fourier approach separates more efficiently ear groups. Calculation time for Fourier approach is few sec per image while its few minutes for hybrid space, with the same operating system.

High pass Fourier filtering gives global satisfying results. Although a close range of settings has been determined, inverse FFT remains a parameter that has to be adjusted according to input image. An empirical value has been found and gives good results for most of images but it could be optimized with an automatic threshold selection such as k-means methods (MacQueen, 1967).

In the context of wheat detection, it has been observed that some ear objects are eliminated after cleaning step. These non-detections mainly correspond to near ground ears or ears massively hidden under leaves, hence, it should be relativized as too low ears may have development problems and may not be considerate within wheat yield. Ears that are located in over or under exposed part of image are not well detected but it is not due to the algorithm but to the quality of the acquisition, which is limited by the natural conditions. Small amount of very big leaves also remain after cleaning step and eliminate these artifacts constitute a further axis of development, including shape analysis in cleaning step.

In the context of wheat ear counting, it is observed that counting error percentage decrease with number of ears in images, hence, best results may be obtained with images representing more area and yet, more ears. Actually, worst error, 5,56%, is obtained with only 36 wheat ears in the image.

It is important to note that in most cases, Fourier approach returns slightly higher counts than manually counts. It should be due to missing detection, such as remaining leaves or over exposed part of images. Counts should be more precise with the including of shape analysis in cleaning step. In all cases, whatever the method used, the only way to obtain a right detection is to use 3-Dimensional information.

4. Proxy-detection image processing for weed-wheat crop discrimination by hyperspectral imagery

In the domain of weed control, since the environmental and economical stakes are particularly great, the herbicides are largely spread in order to assure sufficient yields for the whole field whatever the infestation rate.

The reason is essentially a technological one. Even if some low-cost devices are currently available to assure a localised spraying of herbicides on bare soil (vegetation detection by photoelectric cells), no commercial product allows a reliable and localised post-emergence treatment. Indeed, a such apparatus needs a sophisticated perception system, based on digital vision and allowing to distinguish between weeds and crops. The identification of varieties inside the vegetation is nowadays the principal lock to the development of localized weed control.

The corresponding research are numerous and can be divided into two main approaches (Slaughter et al., 2008):

- the spectral approach, in which the plant reflectance is the main parameter, using hyper- or multispectral images (Feyaerts & Van Gool, 2001; Vrindts, 2002; De Baerdemaeker et al., 2002). The difficulty is then to propose spectral differences sufficiently robusts to lightning conditions
- the spatial approach, based on spatial criteria such as plant morphology (Chi et al., 2003; Manh et al., 2001), plant texture (Burks et al., 2000) ... The main difficulty is tied here to the complexity and the natural variability of the scenes.

The study proposed in this chapter is tied to the previous approach: hyperspectral images of the wheat crop are acquired during the weed control period, and associated to specific pre-processing to avoid illumination conditions. After, the possible spectral discrimination between wheat and dicotyledonous weeds by means of chimiometric tools has been evaluated.

4.1 Spectral pre-processing

The images acquired by the camera (after a first internal processing taking into account the spectral sensitivity of the sensor) are images of luminance, which by definition depend on both the reflectance of the scene and lighting conditions. This is why a reference surface (gray ceramic), for which the reflectance Rc was measured in the laboratory, was systematically placed in each scene (figure 14). The average luminance Lc observed in the image to the reference can then be corrected reflectance for each pixel of vegetation :

$$Rf = Rc * (Lf/Lc) \tag{11}$$

where Lf is the observed luminance for that pixel.

It is important to note that the Rf value obtained is a "apparent" reflectance. Other phenomena are to be considered, namely:

- The inclination of the observed leaf from the direction of the solar source, introducing a multiplicative factor of the light, and hence Rf
- the possibility of specular reflection, adding a component to the spectrum colorless, and resulting in an additive constant of Rf.

These two phenomena are taken into account by applying a standard-centering operation (SNV, or "standard normal variate") on each spectrum of the image (Vigneau et al., 2011).

4.2 Chemometric model

Because hyperspectral images include hundreds of contiguous spectral bands, they have a very high potential for spectral discrimination, compared to classical colour or multispectral images. Unfortunately, in a classification context, these kinds of high-dimensional datasets are difficult to handle and tend to suffer from the problem of the "curse of dimensionality", well known as "Hughes phenomenon" (Hughes, 1968) which causes inaccurate classification (figure 14).

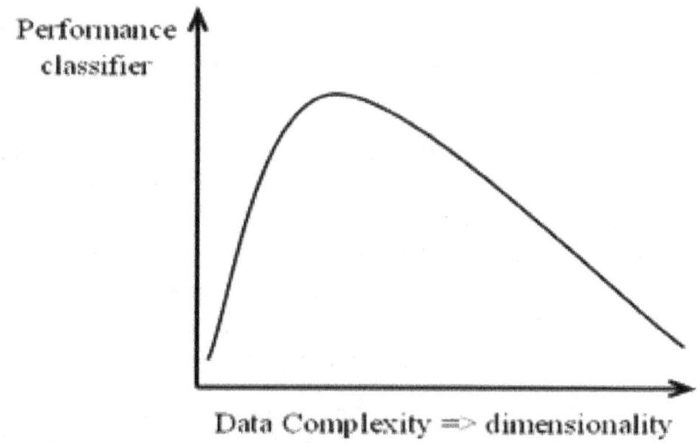

Fig. 14. Illustration of Hughes phenomenon

In this context, it is important to note that some news techniques already exist to overcome this problem. Among all possible methods, many recent works have focused on the area of dimensionnality reduction approaches in order to reduce this high-dimensional datasets while maintaining the relevance of the information contained in the signal.(Journaux et al., 2008). One of these approaches is the Partial Least Square Regression (PLS-R) , which builds a low-dimension subspace by determining a set of spectral "latent variables". Compared to other reduction methods such as Principal Component Analysis (PCA), the PLS takes into account both inputs and outputs to build its subspace, leading to better performances. The PLS-DA (PLS Discriminant Analysis) is an adaptation of PLS-R to discrimination problems.

In our case, a total of 335 spectra was increased step by step on a first reflectance image (figure 15) in three categories, namely wheat (157 spectra), dicotyledons (60 spectra) and ground (118 spectra).

A PLS-DA model was then determined on this data set by cross-validation using commercial software in chemometrics (The Unscrambler v9.7, CAMO Software AS, Oslo, Norway).

The resulting discriminant model, which involves 8 latent variables, was then exported and applied to all pixels in an image test, using a dedicated software developed in C + +.

Fig. 15. Calibration image and sample positions (the reference ceramic can be seen on the left)

4.3 Results

Figure 16 shows the image test and the classification results. We observe an excellent discrimination between wheat and weeds whatever the local conditions of illumination (shadow), with the exception of one type of weed (red circle) which was not present in the calibration image, and was therefore not included in the model.

Fig. 16. Test image and classification results (light gray: ground, red: weed, green: wheat)

These results show the very high potential of hyperspectral image processing to extract pertinent information for agronomic applications.

5. Proxy-detection image processing for seed development and germination performance of chicory achenes by chlorophyll fluorescence

Several aspects of seed quality are tested routinely to minimize the risk of sowing seedlots that do not have the capacity to produce the desired crop. Amongst these tests, seed germination is important since it represents the percentage of pure seeds that have the potential to produce established seedlings in the field. The rate of germination corresponds to the reciprocal of the time needed for a given germination percentage to be reached (Halmer, 2008). Accurate procedure of germination tests performed in laboratory is defined by the International Seed Testing Association (ISTA).

Besides these laboratory tests, seed processing lines includes cleaning machines that remove dust and waste material, and conditioning machines performing dimensional sizing, density sorting, and colour sorting. In conventional colour sorting, discoloured seeds are rejected, on basis of inspecting seeds individually to detect differences in reflected colour.

However, even after several sorting operations, some seed batches can contain a large proportion of viable seeds but still not sufficient for commercial use. These batches are lost because the viable and non-viable seeds cannot be separated using the conventional processing methods. The proportion of immature seeds in these lost batches is unknown. It would therefore be useful to provide a new, non-destructive method of distinguishing immature seeds from mature seeds in order to improve sorting processes.

In this context, the potential of fluorescence imaging (Chen et al., 2002; Nedbal & Whitmarsh, 2004) has been examined. The chlorophyll degrades during fruit ripening and the process of degradation was described by Barry (2009). The chlorophyll is also a highly fluorescent molecule. Fluorescence occurs when some of the light absorbed by the chlorophylls is re-emitted at longer wavelength, typically between 650 and 750 nm. The fluorescent properties of chlorophylls have been used to evaluate the maturity of cabbage seeds (*Brassica oleracea* L.) (Jalink et al., 1998; Jalink et al., 1999). The results showed that the magnitude of the chlorophyll fluorescence (CF) signal was inversely related to the quality of seeds. The relationship between the CF and germination performance was studied for tomato (*Solanum lycopersicum* L.) by Jalink et al. (1999) as in their previous study (Jalink et al., 1998). They concluded that seeds with an intermediate CF level were of the best quality, followed by seeds having a low CF signal. Seeds having a high CF signal were the worst. Konstantinova et al. (2002) measured the CF of barley grains (*Hordeum vulgare* L.) with a SeedScan I Laser Sorter (Satake, Stafford, TX. USA), using the principle developed by Jalink et al. (1998) but including a laser light source instead of a LED. They concluded that sorting a barley seed lot into six subsamples varying in CF values resulted in an optimal quality for the subsamples with low and intermediate CF signals. Suhartanto (2002) thoroughly described the relationships between the fruit CF, seeds CF and germination performance of tomato.

5.1 Image analysis

A specific image analysis code was developed with the GNU Octave language (Ooms & Destain, 2011). After applying background correction (the fluorescence values were divided by the reflectance signal of paper and multiplied by 60), the images were segmented and images of individual seeds were created, each of them being rotated along the main axis of the seed. The pappus side, which is brighter and larger than the radicle tip, was automatically detected on the basis of the mean width of the left half, its mean fluorescence intensity, the right half width and the right half fluorescence intensity. The accuracy of the detection was greater than 98%. The image was thereafter divided into the "pericarp zone" (Pe, 77% of the seed length) and the "pappus zone" (Pa, 23% of the length). The value of 77% was a compromise based on the observation of 100 seed images. The mean fluorescence values of the two zones were recorded for data analysis. The measurement system and image analysis are summarised in figure 17.

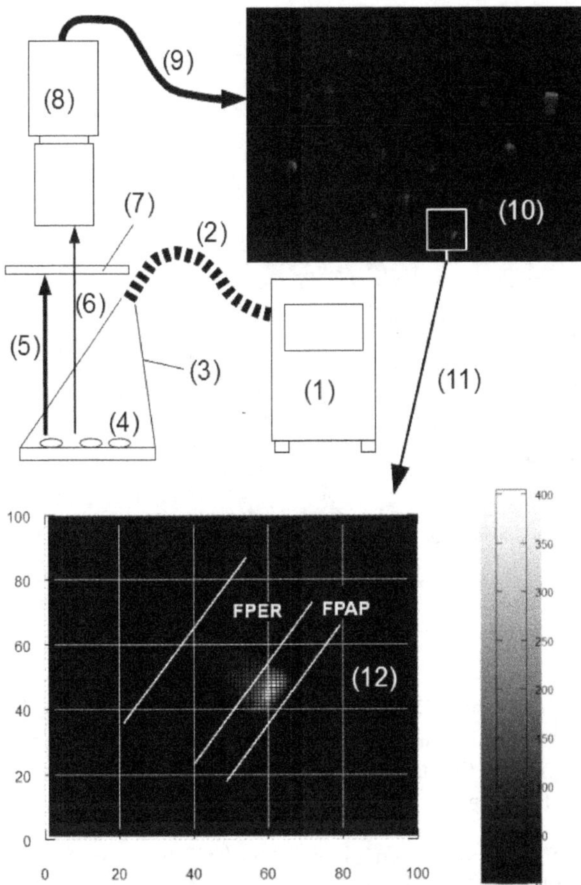

Fig. 17. The components of the chlorophyll imaging device and the main steps of image analysis. (1) Xenon lamp with blue optical bandpass filter (370 - 450 nm), (2) optical fiber, (3) cone of blue light, (4) seeds from one capitulum, (5) blue light reflected by the seeds, (6) chlorophyll fluorescence (650 - 730 nm), (7) highpass optical filter (665 nm), (8) CCD camera with zoom objective, (9) connection to the computer, (10) raw image, (11) background correction, then creation of individual images, (12) automated detection of the pappus using dedicated software and estimation of the levels of fluorescence FPER and FPAP (Ooms & Destain, 2011).

5.2 Results

The observed period corresponds to the phase of reserve deposition in the seed described in Bewley & Black (1994). Figure 18 shows the evolution of CF during this phase on the stalk, the weight parameters (dry weight DW and water content WC) and the germination performance (GP and GR). The dry weight increased, while the water content was still high at the end of the observed period (> 45 % on the stalk at 44 DAF, while the WC of stored seeds is about 6 %).

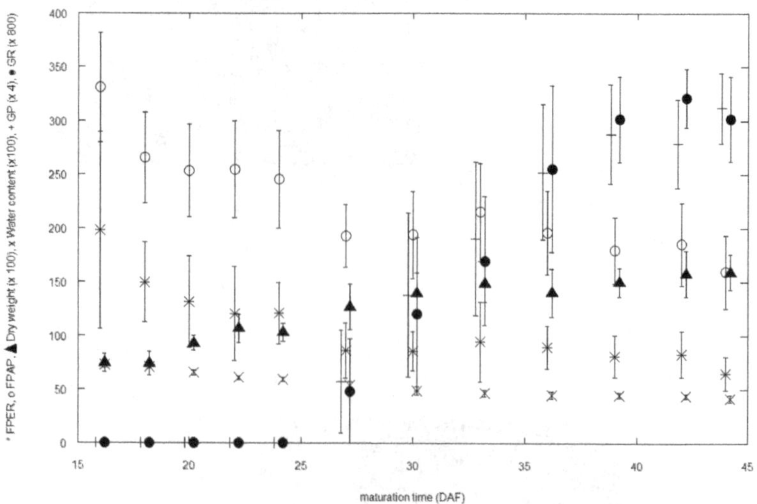

Fig. 18. Evolution of chlorophyll fluorescence (FPER: pericarp, FPAP: pappus), dry weight, water content at harvest, germination percentage and germination rate of chicory seeds at each maturation duration on the stalk from 16 to 44 days after flowering. Means with confidence intervals of the means.

The two following facts are in favour of the use of CF features for the differentiation of immature chicory seeds from mature ones, and as indicators of seed vigour:

- the CF decreased during the filling phase;
- the end of the filling phase corresponded to the physiological maturity, where themaximal germination percentage and vigour is attained (Black et al., 2008).

On the other hand, the efficiency of CF features may be negatively affected by the large variability of individual measurements (random differences between individuals).

Current work aims at estimating the correlations between the CF features, the weight parameters and the germination variables in outdoors and greenhouse environments, and to assess the added value of CF features in comparison to weight, size and density features to distinguish between viable and non-viable seeds using sorting simulations.

6. Conclusion

To predict wheat yield or to determine wheat growth stages, remote sensing is not the only solution. Proxy-detection systems allow to acquire high resolution images to be treated by robust algorithms such as high pass Fourier filtering. For example, to predict yield, the number of wheat ears has to be determined and is the result of a detection and a counting step. Detection has been done by developing specific image acquisition system and by implementing algorithms like colour-texture hybrid space or more classical image processing based on Fourier filtering, according to the frequential information (redundant information) included in the images. In this chapter, we presented the high pass Fourier filtering technique which gives satisfying and robust wheat ear detection with lower computing time. Moreover, we have compared the detection results with those obtained by representation in a hybrid space. Even if satisfying results are obtained for this qualitative experiment, some improvements should be done such as including an automatic threshold determination after the inverse FFT and an efficient shape analysis in order to obtain a finer wheat ear detection and better artefacts elimination.

This work has also revealed the possibility of extending the use of pattern recognition techniques and textural feature analysis to other applications, including the automatic determination of the wheat growth stage, aiming at creating a decision tool for farmers. At this stage, it seems that we should include an agronomic validation in order to propose more specific model to help farmers. In the same context one perspective could be to couple proxy-detection with satellite or aerial images. Moreover using multispectral data would probably improve even more the efficiency of such an approach.

The study of hyperspectral imaging demonstrates the potential to separate weed-culture. A more comprehensive study must now be conducted to assess the robustness and the spectral and spatial minimum necessary resolutions in the context of an operational implementation.

Most fruits and seeds have measurable levels of chlorophyll, respectively in their pericarp or testa, and this chlorophyll degrades with time. In the case of chicory, the commercial seed is a fruit which cannot be hulled and the seed is not observable directly. The low amount of 10 chlorophyll implies the use of a highly sensitive device and because of the presence of a distinct pappus, the imaging of fluorescence is favourable. It would be profitable to identify the factors (variety, season, climate, hydric stress, etc.) influencing the evolution of CF to predict the characteristics of the decrease of CF and to predict if its decrease is always concomitant with the increase of the germination performance.

Finally, combination of several image acquisition systems should give more interesting results and open the door of the detection of other important crop characteristics.

7. References

Aparicio, N., Villegas, D., Casadesus, J., Araus, J.L., & Royo, C. (2000). Spectral vegetation indices as non-destructive tools for determining durum wheat yield. *Agronomy Journal*, Vol. 92, pp. 83–91.

Barry, C.S. (2009). The stay-green revolution: Recent progress in deciphering the mechanisms of chlorophyll next term previous term degradation next term in higher plants. *Plant Science*, 176(3), 325-333.

Bewley J., & Black M. (1994). Seeds: physiology of development and germination. *Plenum*, ISBN 0-306-44748-7.

Black M., Derek Bewley J., & Halmer P. (2008). The Encyclopedia of Seeds. *Science, Technology and Uses*. Cabi, ISBN-13 978-0-85199-723-0.

Burks, F.T., Shearer, S.A., & Payne, F.A. (2000).Classification of weed species using color texture features and discriminant analysis.*Transactions of the ASAE*, Vol. 43, N°2, pp. 441-448.

Chen, Y. R., Chao, K., & Kim, M. S. (2002). Machine vision technology for agricultural applications. *Computers and Electronics in Agriculture*, Vol. 36, pp. 173-191.

Chi, T.Y., Chien, C.F., & Lin, T.T. (2003). Leaf shape modeling and analysis using geometric descriptors derived from Bezier curves. *Transactions of the ASAE*, Vol. 46, N°1, pp. 175-185.

Cointault, F., Journaux, L., &Gouton, P. (2008a). Statistical methods for texture analysis applied to agronomical images.*Proceedings of theIS&T/SPIE 20th Annual Symposium on Electronic Imaging*, 26-31 January, San Jose, CA, USA.

Cointault,F., Guérin, D., Guillemin, J.P., &Chopinet, B. (2008b). In-Field Wheat ears Counting Using Color-Texture Image Analysis. *New Zealand Journal of Crop and Horticultural Science*, Vol. 36, pp. 117–130.

Cooley, J.W., &Tukey, J.W. (1965).An algorithm for the machine calculation of complex Fourier series.*Math.Comput*, Vol. 19, pp. 297-301.

Feyaerts, F., & Van Gool, L. (2001).Multi-spectral vision system for weed detection. *Pattern Recognition Letters*, Vol. 22, N°6-7, pp. 667-674.

Fournier, C., Andrieu, B., Ljutovac, S., & Saint-Jean, S. (2003). ADEL-Wheat: A 3D architectural model of wheat development. *Plant growth modeling, simulation, visualization and their applications*, J. M. Hu B.G., Springer Verlag, pp. 54-63.

Germain, C., Rousseaud, R., & Grenier, G. (1995). Non Destructive Counting of Wheatear with Picture Analysis.*Proceeding of 5th IEE International conference on Image Processing and its Applications*, Edimburg, UK, 1995.

Halmer, P. (2008). Germination rate. In: Black M., Derek Bewley J., Halmer P. (2008). The Encyclopedia of Seeds. *Science, Technology and Uses*. Cabi, ISBN-13 978-0-85199-723-0.

Haralick, R.M., Shanmugam, K., &Dinstein, I. (1973).Textural Features for Image Classification.*IEEE Trans. On Systems, Man and Cybernetics*, Vol. 3, N°6, pp. 610-621.

Hughes, G.F. (1968).On the mean accuracy of statistical pattern recognizers.*IEEE Transactions on Information Theory*,Vol.14, pp. 55-63.

International Seed Testing Association (2005). *International rules for seed testing*. Bassersdorf, Suisse: ISTA.

Jalink H., van der Schoor R., Frandas A., van Pijken J.G., & Bino R.J. (1998). Chlorophyll fluorescence of *Bassica oleracea* seeds as a non-destructive marker for seed maturity and seed performance. *Seed Science Research*, Vol. 89, pp. 437-443.

Jalink, H., Van der Schoor, R., Birnbaum, Y. E., & Bino, R. J. (1999). Seed chlorophyll content as an indicator for seed maturity and seed quality. *Acta Horticulturae*, Vol. 504, pp. 219-227.

Journaux, L., Destain, M.F., Miteran, J., Piron, A., Cointault, F. (2008).Texture Classification with Generalized Fourier Descriptors in Dimensionality

Reduction Context : an Overview Exploration.*Third International Workshop on Artificial Neural Networks in Pattern Recognition,* Pierre & Marie Curie University, Paris, France, IAPR, 2008.

Konstantinova, P., Van der Schoor, R., Van den Bulk, R., & Jalink, H. (2002). Chlorophyll fluorescence sorting as a method for improvement of barley (*Hordeum vulgare* L.) seed health and germination. *Seed Science and Technology,* Vol. 30, pp. 411-421.

MacQueen, J.B. (1967). Some Methods for classification and Analysis of Multivariate Observations.*Proceedings of 5-th Berkeley Symposium on Mathematical Statistics and Probability,* Berkeley, University of California Press, Vol. 1, pp. 281-297.

Manh, A.-G., Rabatel, G., Assémat, L. & Aldon, M.J. (2001). Weed Leaf Image Segmentation by Deformable Templates. *Journal of Agricultural Engineering Research,* Vol. 80, N°2, pp. 139-146.

Martinez de Guerenu C., Germain Ch., Lavialle O., & Grenier G.(1996) Designing an automatic counting system for wheatears, *proc. of AgEng 96, Int. Conf. on Agricultural Engineering, Madrid, Spain, 1996.*

Nedbal, L., & Whitmarsh, J. (2004). Chlorophyll fluorescence imaging of leaves and fruits. In G.C. Papageorgiou, & Govindjee. (Eds.), Chlorophyll a fluorescence: A signature of photosynthesis, pp. 389-407. Dordrecht: Springer.

Ooms, D., & Destain, M. F. (2011). Evaluation of chicory seeds maturity by chlorophyll fluorescence imaging. *Biosystems Engineering,* Vol. 110, pp. 168-177.

Richardson, J., Moskal, L.M., & Kim, S. (2009). Modeling Approaches to Estimate Effective Leaf Area Index from Aerial Discrete-Return LIDAR. *Agricultural and Forest Meteorology,* Vol. 149, pp. 1152-1160.

Robert, C., Fournier, C., Andrieu, B., & Ney, B. (2008).Coupling a 3D virtual wheat plant model with a Septoriatritici epidemic model: a new approach to investigate plant-pathogen interactions linked to canopy architecture.*Functional Plant Biology,* Vol. 35, N°9-10, pp. 997-1013.

Saeys,W., Lenaerts, B., Craessaerts, G., & De Baerdemaeker, J. (2009).Estimation of the crop density of small grains using LiDAR sensors.*Research Paper – Biosystems Engineering,* Vol. 102, pp. 22-30.

Serra, J. (1982). *Image Analysis and Mathematical Morphology.*ISBN 0126372403.

Slaughter, C.D., Giles, D.K., & Downey, D. (2008). Autonomous robotic weed control systems : A review. *Computers and electronics in agriculture,* Vol. 61, N°1, pp. 63-78.

Suhartanto, M. (2002). Chlorophyll in tomato seeds: marker for seed performance? *PhD thesis,* NL:Wageningen Universiteit.

Vigneau, N., Ecarnot, M., Rabatel, G., &Roumet.P. (2011).Potential of field hyperspectral imaging as a non destructive method to assess leaf nitrogen content in Wheat.*Field Crops Research,* Vol. 122, N°1, pp. 25-31.

Vrindts, E., De Baerdemaeker, J., & Ramon, H. (2002). Weed Detection Using Canopy Reflection. *Precision Agriculture,* Vol. 3, N° 1, pp. 63-80.

Zygielbaum, A.I., Gitelson, A.A., Arkebauer, T.J., &Rundquist, D.C. (2009).Non-destructive detection of water stress and estimation of relative water content in maize.*Geophysical Research Letters*, Vol. 36, L12403.

4

Concepts in Crop Rotations

H. Arnold Bruns
*USDA-Agricultural Research Service,
Crop Production Systems Research Unit, Stoneville, MS,
USA*

1. Introduction

1.1 Crop rotations – A historical perspective

Crop rotation is the production of different economically important plant species in recurrent succession on a particular field or group of fields. It is an agricultural practice that has been followed at least since the Middle Ages. During the rule of Charlemagne crop rotation was vital to much of Europe which at that time followed a two-field rotation of seeding one field one year with a crop and leaving another fallow. The following year the fields were reversed (Butt, 2002). Sometime during the Carolingian period the three-field rotation system was introduced. It consisted of planting one field, usually with a winter cereal, a second with a summer annual legume, and leaving a third field fallow. The following year a switch would occur. Sometime during the 17th and /or 18th centuries it was discovered that planting a legume in the field coming out of fallow of the three-field rotation would increase fodder for livestock and improve land quality, which was later found to be due to increased levels of available soil nitrogen (N). During the 16th century Charles Townshend 2nd Viscount Townshend (aka Turnip Townshend) introduced the four-field concept of crop rotation to the Waasland region of England (Ashton, 1948). This system, which consisted of a root crop (turnips (*Brassica rapa* var. rapa)), wheat (*Triticum aestivum* L.), barley (*Hordeum vulgare* L.), and clover (*Trifolium* spp.) followed by fallow. Every third year introduced a fodder crop and grazing crop into the system, allowing livestock production the year-round and thus increased overall agriculture production. Our present day systems of crop rotation have their beginnings traceable to the Norfolk four-year system, developed in Norfolk County England around 1730 (Martin, et al., 1976). This system was similar to that developed by Townshend except barley followed turnips, clover was seeded for the third year and finally wheat on the fourth year. The field would then be seeded to turnips again with no fallow year being part of the rotation.

In the new world, prior to the arrival of European settlers, the indigenous people in what is now the Northeastern United States, practiced slash-and-burn agriculture combined with fishing, hunting, and gathering (Lyng, 2011). Fields were moved often as the soil would become depleted and despite the tale of Native Americans teaching the European settlers to put a fish into the corn hills at planting, there is little or no evidence of the aboriginal people fertilizing their crops. Maize would be planted in hills using crude wooden hoes with gourds and beans (*Phaseolus* spp. L.) being planting alongside and allowed to climb the maize stalks. When an area would become depleted of plant nutrients, it would be abandoned and over time, would recover its natural fertility. Lyng (2011), describes the Native Americans of the northeast as not so much conscience ecologist but rather people with a strong sense of dependences on nature minus the pressure to provide for consumer demands. Plains Indians on the other hand are classified as being of two cultures. There were the nomadic nations that followed the herds of bison that roamed the region and lived mainly on a diet of bison meat and what they might gather in the way of wild berries, fruits, and nuts with very little farming except for some maize and tobacco (*Nicotiana tabacum* L.). There were then the nations that lived on a combination of meat and crops they would raise. These peoples tended to live in established villages and would fish, hunt, and gather wild fruit and berries. The crop farming they practiced again, were maize, beans, and squash (*Cucubina* spp. L.), sometimes referred to as "The Three Sisters" in Native American society (Vivian, 2001). As with the nations in what would become the northeastern United States, the Plains Indians that practiced crop farming would usually clear their garden areas by slash and burn, grow their crops, and then allow a two-year fallow before planting again. Just prior to planting, some villages would carry in brush and other plant debris to burn along with the refuge that grew in the field during fallow to "enrich" the soil for the crops about to be planted.

The early European settlers attempted to raise those crops (wheat, and rye (*Secale cereal* L.)) which they were accustom to, using cultivation methods they had used in the old country. They also, introduced livestock, (cattle, swine, and sheep) which were not found in the New World but that had been a major source of food for them in their native homeland. They soon discovered that clearing fields for planting and pasturing was an arduous task and in order to survive adopted some of the crop production techniques practiced by the indigenous peoples and allowed their livestock to forage open-range (Lyng, 2011). As colonization expanded and available labor increased along with the demand for food, the permanent clearing of arable land increased along with the introduction of more Old World crops and, unfortunately, their pests that continue to demand time and financial resources to contain today.

The first export from the American colonies to England was tobacco. Though not a food crop, tobacco played a pivotal role in helping sustain the Jamestown colony and gave the settlers something to exchange for necessary items to survive. Tobacco is a high cash value, very labor intensive crop. Even as of 2002, with only about 57,000 total farms in the United States being classed as tobacco farms producing an average of 3 hectares of the crop per farm, the average cash value of those 3 hectares was nearly $42,000 (Capehart, 2004). Though tobacco preserved the Virginia colony, within seven years of its cultivation and export, its continued production in the New World would usher in the African slave trade, the darkest part of America's past, and would culminate 200 years later into the American Civil War.

Prior to colonization, a species of cotton, *Gossypium barbadense,* was being grown by the indigenous people of the New World (West, 2004). Columbus received gifts from the Arawaks of balls of cotton thread upon making landfall in 1492. Egyptian cotton (*G. hirsutum* L.) was introduced to the colonies as early as 1607 by the Virginia Company in an attempt to encourage its production and help satisfy the European appetite for the fiber that was currently being exported from India . However, tobacco production and the lucrative prices being paid for it along with the belief that cotton depleted the soil and required too much hand labor, dissuaded the colonist from planting the crop. Even encouragement from the colonial Governors, William Berkley and Edmund Andors could not convince the settlers to switch to cotton. Small hectarages of *G. hirsutum* L. though were grown along the Mid Atlantic colonies for individual household use. The Revolutionary War halted imports of large quantities of cotton to the former colonies from Britain and forced the Americans to grow their own supply. By the mid 1780's production had expanded and the newly formed United States became a net exporter of cotton to Britain.

After the development of the cotton gin by Eli Whitney in 1793 the key to financial success in the southern states was acquiring large hectares of land for cotton production and large numbers of slaves to tend to the crop. Maize, small grains, forages, and food crops were grown only in sufficient quantities to sustain the plantations that had developed. These crops were not grown for the purpose of commerce and were often relegated to some of the marginal lands on the plantation or near the homestead for convenient harvest. The bulk of all cleared fields were devoted to production of tobacco or "King Cotton" as it would become known. From 1800 to 1830 cotton went from making up 7% ($5 million) of exports from the United States to 41% ($30 million) (West, 2004). Tobacco production went from 45.4 million kg at the outbreak of the Revolutionary War to 175.8 million kg prior to the Civil War (Jacobstein, 1907). Crop rotation was not even considered an option with respect to these crops due to the cash value paid for them. By 1835 the top soil of eastern Georgia had eroded away with the remaining clay unsuitable for cotton production. As soils became depleted of nutrients necessary for the crops' production, more wilderness, particularly further west would be cleared and farmed. This resulted in conflicts with the native peoples that resulted in their forced resettlement onto reservations and the spread of slavery westward into newly chartered states in the south. This further deepened political and economic conflicts that would explode into the American Civil War.

1.2 Advent of agricultural education and research

The Morrill Act of 1862 and again 1892 established the American Land-Grant colleges in each state and charged them with the responsibility of teaching the agricultural and mechanical disciplines, along with other responsibilities necessary to an advanced education. The Hatch Act of 1887 then established the Agriculture Experiment Station system which, in most states, is administered by the Land-Grant Universities and was to provide further enhancement of agricultural teaching through experimentation. In 1914 the Smith-Lever Act established the State Cooperative Extension Service which disseminates information to the public of advances in agriculture production discovered by the state agricultural experiment stations. All three of these legislative acts came about because of a need to better understand sound farm management practices, including crop rotations, to improve the nation's farm economy.

The concept of agriculture research stations was not an American idea. The Rothamsted Experiment Station in the United Kingdom is said to be the world's oldest, being established in 1843, while Möcken station in Germany, established in 1850, is said to be the world's oldest state supported agricultural research station. Agricultural research stations can now be found in most all developed countries and even many less developed nations. Research on crop rotations has been and continues to be conducted at virtually all of these stations, with specialization towards the environment and crop species indigenous to their location. Some of these studies have been in existence since the late 19th century (Rothamsted, 2011).

Some of the more famous experiments in the United States that continue to be performed at some of the Land-Grant Universities, and are now designated on the National Register of Historic Places, include The Old Rotation experiment on the Auburn University campus in Alabama, The Morrow Plots on the campus of the University of Illinois, and Sanborn Field at the University of Missouri. Mitchell et al., (2008) published that the Old Rotation experiment in Alabama has shown over the long-term, seeding winter legumes were as effective as fertilizer N in producing high cotton lint yields and increasing soil organic C levels. Rotation schemes with corn or with corn-winter wheat- and soybean (*Glycine max* L. Merr.) produced no yield advantage beyond that associated with soil organic C (Table 1). However, winter legumes and crop rotations contributed to increased soil organic matter and did result in higher lint yields.

	Cotton Lint Yield ($kg\ ha^{-1}$)		
Continous Cotton	1986-1995†	1996-2002†	Soil OM %‡
0 N/no winter legumes	392d	403b	0.8e
winter legumes	952ab	1131a	1.8c
134 kg N ha^{-1}	792c	1154a	1.6d
Cotton-Corn Rotation			
winter legumes	870ab	1120a	1.8c
legumes + 134 kg N ha^{-1}	970a	1276a	2.1b
3-Year Rotation (common-			
winter legumes			
corn-small grain-soybean	850ab	1109a	2.3a

†Values followed by the same letter are not significantly different at P<0.05
‡Recent data show the effect of increasing soil organic matter on cotton productivity.

Table 1. Long-term effects of crop rotations, winter legumes and nitrogen fertilizer on cotton lint yields at the "Old Rotation Experiment" of Auburn University in Alabama. (Mitchell, 2004).

Data from the Morrow Plots in Illinois have shown that yields from continuous corn have always been much less than corn yields from a of corn-oats (*Avena sativa* L.) rotation or a or corn-oats-and hay (clover (*Trifolium* spp.) or alfalfa (*Medicago sativa* L.)) rotation (Aref and Wander, 1998). After the introduction of hybrid corn varieties in 1937, the first plots to show an increase in corn yields due to these varieties were the corn-oats-hay rotation. Yield increases due to hybrids were not noticed in the corn-oat plots until the late 1940's and in the continuous corn plots until the early 1950's. These lower corn yields of the continuous corn and the slower response to corn hybridization in the corn-oat rotation appear to coincide with long-term average levels of soil organic matter and nitrogen observed in the various plots (Table 2).

Rotation	C (g kg-1)	N (g kg-1)	C-N ratio
Continuous corn	19.2a	1.55a	12.36a
Corn-oats	23.0b	1.84b	12.46a
Corn-oats-hay	26.5c	2.12c	12.48a

Table 2. Soil carbon C, nitrogen N, and C-N ration from a crop rotation experiment on the Morrow Plots of the University of Illinois.

Means of samples taken in 1904, 1911, 1913, 1923, 1933, 1943, 1953, 1961, 1973, 1974, 1980, 1986, and 1992. (Aref and Wander, 1998). Values within a column followed different letters are significantly different $P \leq 0.05$.

Corn and wheat yields at Sanborn Field at the University of Missouri have been consistently higher when grown in rotation with each other along with red clover (*Trifolium pratense* L.) inter-seeded into the wheat in late winter for forage the following year (Miles, 1999). Plots of both corn and wheat have been grown continuously since the site's establishment in 1888, some receiving animal manure, some commercial fertilizer, and some no fertility treatment. All have had reduced grain yields compared to those grown in rotation, even with the added manure and/or fertilizer.

Thirty years after Sanborn Field's establishment, its focus began to shift to the study of cropping systems as related to soil erosion and the resulting loss of productivity. An experiment conducted in 1917 by F.L. Duley and M.F. Miller on the campus of the University of Missouri used seven test plots to measure soil erosion resulting from rainfall (Duley and Miller, 1923). This research led to creation of the Soil Conservation Service of the USDA, which in now a component of NRCS-USDA. It led to the establishment of experiment stations throughout the United States dedicated to the study of crop rotations on soil erosion and developing cropping systems to minimize erosion's impact (Weaver and Noll, 1935). Experiments at these stations in Iowa, Missouri, Ohio, Oklahoma, and Texas all showed plots planted to a continuous cropping system had higher surface soil losses and losses of rainfall than plots planted to a forage or in a three or four year rotation (Uhland, 1948).

2. Crop rotations vs. continuous cropping

Crop rotation schemes are, by and large, regional in nature and a specific rotation in one environment may not be applicable in another. Continuous cropping schemes or monocultures for the most part, have fallen out of favor in many farming regions. Roth (1996) published mean corn yields from a 20 year crop rotation experiment in Pennsylvania that included rotation with both soybean and alfalfa showing higher yields with all rotation schemes than continuous corn (Table 3). The extensive use of commercial fertilizers and pesticides has helped mask most of the beneficial effects of crop rotation. But Karlen et al. (1994) has stated" no amount of chemical fertilizer or pesticide can be fully compensated for crop rotation effects". However, economics continues to be the large determining factor into how a field is managed.

Crop Rotation	Yield Mg ha-1
Continuous corn	8.7
Corn/soybean	9.1
Corn/two-year alfalfa	9.6
Corn/corn/three-year alfalfa	9.6†
Corn/corn/three-year alfalfa	9.3‡

†First year corn yield
‡Second year corn yield

Table 3. Mean corn grain yields as influence by crop rotation from 1969 to1989 at Rock Springs, PA. (Roth, 1996).

One primary benefit to crop rotation is the breaking of crop pest cycles. Roth (1996) states that in Pennsylvania, crop rotations help control several of the crop-disease problems common to the area such as gray leaf spot in corn (*Cercospora zeae-maydis)* take-all in wheat (*Gaeumannomyces graminis* var. tritici), and sclerotina in soybean (*Sclerotinia sclerotiorum)*. In corn, corn rootworms (*Diabrotica virgifera* spp.) can be a devastating pest and crop rotation was considered to be the most effect method of control. However, beginning in the late 1980's there was a variant of the Western corn root worm (D. *virgifera virgifera* LeConte) that began egg laying in soybean fields, making larvae present to feed upon first year corn in a soybean-corn rotation (Hammond et al., 2009). Prior to this time the standard method to avoiding rootworm damage was to rotate. However, during the mid-1960's in the Cornbelt there was a movement to engage in growing corn continuously on highly productive soils. Atrazine [2-chloro-4-(ethylamino)-6-(isopropylamino)-s-triazine] was being readily adopted for weed control in corn and a number of insecticides were becoming available for of control corn rootworm and other corn insects. Also sources of nitrogen fertilizer were readily available and relatively inexpensive. Competitive profits for other crops, particularly soybean, and continued research showing tangible benefits to rotations though returned most fields to some sort of rotation scheme. However, there are some producers today who are profitable at growing continuous corn. But, such a system appears to require strict adherence to sound management practices.

Cotton is probably the principle crop that has been grown continuously on many fields, some for over 100 years. The crop was profitable and well suited for production in areas prone to hot summer temperatures and limited rainfall. There was also an infrastructure available in these production regions for processing the lint and seed as well as a social bond that connected the crop to the people who grew it. Corn, hay, and small grains were the "step children" of agronomic crops for generations of southern planters. Corn and winter oats were grown in the Cottonbelt solely as feed grains for the draft animals used to grow cotton and the meat and dairy animals grown for home consumption. There were basically no markets available or facilities to handle some of these crops for commercial trade. Despite being introduced in the 1930's, it wasn't until the early 1950's that soybean became an important crop in the lower Mississippi River Valley (Bowman, 1986). Rice (*Oryza sativa* L.) was introduced to the Mississippi River Delta in 1948 and together these crops provided alternative sources of agronomic income to cotton but did little to encourage crop rotation. Both rice and soybean were relegated to the heavier clay soils of the Mississippi Delta with the sandy loams, silts, and silty clays remaining in cotton. It wasn't until changes in government support programs in the mid-1990's that planters in the Mid South became interested in alternatives to continuous cotton and began to produce corn for commercial sale and rotate it with cotton. Corn hectareage in the states of Arkansas, Louisiana, and Mississippi increased from 161,000 ha in 1990, to 382,000 ha in 2000, to 630,000 ha in 2010 (USDA-NASS, 2011).

Until 2007 research information about corn-cotton rotations were limited. An extensive study on various corn-cotton rotation schemes yielded data on the effects of rotation on yields and reniform nematode (*Rotylenchulus reniformis*) a serious pest to cotton. Bruns, et al. (2007), reported corn grain yields were greater following cotton than in plots of continuous corn. Pettigrew, et al. (2007), noted that cotton plant height increased 10% in plots following one year of corn and 13% following two years of corn when compared to continuous cotton (Table 4). Lint yields increased 13% following two years of corn primarily due to a 13% in bolls per m^2. No other increases were noted however. Stetina, et al., (2007) found that following two years of corn production, reniform nematode populations remained below damaging levels to the cotton plants. However, cotton following just one year of corn would have reniform nematode populations rebound to damaging levels towards the end of the growing season.

Crop	Rotation sequence†	Yield(kg ha^{-1})			
		2000	2001	2002	2003
Cotton	continuous cotton	1101a	1036a	1257	1266b
	corn-cotton-corn-cotton	xxx	1068a	xxx	1353ab
	cotton-corn-corn-cotton	1117a	xxx	xxx	1460a
Corn	continuous corn	10,364a	10,107b	7587a	9032
	corn-cotton-corn-cotton	10,297a	xxx	8157a	xxx
	cotton-corn-corn-cotton	xxx	10,675a	7730a	xxx

†Lint yield for cotton; grain yield at 155 g kg^{-1} seed moisture; all values are means of eight reps averaged across four genotypes.
‡Within each crop and year, means followed by the same letter are not significantly different by lsd ($P \leq 0.05$)

Table 4. Effect of crop rotation sequence on crop yield of corn and cotton from 2000 to 2003 in Stoneville, MS. (Stetina et al., 2007).

3. Rice production and crop rotation

Rice ranks third behind corn and wheat in total tons of grain produced in the world but it is the primary dietary staple for more people than any other cereal (Raun and Johnson, 1999). It is grown on every continent except Antarctica. By the 1990's rice was providing 35% to 59% of the total calories consumed by nearly 2.7 billion people in Asia (Neue, 1993). Peng et al. (1999) quoted that world rice production would need to be at least 600 million tons by 2025, an increase of 266 million tons above 1995 production just to maintain current nutrition levels. This increase will likely not be sufficient to alleviate current malnutrition in many of the rice dependent cultures (Neue, 1993). In areas where it is virtually the sole source of calories it is seldom grown in rotation with other crops. Anders, et al., (2004) stated that producers growing continuous rice will likely experience lower grain yields than those using a rice-soybean rotation.

A common rotation with rice in southern and eastern Asia is a rice-wheat rotation system that occupies an estimated 24 to 27 million hectares (Wassmann, et al., 2004). Lattimore (1994) reviewed the literature pertaining to rice-pasture rotations in southeastern Australia. Annual pastures based on subterranean clover (*Trifolium subterraneum* L.) are well adapted to this part of the world and the rice cropping system. It provides considerable fixed N to the rice crop thus reducing the need level of supplemental N fertilizer as well as breaking weed cycles. It helps sustain a complimentary animal agriculture to use crop residues and provides opportunities for improved farm income. With respect to disease control in rice, both false smut (*Ustilaginoidea virens* (Cooke) Takah) and kernel smut (*Neovossia horrid*a (Takah.) Padwick & A. Khan, syn. *Tilletia barclayana* (Bref.) Sacc. & P. Syd.) two serious fungal pests in rice production areas of the United States, appear to be best controlled when rice is grown in three year rotations with soybean and corn between rice crops (Brooks, 2011). Traditional rotations of rice-soybean, with winter wheat grown between the two summer annuals, were observed to have the highest levels of these diseases especially with high N- fertility levels.

4. The corn-soybean rotation

One of the more widely practiced rotations in the United States involves the corn-soybean rotation scheme used extensively in the North Central and Mid-Atlantic States. Within the 20 year period between 1988 and 2008 nearly 30 publications were known to have been published that compared corn-soybean rotations to continuous corn (Erickson, 2008). Virtually all of this research showed increases in corn grain yields from plots that had been planted to soybean the previous year. A few of these experiments followed soybean with two years of corn and in those studies yields from the second year corn crop were equal to or still greater than those from plots of continuous corn but less than the first year corn crop. One of these studies (Porter et al., 1997) examined the effects of various corn-soybean rotation schemes at three locations in the northern Corn Belt. Data from these studies showed that not only were both corn and soybean yields higher in rotation compared to monocultures of the two crops (Table 5), but that differences between rotations and monocultures were greater in low-yielding environments than in high-yielding conditions.

	Yield (Mg ha^{-1})						
Crop†	1st-yr	2nd-yr	3rd-yr	4th-yr	5th-yr	Cont.	S-C rotation
Corn	9.00a	8.04b	7.90b	7.90b	7.88b	7.81b	8.83a
Soybean	3.26a	2.99b	2.84c	2.82cd	2.8cd	2.77d	3.05b

†Crops were grown under corn or soybean monoculture of the respective crop in the following sequence: 1st-yr, 2nd-yr, 3rd-yr, 4th-yr and 5th-yr corn or soybean after 5 yr of corn or soybean; Cont. (continuous corn or soybean); and S-C (alternating soybean and corn). Values followed by the same letter or letters within each crop species are not significantly different at ($P \leq 0.05$).

Table 5. Corn and soybean yields from 3 locations in Minnesota and Wisconsin representing 29 environments. (Porter, et al., 1997)

Studying the net returns of various crop rotation schemes involving corn, Singer and Cox (1998) calculated a greater net return (\$250 US ha^{-1}) with a corn-soybean rotation than a continuous corn (\$193 US ha^{-1}) or a three year soybean-wheat/red clover rotation (\$133 US ha^{-1}). A recent study reported though, that yield comparisons are not the appropriate basis for decision making on cropping systems but rather economics is most important (Stanger et al., 2008). This report showed that, with the exception of continuous corn grown with 224 kg N ha^{-1}, a corn-soybean rotation was the most stochastically efficient cropping system across a range of N fertility treatments and other rotation schemes.

4.1 Other soybean rotations

Though soybean is one of humankind's oldest crops, it did not really become of significance in the United States until the late 1940's. The species was introduced in Europe from China in the mid 18th century and into the new world in the early 19th century where it was used primarily as a hay crop. The combination of the destruction in China from World War II and the Cultural Revolution removed it as the world's primary supplier of soybeans and opened an opportunity for the United States to develop the crop as a major oil seed (North Carolina Soybean Producers Assn. , 2011). Currently the United States produces about 40% of the world's soybeans followed by Brazil and Argentina combining to produce 50%. Besides corn, soybean is being rotated with rice or cotton in the Mid South and Southeastern States (Anders et al., 2004: Stallcup, 2009). In the eastern Great Plains soybean is often rotated with wheat or grain sorghum (*Sorghum bicolor* L. Moench) as well as corn (Kelley et al., 2003). Kelley et al. (2003), found that in general soybean yields grown in rotations with wheat or grain sorghum produced a 16% greater seed yield than when grown in a monoculture (Table 6). One of these rotations was soybean double-cropped behind winter wheat which is frequently practiced in areas of the United States south of 39° N latitude. This practice does risk failure from drought either causing poor emergence or poor seed set. Above 39° N there is also the risk of early frost terminating growth and above 40° N the practice of double-crop soybean after wheat is not advisable.

Rotation†			
W-S/S	W-Fal/S	GS/S	Cont. S
Yield Mg ha⁻¹‡			
1.91	2.09	1.99	1.68

†W-S/S=Wheat-double crop soybean/soybean; W-F/S= Wheat-fallow/soybean; GS/S= Grain sorghum/soybean; Cont. S= continuous soybean.
‡All means are significantly different by LSD (P≤0.05).

Table 6. 10 Year average 2nd year soybean yields in a two year rotation scheme at Columbus, KS from1980 to1998. (Kelley et al., 2003)

5. Rotations for forage crops

Prior to the extensive production of soybean for seed in the United States, and important rotation scheme in much of the New England, Mid-Atlantic and North Central states was corn-winter wheat-red clover. Frequently the red clover would be over seeded in late winter or early spring in the developing wheat crop. Many times timothy (*Phleum pretense* L.) a cool-season perennial grass would be seeded along with the red clover. The mixture would provide some pasture or hay the first year after the wheat was harvested but was most productive the following year. The two species combined will provide more forage production together than each species separately (Martin and Leonard, 1967). These fields would frequently then be plowed in fall or spring of the second year with the sward providing a green manure crop for the following season. Sometimes only red clover would be over seeded in the wheat solely for the purpose of being used as a green manure crop. Other grass species would occasionally be seeded with the red clover and sometimes the field would remain in red clover-grass for two or more years to provide hay and grazing. Though this rotation is comparatively old, it is still practiced, especially where there are numerous beef cow-calf, dairy, or horse enterprises.

Both beef and dairy cattle farming involves crop management challenges that many tend to overlook when thinking about crop rotations. Many people do not think of pasture swards as being a "crop". But to the cattleman it is a very important source of income and deserves as much attention to management as any other economic plant life. Dairy farms are very dependent on careful management of feed and forage resources in order to be sustainable. Not only is the quantity of feed and forage important to the dairy animal but quality as well, to insure maximum milk production during lactation. Roth et al., (1997), list a number of suggestions to aid dairy farmers in developing long term crop rotation plans that address production issues, such as feed quantities, forage quality, fertility, and pest control issues as it relates to corn silage production. Greater restrictions on pesticide use, are often placed on dairy operations due to concerns over traces of some chemicals carrying over into milk that is consumed by children. Continuous corn silage production carries the risk of corn rootworm damage, and as pointed out earlier, the western corn rootworm has adapted to egg laying in soybean fields and can damage corn following soybean. Roth (1996) points out that western corn rootworm larvae cannot tolerate a rotation to alfalfa. Therefore, seeding alfalfa after corn for silage will not only provide a good source of quality hay but also eliminate the need for a soil insecticide for rootworm control in the following corn crop.

Temporary meadows, which are seeded into a forage species for one to three years and then cultivated for a grain crop are sometimes included in discussions regarding crop rotations. However, a number of dairy farms and beef cow-calf operations occur in areas that include land unsuitable for tillage of any kind and are often used as permanent pastures that should be managed with the same intensity as any other cropland. Without proper attention these areas will often revert to a high proportion of weedy or woody species that are not useable by livestock and low yielding forage that reduces the pasture's carrying capacity or places greater demands on tillable cropland to provide necessary feed to maintain the animal enterprise. Johnson et al., (2007) states that periodic renovation or "renewing" of a pasture is the best way to improve forage yield and animal performance. Pasture renovation is in a way a form of crop rotation if you consider crop rotation as a means of maintaining land productivity. In temperate climates renovation usually begins in the fall by overgrazing the pasture to be renovated to remove excess vegetative material that might interfere with seeding and germination of the new pasture mix (Johnson et al., 2007; Lacefield and Smith, 2009). Seeding usually occurs in mid-winter while the soil is frozen in the first few centimeters. The thawing and refreezing of the soil surface is usually sufficient to allow good contact of the forage seed with the soil for germination. Sometimes a light cultivation with a disk or spike-toothed harrow is done to improve the chances of a good soil-seed contact.

Soil tests are usually acquired prior to seeding to determine nutrient availability. Applications of pulverized limestone are often required to supply Ca and adjust pH levels to facilitate establishment of a forage legume. Many soils in temperate climates tend to be acidic (pH 4.5 to 5.5) and require liming to elevate soil pH levels to 5.7 or higher to be more suitable for legume establishment. Virtually all recommendations for pasture renovation in temperate climates call for the establishment of one or two legumes in the sward (Wheaton and Roberts, 1993; Johnson et al., 2007; Lacefield and Smith, 2009; Teusch and Fike, 2009). Wheaton and Roberts (1993) listed 10 benefits of including a forage legume in a pasture mix. Among them were an increase in animal gain, decrease in herd health problems, an increased conception rate by cows, an increased protein yield per hectare, N being furnished by the legume to the grass, and reduced pasture production costs. Johnson et al., (2007) stated some of the same benefits along with a better seasonal distribution of forage because legumes are generally more productive in mid-summer than cool-season grasses which are the more common grass types grown in temperate climates. Lacefield and Smith (2009), reported that tall fescue (*Festuca arundinacea* Schreb.) growing in conjunction with red clover seeded at 6.7 kg ha^{-1} yielded more dry matter (12,400 kg ha^{-1}) than fescue alone fertilized with 202 kg N ha^{-1} (11,100 kg ha^{-1}). Teusch and Fike (2009) list several legume species to be considered for seeding in pasture renovation. The more common legume species recommended are red clover, ladino clover (*Trifolum repens* L.), annual lespedeza (*Lespedeza striata* Maxim.), alfalfa, and birdsfoot trefoil (*Lotus corniculatus* Cav.). As stated previously, in temperate climates cool season grass species are frequently selected to be seeded along with the legume when renovating pastures. For permanent pastures the species of choice are often tall fescue, orchard grass (*Dactylis glomerata* L.), or smooth brome (*Bromus inermis* L.). Pasture management is an on-going operation and additional seedings may be necessary to maintain a profitable sward. Birdsfoot trefoil has the ability to reseed itself even under grazing and alfalfa is a long-lived legume with individual plants able to survive up to five years. Red clover, though a perennial, will usually last only about three years and will need to be reseeded. Some forage specialists have found that over-seeding the pasture annually with 4.5 kg ha^{-1} of red clover seed in mid-winter will maintain the sward at about 30% legume which is recommended as the proper mix of grass and legume (H.N. Wheaton, 1975, personal communication). Even with good management, most pastures will need renovation about ever four to five years due to weed growth.

Pasture renovation in sub-tropical climates usually involves complete destruction of the old sward and reseeding or sprigging to reestablish the grass. The reasons or renovation in the sub-tropics are usually for weed control, or to reestablish swards lost to insects, over grazing, prolonged drought and in some cases multiple freezes during late winter that kill off the grasses. Renovation may also be done to replace an older grass cultivar with a newer more productive one (Woodruff et al., 2010). One basic difference between temperate pasture renovation and sub-tropical pastures is that in the sub-tropics the old sward is essentially destroyed and a new one established (Verdramini et al., 2010). Sub-tropical pastures are not as apt to include a forage legume due to most of those species being cool-season and may not survive well in hot humid summer months. Also, many of the grass species used are aggressive in nature and effective at crowding out less aggressive species. The more common species grown as pasture grasses in the subtropics are hybrid bermudagrass (*Cynodon dactylon* (L.) Pers.), bahiagrass (*Paspalum notatum*), atra paspalum (*Paspalum atratum* Swallen), digitgrass (*Digitaria decumbens*) (En), limpograss (*Hemarthria altissima*), and Rhodes grass (*Chloris gayana (Kunth)*) (Verdramini et al., 2010). All of these grasses except atra paspalum, bahiagrass and Rhodes grass require vegetative propagation in order to be established. As with temperate pastures, the pasture must first be properly prepared by controlling weeds, insects, removing old or dead vegetative material and fertilizing to soil test recommendations. Burning is often used in preparing bermudagrass pastures for renovation (Stichler and Bade, 2005). Tillage operations are often performed during dry periods in the spring to further control weeds and old sward growth as well as better prepare the land for vegetative propagation with the onset of summer rains (Stichler and Bade, 2005;Verdramini et al., 2010). In both the subtropical and temperate climates warm season annual forages may be included into crop rotations on land used primarily for cultivated crops. This is to provide additional pasture, hay or silage for livestock operations. Species such as pearl millet (*Pennisetum glaucum (L.) R.Br.),* sudangrass *(Sorghum bicolor* subsp. *Drummondii* (Steud.) de Wet ex Davidse), and other forage sorghum and sorghum x sudangrass crosses often fill this role (Hancock, 2009).

6. Summer fallow

Summer fallow, the practice of growing a crop every other year and in most systems, controlling weed growth during the off-year, has been used almost exclusive in semi-arid regions for the production of wheat or other small grains. Its purpose is to accumulate sufficient limited rainfall during the fallow year to grow a crop the following year and to break crop disease cycles. Most of the world's drylands are in developing countries where water resources are usually limited by a lack of rainfall and potential irrigation (Ryan et al., 2008). Stewart et al. (2006) has stated that dry regions worldwide supply about 60% of human food stuffs. Areas where this is a common practice are the Mediterranean, semiarid regions of Africa, Asia, Australia, the Pacific Northwestern United States, and the western regions of the Canadian and United States Great Plains. Many of these areas practice a "clean" summer fallow where the land is kept weed free during the fallow year, usually by cultivation or herbicides. However, in northern Africa a weedy fallow is often employed where palatable weeds and volunteer crop plants are allowed to grow and then be grazed by livestock (Ryan et al., 2008). This allows for an animal agriculture to exist that provides needed calories and returns manure to the land to provide nutrients for the following crop.

In semiarid areas of the Pacific Northwestern United States summer fallow wheat production has been practiced nearly 130 years (Schillinger et al., 2007). In recent years though, studies have been conducted in semiarid regions of the United States and Canada to adapt a continuous cropping system to replace summer fallow. Water storage efficiency in most summer fallow-wheat rotations is usually less than 25% of the rainfall received during a 14 month period using conventional tillage (McGee et al., 1997). Research from Colorado and Nebraska demonstrated that precipitation storage efficiency could be improved to 40% to 60% through minimum or no-till which allows crop residue to remain on the soil surface and soil disturbance is held to a minimum or eliminated (Croissant et al., 2008). However, such practices added little or no increase in wheat yields and it was concluded that the resultant water savings could only be converted to profit by employing intensive cropping systems where fallow time is decreased and summer crops such as maize, grain sorghum, or annual forages and included in the rotation. Under no-till it was determined that it cost more to save the additional water than the value of the added grain yield. Lyon et al. (2004)

reported that winter wheat yields in the central Great Plains were negatively affected by eliminating the 11 to 14 month summer fallow by spring planting a transitional crop before wheat in the fall. However, a spring planted forage crop that was harvested early had a minimum negative impact on wheat yields and that the value of the forage combined with the following wheat yields resulted in greater income than the traditional winter wheat-fallow rotation. Research from the Horse Heaven Hills region of Washington looked at a continuous no-till hard red spring wheat system verses a winter wheat-fallow rotation and found the hard red spring no-till system did not match the winter wheat-fallow system of production in yield or income. However, the continuous no-till system did offer a benefit of providing ground cover that reduced wind erosion and air pollution by dust particles (Young et al., 2000).

Summer fallow is also practiced on fields used to produce castor bean (*Ricinus communis* L.) the previous year. This is not done for accumulating moisture but to rid the field of any volunteer plants. Castor bean contains a very deadly toxin, ricin, which in very small quantities can kill humans and livestock. Summer fallow in this case allows volunteer plants to be destroyed and the field cleaned for future feed and food crops.

7. Organic farming and crop rotations

Organic farming has embraced crop rotations as its backbone to success. Crop rotation is practiced with what appears to be, much more intensity than most conventional farming systems with particular emphasis on sustainability. A crop rotation plan and accompanying records for a field and/or farm are required for certification as an organic farming operation (Johnson and Toensmeier, 2009). Organic farming and its use of crop rotations could be summed up in part, as the employment of proven crop management practices prior to the advent of pesticides and processed fertilizers. This is not to say that improvements on those rotation systems have not occurred. Most have been modified to accommodate mechanization and most other time saving ideas. But, yields may be lower. Maeder et al., (2002) reported on a 21 year study in Europe that crop yields on organic farms were generally 20% lower. However, there was an offsetting decrease of 34% in fertilizer expense, a 53% reduction in energy costs, and a 97% decrease in pesticides. Reganold et al., (1987) reported a comparison of the long-term effects (40 years) of conventional farming to organic systems found that organic farms had significantly higher levels of soil organic matter than conventional systems, greater top soil depth, higher polysaccharide content, and less soil erosion. Clark et al., (1998) reported that over an eight year period of applying organic crop rotation practices that soil organic matter had increased 2% over a comparable field that used conventional practices in a two-year rotation scheme.

A major challenge to organic crop production is control of weeds. Weed management has been identified by producers as the principle problem in organic farming (Walz, 1999). The advent and subsequent extensive use of herbicides after World War II altered crop production practices, especially crop rotations. Conventional crop farming today usually involves two- or three-crop crop rotations as have been previously mentioned with a heavy reliance on herbicides to at least reduce or eliminate weed problems. Regardless of the type of farming system used, conventional or organic, weeds are a constant annual drag on achieving maximum yields of high quality produce. Nave and Wax (1971) reported a reduction in soybean seed yields of between 25% to 30% compared to weed free plots due to the presence of one smooth pigweed (*Amaranthus hybridus* L.) per 30 cm of row and that yield losses from stubble, lodging and stalks were more than double in pigweed and giant foxtail (*Setaria fabric* Herrm.) infested plots compared to weed free plots. Weeds compete with a crop for water, light, and soil nutrients directly reduce yields. Some species have alleopathnogenic effects upon certain crops, reducing their growth and yield or in extreme cases causing their death. Weeds can harbor insect pests or serve as alternate hosts to plant diseases that further reduce yields and produce quality. They can add off-flavors to crop products or in some cases provide toxins to produce, rendering it unhealthy to consume. Weeds present during harvest can also damage the crop being harvested. Ellis et al., (1998) reported increases in damaged soybean seeds by 8.2% to 11.1% with the presences of a plant per m of row of any of five common weed species found in the Mid South. In organic farming, crop rotations are vital to controlling weed growth. Teasdale et al., (2004) evaluated the weed seed dynamics of three organic crop rotations and found that seedbanks of smooth pigweed and common lambsquarter (*Chenopodium album* L.) were usually lower following a hay sward in a four-year rotation or wheat in a three-year rotation than following soybean in a two-year rotation prior to being planted to corn. However, annual grassy weed seedbanks prior to corn planting were generally higher following the hay sward of the four-year rotation than the wheat of three-year rotation or the soybean of the two-year rotation. Porter et al., (2003) also found that weed control in organic corn and soybean were better when they were part of a four-year rotation of corn-soybean-oat-alfalfa compared to a two-year corn-soybean scheme.

Nitrogen is probably the most important of the macro-nutrients in crop production. Libraries at all major agricultural universities have a seemingly endless supply of research articles and texts pointing out the importance of adequate N-fertility in the growth and development of every crop important to humankind. In organic crop farming, and increasingly in conventional cropping systems, the use of crop rotations that supply ample supplies of N to non-leguminous cereals is an important management strategy. Prior to the escalating energy prices and the increasing cost of manufactured N-fertilizer sources, the use of legumes that fix atmospheric N by *Rhizobium*-legume symbiosis, especially clover species, as green manures or cover crops had diminished sharply. Even prior to World War I legumes would almost always be seeded for a green manure crop preceding corn (Heichel and Barnes, 1984). In organic farming, where the emphasis is to refrain from using manufactured N-fertilizer, green manure crops or cover crops are a vital source of N for much of their production. Lupwayi et al., (1998) reported that microbial diversity was greater under wheat preceded by red clover manure or field peas (*Pisum sativum* L.) than under continuous wheat. Long-term crop rotation research in Iowa has shown very little or no increase in corn yields from plots receiving N-fertilizer compared to those following one or two years of an alfalfa- bromegrass- red clover meadow (Voss and Shrader, 1984).

Species selection for organic crop rotations is probably more thought out than for conventional crop rotations. As explained earlier, conventional crop rotations are heavily influenced by market price of the various commodities available to be grown. Though commodity price is important to practitioners of organic farming it is not the primary driving force in making species selections. Baldwin (2006) states that organic farmers face the challenge of practicing crop rotations by defining systems that maintain farm profits while improving soil quality and preserving the environment. Frequently organic farmers will develop rotation schemes and select species based upon the crop's ability to extract nutrients and water from the soil. Some organic producers include vegetable crops into their operations due to these species tendency to extract nutrients and water from shallow depths then follow with a cereal that generally feeds to greater soil depths. Species selection is also made on the basis of weed, insect, and disease control. Delate and Hartzler (2003), states that rye has allopathic properties and is often used as a winter cover crop following corn to aid in weed control in preceding soybean or oat crops. Sustainability through the natural preservation of soil fertility is paramount to the organic farmer and makes the selection of species an important part of the operation. Organic crop production makes as much use of natural pest control, including crop rotations, as is possible.

8. Biofuels production and crop rotations

Interest in using lignocellulosic biomass to produce ethanol is gaining in popularity. Lignocellulosic biomass production mainly involves growing and harvesting plants generally not used for food or feed. Woody species such as willow (*Salix.* spp.) and poplars (*Populus* spp.) (Matthew et al., 2010) and grasses such as switchgrass (*Panicum virgatum*), big bluestem (*Andropogon gerardii),* reed canarygrass (*Phalaris arundinacea*) are several sources of lignocellulosic biomass that have shown to be useful in ethanol production (Hill, 2007). Rotation schemes for growing lignocellulocsic crops are, for the most part, still in development. Production of these materials for biofuels though is being done mostly on land not suitable for extensive corn and soybean production thus relieving pressure to grow more hectares of these crops to satisfy the conventional and biofuels markets. Worldwide it is estimated that about 1% of crop land or about 11-12 million hectares are being used to grow biofuels (de Fraiture et al., 2008). Raghu et al., (2006) points out that some of the traits favorable to producing a lignocellulosic crop such as being a C_4 photosynthetically, lacking pests, rapid early season growth, and long canopy duration can also tend towards them being invasive. This would not work well in a rotation scheme with most conventional crops. Currently lignocellulosic crops contribute little to current U.S. transportation biofuel suppliers but will likely provide the great share of ethanol in the near future.

In the United States, debate is underway concerning the use of corn as a primary source of fuel ethanol. Also, soybean oil is being blended with diesel to extend it. Diversion of these crops for biofuels is believed to increase food prices for consumers as the competing interest of food and fuel vie for the available supply. There is also concern that the increased demands for these grains will negatively impact crop rotations, particularly those that help conserve soil, water, and plant nutrients. One of the primary reasons corn and soybean are currently popular for biofuels over lignocellulosic crops is the comparatively short time to harvest maturity. Most of the lignocellulosic crops require three to five years to reach harvest maturity(Hill, 2007), compared to one year for corn or soybean. Though corn and soybean are regularly rotated with one another, history has shown that a substantial increase in the price received for any crop can encourage monocultures at the expense of proven rotations and their accompanying benefits. The expanded use of corn and soybean as biofuel could diminish the inclusion of small grains and/or forage crops in rotation schemes and the tillage of soils that are not well suited for cultivation.

Besides lignocellulosic crops, the harvesting of crop residues for ethanol production has been considered. Perlack et al., (2006) has reported that nearly 7.0×10^{10} kg of corn stover could be harvested in the U.S. for ethanol production. Worldwide other crops that produce sufficient quantities of residue that could be used to produce ethanol include rice, barley, oat, wheat, sorghum, and sugar cane (*Saccharum officinarum* L.). The use of crop residues for lignocellulosic ethanol production has however run into opposition due to the negative impacts such removals have on C sequestration, soil properties, and nutrient availability for subsequent crops. Wilhelm et al., (2007) reported that between 5.25 and 12.50 Mg ha^{-1} of corn stover are required to maintain soil C at productive levels for subsequent crops. Lal (2004) states that even though the energy acquired from the world's crop residue would be equivalent to 7.5 billion barrels of diesel, a 30% to 40% removal of crop residue would increase soil erosion and its subsequent pollution hazards, deplete soil organic C, and increase CO_2 and other greenhouse gas emissions from the soil. He suggests establishing biofuel plantations of adapted species on marginal lands rather than remove crop residues from land used to grow food and feed grains. Development of such plantations will require more aggressive research into developing crop rotation schemes specific for growing lignocellulosic crops for biofuel.

9. Conservation tillage

Conservation tillage continues to grow in importance in crop production since its inception over 40 years ago. C.M. Woodruff in 1970 was conducting research on strip-planting corn into tall fescue sod on Missouri hillsides with the idea of producing a cash grain/feed crop along with a forage crop for livestock production while maximizing conservation of the soil and rain water (Anonymous, 1970). Anders et al., (2004) found that phosphorus (P) concentrations in run-off water were higher for no-till rice than conventional tilled paddies, most likely due to the P being surface applied. However, total-P concentrations in run-off were lower in no-till because of the reduced loss of soil in no-till and its bound P. Conservation tillage which includes both minimum tillage and no-till practices, are used to grow an array of crops from corn, soybean, cotton, grain sorghum, and several small grain species almost always in some rotation scheme. It is popular not only for the conservation of natural resources as just mentioned, but also for the savings in fuel, time, wear and tear on farm equipment, and the environmentally sustainable attributes of the various practices. Minimum tillage, by reduced soil disturbance, promotes a complex decomposition subsystem that enhances soil system stability and efficiency of nutrient cycling. Basically minimum tillage more closely mimics natural ecosystems than conventional cropping systems (Francis and Clegg, 1990). Tillage has been reported to reduce the diversity of bacteria in the soil by reducing both the substrate richness and evenness (Lupwayi et al., 1998). They found that the influence of tillage on microbial diversity in fields planted to wheat was more prominent at the flag-leaf stage of growth than at seeding and more prominent in bulk soil than in the rhizosphere at the flag-leaf stage.

Kladivko et al., (1986), studied the production of corn and soybean using an array of tillage systems ranging from conventional moldboard plowing and seedbed preparation to no-till on a range of soils differing in organic matter, texture, and slope for seven-year and six-year periods. At one location on a Chalmers silty clay loam (fine-silty, mixed, superactive, mesic *Typic Endoaquolls*) comparisons of a corn-soybean rotation to continuous crops of these two species using various tillage systems was conducted for 10 years (Table 7). Yields of rotated crops tended to be greater than those of the monocultures regardless of tillage. Kladivko et al. (1986) also reported from this research at other locations that conservation tillage systems resulted in increased soil water contents, lower soil temperatures, increased soil organic matter, and more water-stable aggregates near the soil surface with higher bulk densities than conventional tillage. Corn yields were found to be equal to or better than conventional tillage practices when grown on the better drained soils using conservation tillage. Only on the poorly drained soils did corn yields on conservation tilled fields fail to exceed conventional tillage, most likely due to low temperatures and excess wetness in the spring as depicted with the Chalmers silty clay in Table 7.

Tillage	Previous crop	Corn	Soybean	Soybean	Corn
System	Crop	Corn	Corn	Soybean	Soybean
Fall plow		10.7	11.6	3.6	3.8
Fall chisel		10.3	11.4	3.3	3.6
Ridge till		10.4	11.6	3.4	3.6
No-till		9.1	11.2	3.2	3.3

Table 7. Mean corn and soybean yields (Mg ha^{-1}) in response to tillage system and crop rotation on a Chalmers silty clay loam in Indiana in 1980-1984 (data is from 6th to 10th year of the study). (Kladivko et al., 1986).

Roth (1996) presents several crop rotation schemes to use in no-till farming on Pennsylvania dairy farms. One of the more popular is an alfalfa-grass sward for hay followed by no-till corn. This involves killing the sod in the fall with herbicides to control weeds and reduce residue by early spring to facilitate corn planting. This rotation seems to work best where hay production is limited to three years. Alfalfa is also successfully no-tilled into fields that have just been harvested for corn silage or following the harvest of a spring seeded sorghum sudangrass. Lafond et al., (1992), evaluated no-till, minimum till (one pre-seeding tillage operation) and conventional till (fall and spring pre-seeding tillage operations) on a four-year crop rotation study. The rotations were fallow-spring wheat- spring wheat-winter wheat, spring wheat-spring wheat-flax (*Linum usitartissimum* L.)-winter wheat, and spring wheat-flax-winter wheat- field pea. Tillage systems did not affect the amount of water conserved during fallow. However, no-till and minimum till did result in an increase in soil water from the surface to 120 cm in depth over conventional till. All three crops in the study had greater yields in the no-till and minimum till treatments than in the conventional till. In an experiment using conservation tillage practices (strip-till or no-till) in combination with a corn-soybean rotation, both full-season soybean or double-crop soybean following wheat had the most consistent increase in seed yields (Edwards et al., 1987).

In recent years there has been considerable interest in various tillage practices and their influence on the sequestration of atmospheric CO_2 as a partial means of mitigating its current increase and subsequent impact on climate change. Sampson and Scholes (2000), state that the optimization of crop management to facilitate accumulation of soil organic matter could help sequester atmospheric CO_2 and lower the rate of its increase. West and Post (2002), found that, excluding a change to no-till in wheat-fallow rotations, a change from conventional tillage to no-till can sequester between 43 to 71 g C m^{-2} yr^{-1}. These values are within the upper range (10 to 60 g C m^2 yr^{-1}) of those reported in a review by Follet (2001). West and Post (2002) also stated that enhanced crop rotation complexity can sequester an average of 8 to 32 g C m^{-2} yr^{-1} which is similar to an average of 20 g C m^{-2} yr^{-1}

estimated by Lal et al. (1998; 1999) resulting from an improvement in rotation management.

Conservation tillage can present pest control problems that are different from those found in conventional systems, particularly weeds. Weed species composition and abundance often change in response to crop and soil management practices (Cardina et al., 2002). Buhler (1995), wrote that most conservation tillage practices rely heavily on increased herbicide use and that reduced herbicide efficacy has slowed the adoption of conservation tillage practices. Weed populations have tended to shift more towards perennials, summer annual grasses, biennials and winter annual species in conservation tillage systems. Moyer et al.,(1994) stated that successful conservation tillage systems usually involve crop rotations of three or more species and several different herbicides. Legere et al., 1997 concluded that conservation tillage has the potential to produce sustained yields of spring barley in Quebec, provided attention is given to critical aspects such as crop establishment and weed management. With respect to plant diseases, Peters et al., (2003) determined that soil agroecosystems can be modified by crop rotation and conservation tillage to increase disease suppression by enhanced antibiosis abilities of endophytic and root zone bacteria in spring barley and potato (*Solanum tuberosum* L.).

10. Conclusions

Crop rotation is very likely to continue to be an important management practice, especially in the developed part of the world. However, economics is always going to have a major say in the decision making process of what gets planted where and by how much. Our ever increasing world population is going to place greater pressure on getting more production from our shrinking areas of arable land and potable water needed for drinking, personal use, and growing crops. Climate change is, despite all of the predictive computer models, a great unknown, not from the standpoint of whether or not it is occurring but as to just what can be realistically done, if anything, to curb it.

Lands that include a fallow period and/or irrigation to produce crops may eventually be taken out of the food and fiber production equation because of both shifts in the climate and the loss of water for irrigation. Production areas of various crop species may shift due to climate change. Changes in crop genetics through biotech and genetic engineering may contribute some relief but improvements in water use efficiency are not going to eliminate the need for irrigation. Also, genetically engineered crops in many areas of the world are not being well received based on fears, real or imaginary, and increased costs of such seed stocks are often prohibitive to many producers. A nearly 500% increase in energy prices and a shift from food, feed, and fiber crops to renewable energy crops will have an increasing impact on the land available to grow all crops and the rotations to produce them. Given the history of crop rotations and the overall benefits that appear to be gained from them, it is virtually assured that they will continue to be an important practice in food production. Despite the difficulties associated with conducting crop rotation research, it will be beneficial to society as a whole to support efforts in this area and to adequately reward scientists willing to dedicate their professional efforts in such endeavors.

11. References

Anders, M., Olk, M.D., Harper, T., Daniel, T, & Holzauer, J. (2004). The effect of rotation, tillage, and fertility on rice grain yields and nutrient flows. *Proceedings 26th Southern Conservation Tillage Conference for Sustainable Agriculture.* 8-9 June. North Carolina State University, Raleigh, NC. pp 26-33.

Anonymous. (1970). Strip planting corn in fescue sod. *The Southeastern Missourian,* 10 Sept., 1970. Cape Girardeau, MO. p 14.

Aref,S. and Wander, M.M. (1998). Long-term trends of corn yield and soil organic matter in different crop sequences and soil fertility treatments on the Morrow Pltots. *Adv. Agron.* Vol.62:153-197.

Ashton, T.S. (1948). *The Industrial Revolution.* (3rd printing 1965 ed.) Oxford University Press, New York, NY. p 21.

Baldwin, K.R. (2006). Crop rotations on organic farms. The Organic Production publication series. *Center for Environmental Farming Systems. North Carolina Coop. Ext. Service, Raleigh, NC.* URL: http://www.cefs.ncsu.edu/resources/organicproductionguide/croprotationsfinaljan09.pdf .

Bowman, D.H. (1986). A history of the Delta Branch Experiment Station. *Mississippi Agric. & Forestry Exp. Sta., Special Bull. 86-2.* Mississippi State Univ., Mississippi State, MS.

Brooks, S.A. (2011). Influences from long-term crop rotation, soil tillage, and fertility on the severity of rice grain smuts. *Plant Disease* Vol.95:990-996.

Bruns, H.A., Pettigrew, W.T., Meredith, W.R., & Stetina, S.R.. (2007). Corn yields benefit in rotation with cotton. *Crop Management.* URL: doi:10.1094/CM-2007-0424-01-RS.

Buhler, D.D.. (1995). Influence of tillage systems on weed population dynamics and management in corn and soybean in the central USA. *Crop Sci.* Vol.35:1247-1258.

Butt, J.J. (2002). Daily life in the age of Charlemagne p. 82-83. *Greenwood Publishing Group,* Westport, CT., USA

Capehart, T. (2004). Trends in U.S. tobacco farming. *Electronic outlook report from the Economic Research Service. United States Department of Agriculture.* URL: http://www.ers.usda.gov/publications/tbs/nov04/tbs25702/tbs25702 .

Cardina, J. Herms, C.P., & Doohan, D.J. (2002). Crop rotation and tillage systems effects on weed seedbanks. *Weed Sci.*Vol. 50:448-460.

Clark, M.S., Horwath, W.R., Shennan, C., & Scow. K.M. (1998). Changes in soil chemical properties resulting from organic and low-input farming practices. *Agron. J.* Vol. 90:662-671.

Croissant, R.L. Peterson, G.A., & Westfall. D.G. (2008). Dryland cropping systems. *Crop Production Series* no. 0.516. *Colorado State University Extension.* Fort Collins, CO.

de Friature, C., Giordano, M., & Laio, Y. (2008). Biofuels and implications for agricultural water use: blue impacts
of green energy. *Water Policy 10 Suppl.* Vol. 1:67-81.

Delate, K. & Hartzler R. (2003). Weed management for organic farms. *Organic Agriculture Series. Iowa State Univ. Univ. Extension. Ames, IA.* PM 1883 URL: http://www.extension.iastate.edu/Publications/PM1883.pdf .

Duley, F. L., & Miller, M. F. (1923). Erosion and surface run-off under different soil conditions. *Mo. Agr. Exp. Sta.* Res. Bull. 63.

Edwards, J.H., Thrulow, D.L., & Eason, J.T. (1987). Influence of tillage and crop rotation on yields of corn, soybean, and wheat. *Agron. J.* Vol. 80:76-80.

Ellis, J.M., Shaw, D.R., & Barrentine, W.L. (1998). Soybean (*Glycine max*) seed quality and harvesting efficiency as affected by low weed densities. *Weed Tech.* Vol. 12:166-173.

Erickson, B. (2008). Corn/soybean rotation literature summary. URL: http://www.agecon.purdue.edu/pdf/Crop_Rotation_Lit_Review.pdf verified 2 August 2011.

Follett, R.F. (2001). Soil management concepts and carbon sequestration in cropland soils. *Soil Tillage Res.* Vol. 61:77-92.

Francis, C.A. & Clegg, M.D. (1990). Crop rotations in sustainable production systems. *In* C.A. Edwards, R. Lal, P. Madden, R.H. Miller, and G. House. (eds.) *Sustainable Agricultural Systems.* Soil and Water Cons. Soc. CRC Press New York, NY. p 130.

Hammond, R. B., Michel, A., & Eisley, J.B. (2009). Corn rootworm management. *The Ohio State University Extension* Fact Sheet FC-ENT-0016-09. Agriculture and Natural Resources. Columbus, OH.

Hancock, D.W. (2009). Planting warm season annual grasses. *Univ. of Georgia Coop Ext.* CSS-F010. URL: http://www.caes.uga.edu/commodities/fieldcrops/forages/documents/PlantingWarmSeasonAnnualGrasses.pdf. verified 14 Sept., 2011.

Heichel, G.H., & Barnes, D.K. (1984). Opportunities for meeting crop nitrogen needs from symbiotic nitrogen fixation. *In* D.M. Kral editor. *Organic farming: Current technology and its role in a sustainable agriculture.* American Society of Agronomy Spec. Publ. No. 46. ASA, CSSA, SSSA. Madison, WI. pp 49-59.

Hill, J. (2007). Environmental costs and benefits of transportation biofuel production from food- and lignocelluloses-based energy crops. A review. *Agron. Sustain. Dev.* Vol. 27:1-12.

Jacobstein, M. (1907). The Tobacco Industry in the United States. *New York: Columbia University Press,* New York, AMS, Reprint, 1968.

Johnson, K.D., Rhykerd, C.l., Hertel, J.M., & Hendrix, K.S. (2007). Improving pastures by renovation. Forage Information. *Agronomy Ext. Purdue Univ. West Lafayette, IN* URL: http://www.agry.purdue.edu/ext/forages/publications/ay251.htm verified 13 Sept. 2011.

Johnson, S.E. & Toensmeier, E. (2009). How expert organic farmers manage crop rotations. p 3. *In* C.L. Mohler, and S.E. Johnson (eds.) *Crop Rotation on Organic Farms a Planning Manual.* Natural Resource, Agriculture and Engineering Service. Cooperative Extension. Ithaca, NY.

Karlen, A.H., Varvel, G.E., Bullock, D.G., & Cruse, R.H. (1994). Crop rotations for the 21st century. *Adv. Agron.* Vol. 53:1-45.

Kelley, K. W., Long Jr. J.H., & Todd, T.C. (2003). Long-term crop rotations affect soybean yield, seed weight, and soil chemical properties. *Field Crops Res.* Vol. 83:41-50.

Kladivko, E.J., Griffith, D.R., & Mannering, J. V. (1986). Conservation tillage effects on soil properties and yield of corn and soya beans in Indiana. *Soil Tillage Res.* Vol. 8:277-287.

Lacefield, G.D. & Smith, S.R. (2009). Renovating hay and pasture fields. *AGR-26. Coop Ext. Serv. Univ. of Kentucky* College of Agric. Lexington, KY.

Lafond, G.P., Loeppky, H., & Derksen, D.A. (1992). The effects of tillage systems and crop rotations on soil water conservation, seedling establishment and crop yield. *Can. J. Plant Sci.* Vol. 72:103-115.

Lal, R. (2004). World crop residues production and implications of its use as a biofuels. *Environ. Intern.* Vol. 31: 575– 584.

Lal, R., Kimble, J.M., & Follett, R.F. (1997). Soil quality management for carbon sequestration. *In* R. Lal et al. (ed.) *Soil Properties and Their Management for Carbon Sequestration.* USDA,-NRCS, National Soil Survey Center, Lincoln, NE. pp 1-8.

Lal, R., Kimble, J.M., & Follett, R.F., and Cole, C.V. (1998). *The potential of U.S. cropland to sequester carbon and mitigate the greenhouse effect.* Sleeping Bear Press, Chelsea, MI.

Lattimore, M.E. (1994). Pastures in temperate rice rotations of south-eastern Australia. *Australian J. Exp. Agric.* Vo. 34:959-965.

Legere, A., Samson, N., Rioux, R., Angers, D.A., & Simard, R.R. (1997). Response of spring barley to crop rotation, conservation tillage, and weed management intensity. *Agron. J.* Vol. 89:628-638.

Lupwayi, N.Z., Rice, W.A., & Clayton, G.W. (1998). Soil microbial diversity and community structure under wheat as influenced by tillage and crop rotation. *Soil Biol. Biochem.* Vol. 30:1733-1741.

Lyng, R.E. (2011). Transformation of the land in colonial America. *Health Guidance.* URL: http://www.healthguidance.org/entry/6937/1/Transformation-of-the-Land-in-Colonial -America.html .

Lyon, D.J., Baltensperger, D.D., Blumenthal, J.M., Burgener, P.A., & Harveson, R.M. (2004). Eliminating summer fallow reduces winter wheat yields, but not necessarily system profitability. *Crop Sci.* Vol. 44:855-860.

Maeder, P., Fliessbach, A., Dubois, D., Gunst, L., Fried, P., & Niggli, U. (2002). Soil fertility and biodiversity in organic farming. *Sci.* Vol. 296 (No.) 5573:1694-1697.

Martin, J.H. & Leonard, W.H. (1967). Perennial forage grasses. In *Principles of Field Crop Production 2rd Edition.* The Macmillan Co. New York, NY. p 536.

Martin, J.H., Leonard, W.H., & Stamp, D.L. (1976). Fertilizer, green manuring, and rotation practices. In *Principles of Field Crop Production 3rd Edition..* The Macmillan Co. New York, NY. p 166.

Matthew, J.A., Casella, E., Farrall, K., & Taylor, G.. (2010). Estimating the supply of biomass form short-rotation coppice in England, given social, economic and environmental constraints to land availability. *Biofuels* Vol. 1:719-727.

McGee, A.E., Peterson, G.A., & Westfall, D.G. (1997). Water storage efficiency in no-till dryland cropping systems. *J. Soil Water Cons.* Vol. 52:131-136.

Miles, R. (1999). The rotation systems. Sanborn Field Agric. Exp. Station. *Univ. of Missouri. CAFNR.* URL: http://aes.missouri.edu/sanborn/research/anlrpt/rotation.stm .

Mitchell, C.C. (2004). Long-term agronomic experiments. *College of Agriculture Agronomy Crops & Soils* URL: http://www.ag.auburn.edu/agrn/longterm.htm#Additional .

Mitchell, C.C., Delaney, D.P., & Balkcom, K.S. (2008). A historical summary of Alabama's Old Rotation (circa 1896): The world's oldest, continuous cotton experiment. *Agron. J.* Vol. 100:1493-1498.

Moyer, J.R., Roman, E.S., Lindwall, C.W., & Blackshaw, R.E. (1994). Weed management in conservation tillage systems for wheat production in North and South America. *Crop Protection* Vol. 13:243-259.

Nave, W.R. & Wax, L.M. (1971). Effect of weeds on soybean yield and harvesting efficiency. *Weed Sci.* 19:533-535.

Neue, H. (1993). Methane emission from rice fields: Wetland rice fields may make a major contribution to global warming. *Bioscience* Vol. 43:466-473.

North Carolina Soybean Producers Association. (2011). *The history of soybeans* URL: http://www.ncsoy.org/ABOUT-SOYBEANS/History-of-Soybeans.aspx.

Peng, S., Cassman, K.G., Virmani, S.S. Sheey J., and Khush G.S. (1999). Yield potential trends of tropical rice since the release of IR8 and the challenge of increasing rice yield potential. *Crop Sci.* Vol 39:1552-1559.

Perlack, R.D., Wright, L.L., Turhollow, A.F., Graham, R.I. Stokes, B.J., & Erbach, D.C. (2005). Biomass as feedstock for a bioenergy and bioproducts industry. The technical feasibility of a billion-ton annual supply. *ODE/GO-102995-2135. ORNL/TM-2005/66.* Oak Ridge National Laboratory, Oak Ridge, TN.

Peters, R.D., Sturz, A.V., Carter, M.R., & Sanderson, J.B. (1999). Developing disease-suppressive soils through crop rotation and tillage management practices. *Soil and Tillage Res.* Vol. 72:181-192.

Pettigrew, W.T., Meredith, W.R., Bruns, H.A., & Stetina, S.R. (2006). Effects of a short-term corn rotation on cotton dry matter partitioning, lint yield, and fiber quality production. *J. Cotton Sci.* Vol. 10:244-251.

Porter, P.M., Lauer, J. G., Lueschen, W.E., Ford, J. H., Hoverstad, T. R., Oplinger, E S., & Crookston, R. K.. (1997). Environment Affects the Corn and Soybean Rotation Effect. *Agron. J.* Vol. 89:441-448.

Porter, P.M., Huggins, D.R., Perillo, C.A., Quiring, S.R., & Crookston, R.K. (2003). Organic and other management strategies with two- and four-year crop rotations in Minnesota. *Agron. J.* Vol. 95:233-244.

Raghu, S., Anderson, R.C. Daehler, C.C., Davis, A.S., Wiedenmann, R.N., & Simberlotf, D. (2006). Adding biofuels to the invasive species fire? Sci. Vol. 313:1742.

Raun, W.R. and Johnson, G.V. (1999). Improving nitrogen use efficiency for cereal production. *Agron. J.* Vol. 91:357-363.

Reganold, J.P., Elliott, L.F., & Unger, Y.L. (1987). Long-term effects of organic and conventional farming on soil erosion. *Nature* Vol. 330: 370-372.

Rodhe, H. (1990). A comparison of the contribution various gases to the greenhouse effect. *Sci.* Vol. 248:1217-1219.

Roth, G.W. (1996). Crop rotations and conservation tillage. *Conservation tillage series* No. 1 5M696ps7071. . College of Agriculture, Coop. Ext. University Park, PA.

Roth, G.W., Harper J., & Kyper, R. (1997). Crop rotation planning for dairy farms. *Agronomy Facts* 57. 5M1097PS. College of Agriculture, Coop. Ext. University Park, PA. URL: http://cropsoil.psu.edu/extension/facts/agronomy-facts-57.

Rothamsted. (2011). e-RA: *the electronic Rothamsted Archive*. URL: http://www.era.rothamsted.ac.uk/.

Ryan, J., Singh, M., & Pala, M. (2008). Long-term cereal-based rotation trials in the Mediterranean region: Implications for cropping sustainability. *Adv. Agron.*Vol. 97:273-319.

Sampson, R.N., & Scholes, R.J. (2000). Additional human-induced activities. Article 3.4 *In* R.T. Watson et al. (ed.) *Land Use, Land-Use Change, and Forestry: A Special Report of*

The Intergovernmental Panel on Climate Change. Cambridge University Press, New York, NY. pp 181-281.

Schillinger, W.F., Kennedy, A.C., & Young, D.L. (2007). Eight years of annual no-till cropping in Washington's winter wheat - summer fallow region. *Agric. Eco. Env.* Vol. 120:345-358.

Singer, J.W. & Cox, W.J. (1998). Economics of different crop rotations in New York. *J. Prod. Agric.* Vol. 11:447-451.

Stallcup, L. (2009). Rotation boosts yield, improves soil fertility. *Delta Farm Press*. Penton Media Co. URL: http://deltafarmpress.com/soybeans/rotation-boosts-yield-improves-soil-fertility.

Stanger, T.F., Lauer, J.G., & Chavas, J.P. (2008). The profitability and risk of long-term cropping systems featuring different rotations and nitrogen rates. *Agron. J.* Vol. 100:105-113.

Stetina, S.R., Young, L.D., Pettigrew, W.T., & Bruns, H.A. (2007). Impact of Corn-Cotton Rotations on Reniform Nematode Populations and Crop Yield. *J. Nematropica* Vol. 37:237-248.

Stewart, B.A., Koohafkan, P., & Ramamoorthy, K. (2006). Dryland agriculture defined and its importance to the world *In* G.A. Peterson, P.W. Unger, and W.A. Payne. Eds. *Dryland Agriculture Agronomy Monograph No. 23* American Society of Agronomy, Crop Science Society of American, Madison, WI. pp 1-26.

Stichler, C., & Bade, D. (2005). Forage bermudagrass: Selection, establishment, and management. E-179 *Texas Coop. Ext.* Texas A&M Uni.. College Station, TX.

Teasdale, J.R., Mangum, R.W., Radhakrishnan, J., and Cavigelli, M.A. (2004). Weed seedbank dynamics in three organic farming crop rotations. *Agron. J.* Vol. 96:1429-1435.

Teutsch, C.D. & Fike, J.H. (2009). Virginia's horse pastures: Renovating old pastures. *Virginia Coop. Ext. Pub. 418-104*. Virginia Polytech. Inst. And State Univ., Blacksburg, VA.

[USDA-NASS] USDA National Agricultural Statistics Service. (2011). *Grain crops national and state data*. NASS, Washington, DC. URL: http://www.nass.usda.gov/QuickStats/PullData_US.jsp .

Uhland, R.E. (1948). Rotations in conservation. *Yearbook of Agriculture 1943-1947 Science in Farming*. U.S Govt. Printing Office. Washington D.C. pp:527-536.

Vendramini, J., Newman, Y., Blout, A., Adjei, M.B., & Mislevy, P. (2010). Five basic steps to successful perennial pasture grass establishment from vegetative cuttings on south Florida flatwoods. SSAGR24. *IFAS Extension*. Univ. of Florida, Gainesville, FL.

Vivian, J. (2001). The three sisters, the nutritional balancing act of the Americas. *Mother Earth News* URL: http://www.motherearthnews.com/Nature-Community/2001-02-01/The-Three-Sisters.aspx .

Voss, R.D. & Shrader, W.D. (1984). Rotation effects and legume sources of nitrogen for corn. *In* D.M. Kral editor. *Organic Farming: Current Technology and Its Role in a Sustainable Agriculture*. American Society of Agronomy Spec. Publ. No. 46. ASA, CSSA, SSSA. Madison, WI. pp 61-68.

Walz, E.. (1999). Final results of the Third Biennial National Organic Farmers' Survey. *Organic Farming Research Foundation*, Santa Cruz, CA.

Wassmann, R., Neue, H.U., Ladha, J.K., & Aulakh, M.S. (2004). Mitigating greenhouse gas emissions from rice-wheat cropping systems in Asia. *Environ. Devel. Sust.* Vol. 6:65-90.

Wheaton, H.N. & Roberts, C.A. (1993). Renovating grass sods with legumes. *Univ. of Missouri Ext. G4651*. University of Missouri. Columbia, MO.

Weaver, J.E. & Noll, W. (1935). Measurement of run-off and soil erosion by a single investigator. *Ecol.* Vol. 16:1-12.

West, T.O., & Post, W.M. (2002). Soil organic carbon sequestration rates by tillage and crop rotation; a global data analysis. *Soil Sci. Soc. Amer. J.* Vol. 66:1930-1946.

West, J.M. (2004). King cotton: The fiber of slavery. *Slavery in America.* URL: http://www.slaveryinamerica.org/history/hs_es_cotton.htm .

Wilhelm, W.W., Johnson, J.M.F., Karlen, D.L., & Lightle, D.T. (2007). Corn stover to sustain soil organic carbon further constrains biomass supply. *Agron. J.* Vol. 99:1665-1667.

Woodruff, J.M., Durham, R.G. & Hancock, D.W. (2010). Forage establishment guidelines. *Univ. of Georgia College of Agriculture & Environmental Sciences.* URL: http://www.caes.uga.edu/commodities/fieldcrops/forages/establishment.html#pasture.

Young, D.L., Hinman, H.R., & Schillinger, W. F. (2000). Economics of winter wheat-summer fallow vs. continuous no-till spring wheat in the Horse Heaven Hills, Washington. Farm business management reports. EB1907. *Cooperative Extension Service,* Washington State University. Pullman, WA.

Section 3

Crop Response to Water and Nutrients

5

Long-Term Mineral Fertilization and Soil Fertility

Margarita Nankova
Dobrudzha Agricultural Institute – General Toshevo
Bulgaria

1. Introduction

Long-term experiments are very important in studying the changes of soil fertility and environmental conditions as well as in analyzing the stability and quality of crop production. Such experiments give us more information how to use the good agronomic practices and how to protect the nature. Probably the oldest still-running arable crop fertilizer experiment is the Broadbalk Experiment established by John B. Lawes in Rothamsted (UK) in 1843 (Goulding et al., 2000). Thanks to this experiment many other long-term fertilizer experiments were established worldwide (Sims, 2006; Khan et al., 2007; Takahashi&Anwar, 2007; Kunzova&Hejcman, 2009).

In Bulgaria also have investigations on such long-term fertilizer trails (Koteva, 2010; Panayotova, 2005; Nankova et al., 1994 & 2005; Nankova, 2010).

Dobrudzha Agricultural Institute-General Toshevo is situated in North-Eastern part of Bulgaria on black earth zone (Picture 1). The main soil type is chernozem (Haplic Chernozems WRBSS, 2006).

The aim of this investigation was to follow the effect of the long-term agronomy practices and especially fertilization on the nutrition regime of slightly leached chernozem soil in the region of South Dobrudzha after 40 years mineral fertilization with different norm and combination between nitrogen, phosphorus and potassium.

A long-term fertilizer experiment , which was established in 1967 is still running. In two field crop rotation (wheat-maize) four nitrogen and phosphorus and three potassium norms were tested - 0, 60, 120, 180 and 0, 60, 120 kg/ha respectively. The experiment was designed according to the method of the "net square", applying the full version of the design in four replications. The experiment was designed by the method of the "net square", applying the full version of the design (4 x 4 x 3 = 48) in four replications. On the 40th year from the beginning of the trial (2007) after wheat harvest, soil samples were taken every 20 cm down the soil profile till depth 400 cm. A motor-driven portable soil sampler was used (Iliev&Nankova, 1994; Iliev, 2000). The changes of some agrochemical characteristics were determined in selected variants with high average 40th year productivity.

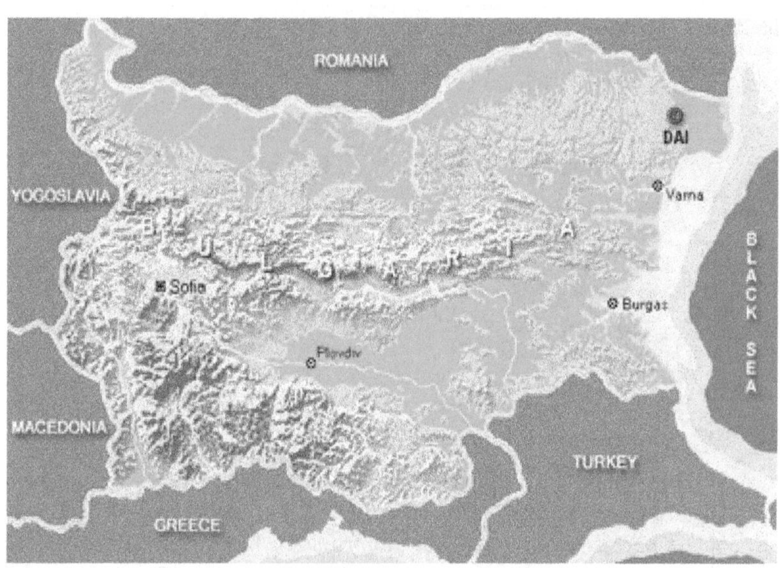

Picture 1. Position of Dobrudzha Agriculture Institute on Bulgaria map (43° 40′ northen latitude and 28° 10′ eastern longitude)

The soil acidity forms were determined by Ganev&Arsova (1980).

The potential nitrogen-supplying ability of soil was determined through incubation under constant temperature of 30° C at 60 % humidity from its total moisture absorption capacity in order to develop optimal conditions for nitrification. Incubation was done in thermostate to investigate its dynamics at the 14th, 28th and 56th day. The samples were analyzed to determine the amount of nitrate nitrogen in 1 % K_2SO_4 extract. The ability of NO_3-N to form intensive yellow coloration when interacting with disulphurphenoloc acid [$C_6H_3OH(HSO_3)_2$] in alkali media was used.

Carbon contend was valuated using the Tyurin modification (oxidizing with $K_2Cr_2O_7/H_2SO_4$ solution in thermostate at 125°C, 45 min, at presence of Ag_2SO_4 and titration with $(NH_4)_2SO_4.FeSO_4.6\ H_2O$ (Kononova&Belchikova,1961; Spiege at al, 2007; Hegymegi at al, 2007). Composition of soil organic matter was determined by Konnova (1963) and Filcheva&Tsadilas (2002).

Data were analysed with Excel and SPSS 16.0 (2007) and means separated by the Waller-Duncan test ($P<0,05$).

2. Influence of long-term mineral fertilization on some agrochemical characteristics of slightly leached chernozems (Haplic Chernozems)

One of the main degenerative processes in soil is the so called acidification. Various acid complexes are formed in soil as a result from soil formation processes on the one hand (erosion, humification, leaching, podzolization), and on the other – as a result from the activity of micro organisms and plants. Soils also possess buffer systems to counteract the acidification, which differ by their capacity.

2.1 Changes in soil acidity forms

2.1.1 Soil acidity forms for the 0-400 cm profile

The soil acidity forms, averaged for the investigated depth of the 0-400 cm profile, were significantly affected by the type of fertilizer combination. The independent effect of the factor mineral fertilization was higher on exchangeable Al^{3+}, Ca^{2+} and the sum of Ca^{2+} and Mg^{2+}, and significantly lower - on the values of residual hydrolytic acidity and the rate of alkali saturation. The depth of the investigated profile was the factor with decisive effect on all forms of soil acidity. Its effect on the pH values, the residual hydrolytic acidity and the alkali saturation degree was over 90 %. Significantly lower was its influence on the exchangeable Al^{3+} (22.5%) and the sum of exchangeable Ca and Mg (45.8%).

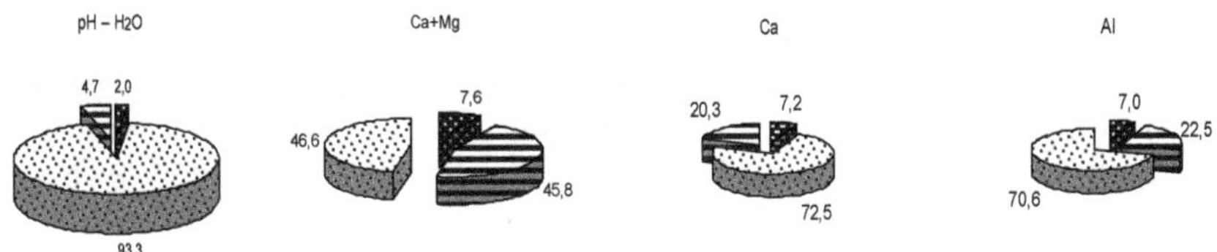

Fig. 1. Power of factors influence

In spite of the maximum degree of significance of the effect of mineral fertilization on the forms of soil acidity, the amplitude of variation of the separate indices was not so well expressed as in the separate soil layers up to 400 cm down the soil profile. Averaged for the fertilization variants, pH varied from 6.35 (10-20 cm) to 8.53 (260 - 300 cm). Soil reaction increased down the soil profile and at the 4th meter there was well expressed correlation between the soil layers forming it. It, however, showed similarities to layers 160-180, 180-200 and 200-220 cm. The layers from 220 to 300 cm possessed higher pH values in comparison to the layers of the 4th meter.

The amount of exchangeable Ca^{2+} showed a gradual tendency toward decreasing down the depth profile. Amplitude of variation was from 28.49 $cmol_c kg^{-1}$ (60-80 cm) to 18.79 $cmol_c kg^{-1}$ (380-400 cm). The surface layers 0-10 and 10-20 cm had lower content of exchangeable Ca^{2+} in comparison to the layers under them up to depth of 100 cm, being more similar to the amounts found in the 2nd meter. Highest amounts were detected in layers 60-80 cm and 80-100 cm.

The amount of exchangeable Mg^{2+} had a clear tendency toward increasing down the soil profile, being highest in the 340-360 cm layer (8.10 $cmol_c kg^{-1}$). In the trial field, layers 80-100, 120-140 and 60-80 cm had lowest content of exchangeable Mg^{2+} - about 1-2 $cmol_c kg^{-1}$. The surface layers within the 1st meter were comparatively richer in it, but their content considerably conceded to the content in the deeper layers of the 3rd and 4th meter.

The sum of the two exchangeable cations down the profile varied from 25.38 $cmol_c kg^{-1}$ (120-140 cm) to 30.51 $cmol_c kg^{-1}$ (60-80 cm). The surface layers (0-10 cm and 10-20 cm) had lower sorption capacity, $\sum$Ca+Mg and degree of saturation with bases than the 0-20 cm layer according to the trial beginnig. According to Nankova (2005, Personal Communication) at the start of this long-term experiment the values of these parameters were 34,44 $cmol_c kg^{-1}$, 30,80 $cmol_c kg^{-1}$ and 91,2% respectively. Further down the profile the sorption capacity decreased. What is very impressive is that it significantly increased in the 4th meter regardless of the occurrence of Ha in the 360-380 and 380-400 cm layers. The main reason for this is the higher amount of exchangeable Mg^{2+} in the 4th meter, which makes it very distinctive from the layers above it.

The degree of saturation with bases was lowest in the surface layers 0-10, 10-20 and 20-40 cm due to the intensive anthropogenic activity on the one hand, and on the other - due to the presence of plants. From the 40-60 cm layer the values of this index increased. In the entire 2nd meter the degree of saturation with bases was more than 99 %, and in the third meter it was 100 %. This value remained the same in the upper part of the 4th meter but in the layers 360-380 and 380-400 cm decreased slightly and was closer to the values registered in the 2nd meter. The reason for this is the occurrence of residual hydrolytic acid in the lower part of the 4th meter.

The separate meters up to depth 400 cm, as well as the sub-layers (horizons) at each meter (every 20 cm) affected to a maximum degree of significance the investigated indices characterizing the soil acidity forms in the investigated fertilization variants. The comparison of the results for the indices characterizing soil acidity revealed clear differentiation by each meter down the investigated soil profile.

Depth, cm	pH/ KCl	pH/H_2O	$T_{8.2}$, Sorption capacity	Exchangeable cations					Degree of saturation with bases, V%
				$H_{8.2}$	Al^{3+}	Ca^{2+}	Mg^{2+}	∑Ca+Mg	
			In $cmol_c kg^{-1}$. soil						
0-10	5,54 b	6,46 b	33,21 p	5,65 k	,17 b	23,80 k	3,60 h	27,40 h	82,99 b
10-20	5,38 a	6,35 a	33,06 o	5,90 l	,26 c	23,27 i	3,62 h	26,89 f	82,17 a
20-40	5,72 c	6,74 c	33,06 o	4,65 j	,00 a	25,09 m	3,33 g	28,42 l	85,93 c
40-60	6,45 d	7,48 d	32,83 n	2,88 i	,00 a	27,44 n	2,52 e	29,95 n	91,17 d
60-80	6,80 e	7,89 e	32,13 m	1,62 h	,00 a	28,49 p	2,02 c	30,51 o	94,91 e
80-100	7,15 f	8,21 f	30,83 l	,97 g	,00 a	28,28 o	1,58 a	29,86 m	96,89 f
100-120	7,35 g	8,30 g	28,37 k	,47 f	,00 a	24,97 l	2,94 f	27,90 j	98,38 g
120-140	7,43 jk	8,34 h	25,56 a	,19 e	,00 a	23,64 j	1,74 b	25,38 a	99,29 h
140-160	7,41 i	8,40 i	26,00 d	,09 d	,00 a	23,67 j	2,25 d	25,91 c	99,67 i
160-180	7,39 h	8,48 jk	26,54 f	,03 b	,00 a	23,58 j	2,94 f	26,51 e	99,91 k
180-200	7,41 i	8,51 klm	26,53 f	,04 b	,00 a	23,08 h	3,41 g	26,49 e	99,86 k
200-220	7,42 ij	8,47 j	26,97 g	,00 a	,00 a	23,11 h	3,86 i	26,97 f	100,00 l
220-240	7,43 k	8,52 m	26,51 f	,00 a	,00 a	22,05 g	4,46 k	26,51 e	100,00 l
240-260	7,49 m	8,52 lm	25,88 c	,00 a	,00 a	21,62 f	4,25 j	25,88 c	100,00 l
260-280	7,47 l	8,53 m	25,88 c	,00 a	,00 a	21,55 f	4,32 j	25,88 c	100,00 l
280-300	7,47 l	8,53 m	25,70 b	,00 a	,00 a	19,88 b	5,82 l	25,70 b	100,00 l
300-320	7,49 m	8,48 j	27,24 h	,00 a	,00 a	21,09 e	6,15 m	27,24 g	100,00 l
320-340	7,49 m	8,48 jk	28,25 j	,00 a	,00 a	20,79 d	7,46 n	28,25 k	100,00 l
340-360	7,52 n	8,49 jkl	28,33 jk	,00 a	,00 a	20,23 c	8,10 p	28,33 k	100,00 l
360-380	7,51 n	8,49 jkl	27,75 i	,06 c	,00 a	19,98 b	7,71 o	27,69 i	99,79 j
380-400	7,53 o	8,49 jkl	26,36 e	,06 c	,00 a	18,79 a	7,51 n	26,30 d	99,79 j

Table 1. Sorption capacity ($T_{8.2}$), exchangeable cations and degree of saturation with bases down the soil profile

Depth, cm	pH/ KCl	pH/ H_2O	$T_{8.2}$	Exchangeable cations					Degree of saturation with bases
				$H_{8.2}$	Al^{3+}	Ca^{2+}	Mg^{2+}	∑Ca+Mg	
			In $cmol_c kg^{-1}$. soil						
0-100	6,32 a	7,35 a	32,40 d	3,18 d	,043 b	26,57 d	2,61 a	29,186 d	90,30 a
100-200	7,40 b	8,41 b	26,60 b	,16 c	,000 a	23,79 c	2,65 a	26,44 b	99,42 b
200-300	7,46c	8,51 d	26,19 a	,00 a	,000 a	21,64 b	4,54 b	26,19 a	100,00 d
300-400	7,51 d	8,48 c	27,59 c	,025 b	,00 a	20,18 a	7,39 c	27,56 c	99,91 c

Table 2. Sorption capacity ($T_{8.2}$), exchangeable cations and degree of saturation with bases by meter down the soil profile

Averaged for the investigated fertilization variants, the 1st meter had lowest pH values, exchangeable Mg^{2+} and degree of saturation with bases. This meter, at the end of the 40-year period of investigation, showed harmful exchangeable acidity untypical for the natural status of this soil type. The first meter was also characterized by significantly higher content of residual hydrolytic acidity and higher sorption capacity in comparison to the other depths down the profile. In the 2nd and 3rd meter, with the exception of pH and the degree of saturation with bases, all other indices decreased their values. The third meter was characterized with complete absence of residual hydrolytic acidity, 100 % saturation with bases and increased content of exchangeable Mg^{2+} - with 73.9 % more above the 1st meter and with 71.2 % more above what has been established in the 2nd meter. What was typical for the separate layers of the 4th meter, besides the visually distinct coloration of these layers, was once again the occurrence of residual hydrolytic acidity, comparatively high increase of the content of exchangeable Mg^{2+} in comparison to the 3rd meter (62.6 %). Averaged for this depth, a higher sum of exchangeable bases was determined in the trial field in comparison to the 2nd and the 3rd meter, as well as higher sorption capacity of soil.

In spite of the average data for 400 cm, 40 years mineral fertilization caused a big difference on the forms of soil acidity accorfing to kind of fertilizer variant. The differentiation between variants of fertilization is very well expressed (Table 3). Soil reaction varied in narrow limits, but in spite of this it was established decreasing of pH in variants $N_{180}P_{60}K_{60}$ и $N_{60}P_{180}K_0$ according to the control variant. By the Waller-Duncan test there were established a very well expressed differences in all soil acidity forms, as well as in degree of saturation.

The lowest values of residual hydrolytic acidity ($H_{8,2}$) were registered in the variant with independent fertilization with 180 kg P_2O_5/ha (0.87 $cmol_c kg^{-1}$ soil). Residual hydrolytic acidity is one of the forms strongly affected by the long-term mineral fertilization, especially in the variant with $N_{180}P_{60}K_{60}$ (1.36 $cmol_c kg^{-1}$ soil). High amplitude of variation was determined for residual hydrolytic acidity down the soil profile: from its complete lack to 5.90 $cmol_c kg^{-1}$ soil. As was shown the highest values were established in the surface layers, 0-10 and 10-20 cm, which are most influenced by the agronomy practices fertilization and tillage.

The strongest evidence for the high effect of the long-term mineral fertilization with various norms and ratios on the agrochemical condition of the slightly leached chernozem soil in the trial field was the occurrence of exchangeable Al^{3+} in the soil absorption complex. It was detected in the surface layers (0-10 and 10-20 cm) in the variants $N_{180}P_0K_0$ and $N_{180}P_{60}K_{60}$. It was not present in the soil absorption complex further down the profile.

Fertilizer variants	pH/ KCl	pH/ H_2O	$T_{8.2}$	Exchangeable					Degree of saturation with bases – V %
				$H_{8.2}$	Al	Ca	Mg	Ca+Mg	
			cmolckg-1. почва						
$N_0P_0K_0$	7,12 c	8,14 d	28,55 d	1,03 c	,00 a	24,60 f	2,92 a	27,52 d	96,83 f
$N_{60}P_0K_0$	7,18 f	8,13 d	28,71 e	1,13 f	,00 a	23,16 d	4,41 d	27,58 e	96,50 c
$N_{120}P_0K_0$	7,13 d	8,15 d	28,96 g	,89 b	,00 a	23,79 e	4,29 c	28,07 h	97,30 g
$N_{180}P_0K_0$	7,13 d	8,23 e	27,45 a	1,11 e	,07 b	21,63 a	4,64 e	26,27 a	96,59 d
$N_0P_{180}K_0$	7,15 e	8,21 e	28,72 e	,87 a	,00 a	23,19 d	4,65e	27,84 g	97,39 h
$N_{60}P_{180}K_0$	6,98 a	7,96 b	27,94 b	1,16 g	,00 a	22,92 c	3,87 b	26,78 b	96,41 b
$N_{120}P_{120}K_{120}$	7,04 b	8,06 c	28,79 f	1,05 d	,00 a	22,88 c	4,86 f	27,74 f	96,69 e
$N_{180}P_{60}K_{60}$	6,98 a	7,93 a	28,30 c	1,36 h	,10 c	22,36 b	4,48 d	26,84 c	95,89 a

Table 3. Sorption capacity ($T_{8.2}$), exchangeable cations and degree of saturation with bases according to variants of fertilization, average for 0-400 cm depth

The variant with independent nitrogen fertilization with 180 kg/ha has the lowest $\sum$ Ca+Mg and the lowest value of sorption capacity. Intensive mineral fertilization with $N_{180}P_{60}K_{60}$ caused decreasing of degree of saturation with bases.

2.1.2 Compare the changes of soil acidity forms after 30th and 40th years mineral fertilization

For the first time in Bulgaria in such long-term trial we compared the obtained results for the individual indices which characterize the forms of soil acidity for the end of the 30th and the end of the 40th year since the initiation of the experiment in some of the variants to 60 cm depth.

At the end of the 40th year, the tendency towards lower values of pH and sorption capacity of soil became more prominent to various degrees according to the type of fertilization variant . This tendency was most evident in the variant with $N_{120}P_{120}K_{120}$, where a decrease with 17.2 % according to the end of the 30th year was determined.

A serious change was observed also towards decrease of the values of acidity on the strongly acid positions (T_{CA}) of the soil adsorbent with the increase of the duration of mineral fertilization in the above variants, and respective significant increase of this acidity, but this time on the slightly acid positions of the soil adsorbent. The indicated change led also to decrease of the rate of alkali saturation. The decrease varied from 4.4% ($N_{120}P_0K_0$) to 8.1% ($N_{120}P_{120}K_{120}$).

Fertilizer variants	pH/H_2O		$T_{8.2}$ Sorption capacity		T_{CA} Strongly acid positions		T_A Slightly acid positions		Degree of saturation with bases	
	30th	40th	30th	40th	30th	40th	30th	40th	30th	40th
$N_0P_0K_0$	7,60 c	6,85 a	34,87 b	32,93 b	32,33 b	23,92 b	2,53 ab	9,01 b	92,63 b	85,65 a
$N_{120}P_0K_0$	7,48 b	7,09 c	35,17 b	33,15 c	32,55 b	24,11 b	2,62 b	9,03 b	92,59 b	88,55 c
$N_{180}P_0K_0$	7,14 a	6,99 b	32,74 a	32,92 b	29,57 a	24,83 c	3,18 c	8,09 a	91,09 a	86,84 b
$N_{120}P_{120}K_{120}$	7,44 b	6,84 a	38,32 c	31,71 a	35,87 c	22,23 a	2,45 a	9,48 c	93,15 c	85,65 a

Table 4. Comparison of the soil acidity forms at the end of the 30th and at the end of the 40th year from the initiation of the trial according to the fertilization variant applied

Furthermore, during the last decade the balanced treatment with NPK at norm 120 kg/ha was the variant with lower values of $T_{8,2}$ and T_{CA}, while at the end of the 30th year of this trial the values of the above indices were higher than the values of the check variant and the values of the variants with independent nitrogen fertilization with increasing norms. The negative changes in the absolute values of these indices occurred also when determining the percent of acidity on the highly acid positions of the soil adsorbent (Fig. 2).

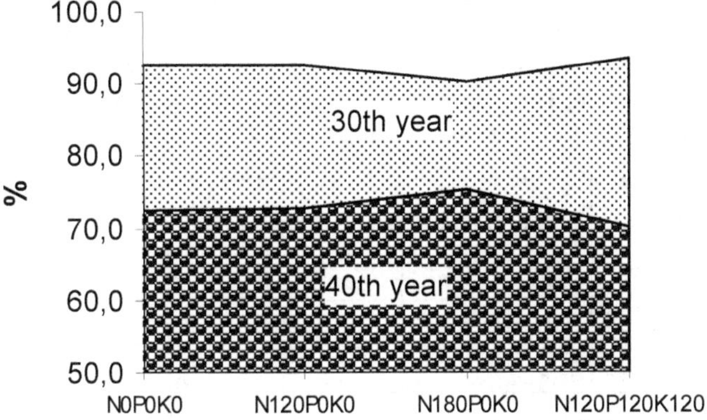

Fig. 2. Acidity on the highly acid positions of the soil adsorbent as percent of $T_{8.2}$ according to the fertilization variant at the end of the 30th and the 40th year of the trial.

The changes caused by the long-term agricultural usage of the land concerned also changeable Ca and Mg and their respective sum (Table 5). During the last investigated period changeable Ca^{2+} decreased in all tested variants, most strongly in the variants treated annually with $N_{120}P_{120}K_{120}$. The check variant ($N_0P_0K_0$), as well the variants with independent nitrogen fertilization, were less affected by this process. The amount of exchangeable Ca^{2+} at the end of the 40th year averaged for the investigated variants was 88.76 % from the amount at the end of the 30th year. A tendency was found towards lower amounts of exchangeable Mg^{2+} in the check variant and the variants with independent nitrogen fertilization, and towards significant increase in the variant with systematic balanced introduction of the main macro elements ($N_{120}P_{120}K_{120}$). At the end of the 30th year the investigated variants were characterized with a mean content of exchangeable Mg^{2+} of 3.66 $cmol_c kg^{-1}$ soil, and at the end of the 40th year – of 3,05 $cmol_c kg^{-1}$ soil. The comparison of the results for ∑ Ca+Mg revealed their decrease with averagely 12 % according to the data from the end of the 30th year, the decrease being greatest in the variant $N_{120}P_{120}K_{120}$.

Fertilizer variants	Exchangeable cations							
	Ha		Ca^{2+}		Mg^{2+}		∑Ca+Mg	
	30th	40th	30th	40th	30th	40th	30th	40th
$N_0P_0K_0$	2,57 a	4,73 d	28,63 b	25,12 b	3,63 b	3,00 ab	32,27 b	28,12 b
$N_{120}P_0K_0$	2,58 a	3,80 a	28,23 b	26,98 c	4,33 c	2,38 a	32,57c	29,36 d
$N_{180}P_0K_0$	2,60 a	4,33 b	25,82 a	25,49 b	3,78 b	3,11 bc	29,60 a	28,60 c
$N_{120}P_{120}K_{120}$	2,92 b	4,56 c	31,17 c	23,43 a	2,88 a	3,72 c	34,05 d	27,15 a

Table 5. Comparison of the changes in the exchangeable cations at the end of the 30th year and at the end of the 40th year from the trial depending on the type of the fertilization variant

At the end of the 40th year the acidity on the highly acid positions averaged for depth 0-60 cm was 23.77 $cmol_c kg^{-1}$ soil, compared to 32.58 $cmol_c kg^{-1}$ soil at the end of the 30th year, i.e. there was a decrease with 27.04 % (Table 5). pH variations affected most strongly the 20-40 cm layer. The established negative tendencies affected the 40-60 cm layer as well, where a significant decrease of the sorption capacity of soil was determined: with 9.6 % according to the values at the end of the 30th year.

At the same time the acidity on the slightly acid positions strongly increased in all three layers, the mean increase being almost three times higher, and affected mostly the 40-60 cm layer. These results also concern the rate of alkali saturation, which, too, demonstrated a tendency towards decrease. The decrease was highest in the surface 0-20 cm layer (8.1 %), in the 20-40 cm layer (6.5 %) and in the 40-60 cm layer (4.1 %).

Soil depth, cm	pH/H_2O		$T_{8.2}$ Sorption capacity		T_{CA} Strongly acid positions		T_A Slightly acid positions		Degree of saturation with bases	
	30th	40th	30th	40th	30th	40th	30th	40th	30th	40th
0-20	6,99 a	6,42 a	34,77 a	32,69 b	31,21 a	22,89 a	3,56 c	9,80 c	89,49 a	82,29 a
20-40	7,44 b	6,86 b	35,24 b	32,96 c	32,39 b	24,01 b	2,85 b	8,96 b	92,51 b	86,49 b
40-60	7,81 c	7,55 c	35,81 c	32,38 a	34,14 c	24,42 c	1,67 a	7,96 a	95,11 c	91,25 c

Table 6. Comparison of the soil acidity forms at the end of the 30th year and at the end of the 40th year of the trial depending on the depth of the layer

The changes which occurred down the investigated profile confirmed the established tendencies towards change in the values of the exchangeable cations during the respective periods of investigation depending on fertilization. The increase of the values of residual hydrolytic acidity affected most the surface 0-20 cm layer (Table 6). This layer was characterized with highest decrease of the changeable Ca^{2+} values, the amount of exchangeable Mg^{2+} remaining practically the same. Within both periods of analysis the sum of exchangeable alkali increased down the profile due to the exchangeable Ca^{2+}.

Soil depth, cm	Exchangeable cations							
	Ha		Ca		Mg		Ca+Mg	
	30th	40th	30th	40th	30th	40th	30th	40th
0-20	3,65 c	5,79 c	27,25 a	22,96 a	3,76 b	3,90 c	31,01 a	26,86 a
20-40	2,62 b	4,45 b	27,46 a	25,52 b	4,61 c	2,99 b	32,08 b	28,52 b
40-60	1,74 a	2,83 a	30,68 b	27,28 c	2,60 a	2,26 a	33,28 c	29,54 c

Table 7. Comparison of changes in the exchangeable cations between the 30th and the 40th year from the trial depending on the depth of soil layer

2.2 Changes of the soil mineralization ability after 40-years mineral fertilization

Transportation, redistribution and transformation of nitrogen down the soil profile was affected by a number of factors such as the structure of soil units, aeration, macro pores, composition, amount and depth of post harvest residue incorporation, mineral fertilization and nitrogen norm, mineralization of organic substance, leaching, productive moisture, etc (Goldbi et al., 1995; Karlen et al., 1998). The size of the nitrogen norm is significant for agricultural production under moist, semi-dry and dry conditions to obtain acceptable balance between economic and non-economic part of the produce and avoid possible losses (Cantero-Martinez et al.,1995). It is well known that the availability of the nitrogen from the mineral fertilizers depends strongly on the type of the nitrogen source, the soil type, the crop, the fertilization norm, etc. Many farmers tend to apply higher nitrogen norms to ensure higher yields (Franzluebbers et al., 1999). This in many cases is not necessary due to changes in the distribution of the nitrogen in the surface of the soil profile and its improved mobility (Rice et al., 1986).

The ability of soil to nitrify nitrogen under optimal conditions was significantly affected by the mineral fertilization and the investigated layer up to depth 400 cm (Table 7). During all three investigated periods of increasing incubation, this effect was significant to a maximum degree both under the independent influence of the investigated factors and under their interaction.

Source	Dependent Variable	df	Mean Square	F	Sig.
Fertilizer variants (A)	14 days	7	1201,997	1071,012	,000
	28 days	7	895,323	644,004	,000
	56 days	7	1760,257	808,230	,000
Soil depth (B)	14 days	20	3566,615	3177,951	,000
	28 days	20	9128,237	6565,923	,000
	56 days	20	14022,587	6438,532	,000
A x B	14 days	140	84,304	75,117	,000
	28 days	140	63,912	45,972	,000
	56 days	140	113,244	51,997	,000

Table 8. Variance analysis of the mineralization ability during a 40-year period of investigation

The depth of the soil layer was the factor with higher effect on the values of the soil's mineralization ability in comparison to mineral fertilization during all three incubation periods (Figure 1). The longer the period of incubation was, the higher its effect, reaching a maximum at 28-day incubation. Regardless of a slight decrease in the effect of this factor at 56-day incubation, the longer incubation had higher effect on the obtained results in comparison to 14-day incubation. The effect of mineral fertilization was lowest in the second incubation period and slightly increased in the third incubation period. The long-term mineral fertilization affected the amount of the established NO_3-N to a highest degree at 14-day incubation. The same was valid for the interaction between the two factors.

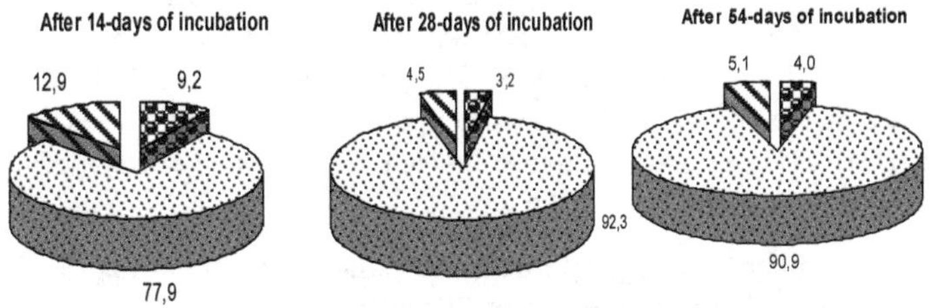

Fig. 3. Effect of factors according to the period of incubation, %

The distribution of the amount of nitrified nitrogen averaged for the periods of incubation by meters down the soil profile showed interesting results (Table 8). The soil layers of the 1st meter had highest potential nitrogen-supplying capacity. The layers forming the 2nd and the 3rd meter (loess horizon) had lowest nitrification capacity regardless of the favorable conditions for this process. The Waller-Duncan test did not reveal differences between them. It, however, differentiated the results obtained for the content of NO_3-N averaged for the 4th meter in a separate group after what was established in the 1st meter.

Meters	Value	Group
2	3,59	a
3	3,74	a
4	5,08	b
1	32,92	c

Table 9. NO_3-N content by meters down the soil profile (mg NO_3-N/1000 g soil)

The effect of mineral fertilization of the different variants averaged for depth 0-400 cm and the incubation periods was strongly expressed depending on the norms and ratios between the three macro elements (Table 9).

Fertilizer variants	Value	Group
$N_0P_0K_0$	5,95	a
$N_{120}P_{120}K_{120}$	8,64	b
$N_0P_{180}K_0$	9,17	c
$N_{60}P_0K_0$	10,28	d
$N_{60}P_{180}K_0$	10,71	d
$N_{120}P_0K_0$	11,36	e
$N_{180}P_0K_0$	12,93	f
$N_{180}P_{60}K_{60}$	21,63	g

Table 10. Mean content of NO_3-N according to the type of fertilization variant (mg NO_3-N/1000 g soil)

The check variant ($N_0P_0K_0$) reflected the natural fertility of *Haplic Chernozems* in the trial field after its long-term cultivation. The check variant had lowest content of NO_3-N after incubation among all tested variants. The fertilization variants were well differentiated on the basis of the total amount of NO_3-N after incubation. The independent nitrogen fertilization with increasing norms was accompanied with proportional increase of the amount of nitrified nitrogen, the values of which fell within separate groups of higher orders, compared one to another.

When combining nitrogen with phosphorus and potassium depending on the norms and ratios between the three elements, the 4-m soil layer had variable capacity to supply nitrates as a result from incubation. Highest amounts of this inorganic nitrogen form were found after systematic application of $N_{180}P_{60}K_{60}$ - 21.63 mg NO_3-N/1000 g soil. The systematic application of $N_{120}P_{120}K_{120}$ for a period of 40 years showed lowest amounts of nitrified nitrogen following the check variant. Averaged for the 4-m soil profile, they were lower than the amounts after independent application of the same nitrogen norm. The main reason for this was that after this type of fertilization the highest yields from wheat were obtained, averaged for the 40-year period of investigation, which, on its part, was an indication for their uptake and respective realization into cash crop. The results with regard to the nitrification capacity from the analysis of this fertilization variant revealed considerable similarities to that of the check variant. The comparatively low amounts of nitrified nitrogen after systematic fertilization with $N_{120}P_{120}K_{120}$, combined with high agronomy effect, showed that this fertilization combination can not lead to accumulation of inorganic nitrogen in soil (in nitrate form) down the soil profile.

The incubation periods of soil under conditions favorable for the nitrification process also significantly affected the values of nitrified nitrogen (Table 10). With the longer incubation periods, the mean total amount of nitrified nitrogen increased with increasing the days of incubation. This lead to clear differentiation of the incubation periods and to formation of the results into separate groups.

Days of incubation	Value	Group
14	7,2081	a
28	11,0113	b
56	15,7811	c

Table 11. Mean content of NO_3-N according to the incubation period (mg NO_3-N/1000 g soil)

2.3 Changes of the soil organic matter after long-term mineral fertilization

2.3.1 Carbon concentration along the soil profile to 400 cm depth

Systematic mineral fertilization carried out for 40 years in two-field crop rotation (wheat - maize) affected the content of Ctotal deep down the profile of the slightly leached chernozem soil. Annual fertilization with N180P60K60 during 40 years contributed most for the increase of its content at average depth 0-400 cm. Independent nitrogen fertilization with increasing norms, especially with 120 and 180 kg N/ha, had low effect on the content of Ctotal, averaged for depth 0-400 m (Fig 4). This type of fertilization contributed less to the total carbon reserves in soil, averaged for the 60 cm layer. The fertilization variants involving phosphorus and phosphorus plus potassium in the norms and ratios investigated in this study had more significant effect on the increase of these reserves; in this case there was an average increase with 18.7 % in comparison to the check variant without fertilization.

Along the soil profile, the sub-depths forming the 3rd meter had lowest Ctotal (respectively humus). No differentiation affected by the fertilization variant was found in this zone. The layers comprising the 4th meter had higher Ctotal content in comparison to the 3rd meter, and the differentiation in its content depended on the applied fertilization variant.

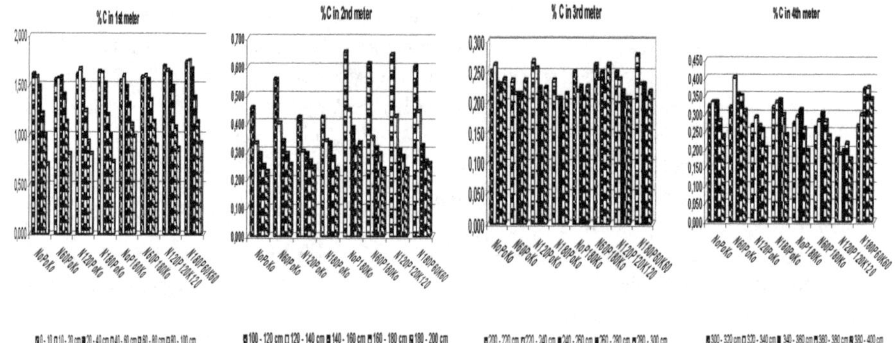

Fig. 4. Content of C_{total} (%) by layers for every meter up to 400 cm averaged for the fertilization variants

2.3.2 Soil organic mater composition along the soil profile to 400 cm depth

The variance analysis of the composition of the soil organic substance down the layers of the soil profile up to depth 400 cm revealed the dynamics in the degree of significance of the changes of C in the respective groups and fractions as a result from the long-term mineral fertilization. Although the indices of variations of the respective fertilization variants were not significant, the established differences between the investigated fertilization combinations were significant to a maximum degree, averaged for depth 0-400 cm.

Depth cm	$C_{organic}$	Humic acides (HA)	Fulvic acides (FA)	HA/FA	HA-Ca	HA-R_2O_3	Humin	FA in H_2SO_4
0-20	,000	,000	,000	,000	,000	,000	,000	,001
20-40	,000	,000	,000	,000	,000	,055	,023	,006
40-60	,000	,102	,008	,001	,052	,000	,000	,000
60-80	,000	,001	,006	,000	,001	,000	,000	,006
80-100	,000	,000	,001	,000	,000	,013	,002	,001
100-120	,001	,003	,001	,001	,003	,001	,001	,000
120-140	,002	,004	,006	,000	,003	,001	,001	,001
140-160	,001	,000	,000	,000	,000	,001	,001	,001
160-180	,000	,000	,001	,001	,000	,000	,000	,008
180-200	,002	,000	,005	,000	,000	,000	,007	,239
200-220	,003	,000	,381	,073	,000	,000	,009	,003
220-240	,001	,000	,004	,000	,000	,000	,107	,001
240-260	,000	,000	,022	,069	,000	,000	,011	,000
260-280	,008	,003	,008	,023	,001	,003	,007	,000
280-300	,002	,000	,012	,022	,000	,000	,032	,001
300-320	,004	,003	,001	,009	,007	,000	,074	,003
320-340	,000	,000	,000	,000	,000	,000	,000	,008
340-360	,000	,000	,000	,000	,000	,026	,001	,000
360-380	,000	,000	,000	,000	,000	,007	,002	,001
380-400	,000	,000	,000	,000	,000	,000	,000	,000
0-400 cm	**,000**	**,000**	**,000**	**,000**	**,000**	**,000**	**,000**	**,000**

Table 12. Variance analysis of the soil organic matter composition after a 40-year mineral fertilisation

Organic C of soil was also subjected to dynamic changes averaged for the entire 4 m depth. In this index the differentiation between the variants was less expressed in comparison to total C. Its amount was lowest in the untreated check variant.

Highest differentiation in the content of organic C according to the type of the fertilization variant was established in layers 40-60 cm and 60-80 cm, and lowest variation was found in the 260- 280 cm layer.

The independent nitrogen fertilization with 180 kg N/ha contributed most significantly to the increased amount of $C_{organic}$ averaged for a considerable depth down the profile. In this case, however, C_{humin} had lowest values. A similar tendency was found in the independent nitrogen fertilization with 120 kg N/ha, as well. In these two variants the amount of C_{humin}, also called "guard of humus", was below the level of the check variant. The independent nitrogen fertilization with 60 kg N/ha, the independent phosphorus fertilization with 180 kg P_2O_5/ha, the combination between them and the systematic balanced introduction of NPK at norm 120 kg/ha did not in practice affect the insoluble fraction of organic substance of soil under systematic agricultural cultivation of the land. Not only in the respective layers, but also in the entire 0-400 cm depth, the long-term independent nitrogen fertilization with 120 and 180 kg/ha lead to lower amounts of the insoluble residue. This is valid to a higher degree for the norm 180 kg/ha. Lowest differentiation in the values of Cresidue between the fertilization variants was determined in the 320-340 cm layer. Highest variations between the fertilization variants were established in the 0-20 cm, 60-80 cm and 380-400 cm layers.

The systematic introduction of $N_{180}P_{60}K_{60}$ had most significant contribution for $C_{residue}$ increase average for 0-400 cm profile.

Depth cm	$N_0P_0K_0$	$N_{60}P_0K_0$	$N_{120}P_0K_0$	$N_{180}P_0K_0$	$N_0P_{180}K_0$	$N_{60}P_{180}K_0$	$N_{120}P_{120}K_{120}$	$N_{180}P_{60}K_{60}$
0-20	,7969 a	,8940 c	,7868 a	,8962 c	,8284 b	,8827 c	,8949 c	,9461 d
20-40	,7824 a	,8209 b	,7625 a	,8962 d	,7782 a	,8705 c	,8583 c	,8660 c
40-60	,6539 b	,7235 d	,5919 a	,6954 c	,7406 de	,7730 f	,7486 e	,7479 e
60-80	,5172 a	,5773 cd	,5311 ab	,5448 abc	,7782 f	,5901 de	,5657 bcd	,6245 e
80-100	,3556 a	,3947 b	,3971 b	,4694 c	,5021 cd	,5047 d	,5047 d	,4818 cd
100-120	,2312 ab	,3244 d	,2498 bc	,2059 a	,3264 d	,3219 de	,2853 cd	,3026 de
120-140	,1318 a	,1632 ab	,2498 d	,2508 d	,2385 cd	,1951 bc	,2121 cd	,2227 cd
140-160	,1572 ab	,1617 abc	,2498 d	,1933 bc	,2008 c	,2438 d	,1268 a	,1424 a
160-180	,1697 b	,2348 cd	,2583 d	,2222 c	,1883 b	,2926 e	,1268 a	,1172 a
180-200	,1074 a	,2348 d	,1249 a	,2008 cd	,1757 bc	,1951 cd	,1390 ab	,0990 a
200-220	,1499 c	,1349 bc	,1413 c	,2008 d	,1255 bc	,0834 a	,1146 abc	,0990 ab
220-240	,1324 d	,1267 cd	,1357 d	,1255 cd	,0628 a	,0771 a	,0866 ab	,1053 bc
240-260	,0949 a	,0780 a	,1385 b	,2008 c	,0753 a	,0992 a	,0829 a	,0743 a
260-280	,0824 a	,1125 ab	,1520 b	,1506 b	,1004 a	,0708 a	,0829 a	,0806 a
280-300	,0999 ab	,1267 b	,1291 b	,1933 c	,1004 ab	,1244 b	,0829 a	,0619 a
300-320	,1199 ab	1389 bc	,1745 cd	,2008 d	,1255 ab	,0975 ab	,0902 a	,1231 ab
320-340	,1449 b	,2099 c	,0804 a	,3306 d	,1255 b	,1146 ab	,1073 ab	,1261 b
340-360	,1449 bc	,1876 d	,1291 abc	,3393 e	,1506 c	,1146 a	,1317 abc	,1240 ab
360-380	,1449 cd	,1632 d	,1269 bc	,2987 e	,1506 cd	,1204 ab	,1024 a	,1177 ab
380-400	,1574 c	,1369 bc	,1047 ab	,2436 d	,1506 c	,0975 a	,1354 bc	,1055 ab
0-400 cm	**,2585 a**	**,2972 c**	**,2757 b**	**,3429 d**	**,2962 c**	**,2934 c**	**,2740 b**	**,2784 b**

Table 13. Content of $C_{organic}$ by layers up to 400 cm according to the fertilization variants

Depth cm	$N_0P_0K_0$	$N_{60}P_0K_0$	$N_{120}P_0K_0$	$N_{180}P_0K_0$	$N_0P_{180}K_0$	$N_{60}P_{180}K_0$	$N_{120}P_{120}K_{120}$	$N_{180}P_{60}K_{60}$
0-20	,8055 d	,6837 a	,8779 e	,7366 bc	,7667 c	,7124 ab	,7553 c	,8085 d
20-40	,7490 bc	,7655 bc	,7892 c	,6235 a	,7154 abc	,6840 ab	,7716 bc	,8045 c
40-60	,5700 b	,6889 d	,6494 cd	,5053 a	,5762 b	,5785 b	,7479 e	,6384 c
60-80	,4805 c	,5654 e	,4028 b	,4616 c	,3326 a	,5526 de	,5132 cde	,5096 d
80-100	,3492 b	,4146 bc	,4150 bc	,2585 a	,4840 d	,3856 bc	,3653 b	,4376 cd
100-120	,2241 b	,2325 b	,1707 a	,2147 ab	,3262 cd	,2901 c	,3586 d	,2979 c
120-140	,1989 bc	,2370 c	,0489 a	,0885 a	,2082 bc	,1558 b	,2142 c	,2211 c
140-160	,1329 b	,1777 bc	,0402 a	,1344 bc	,1820 c	,0694 a	,1748 bc	,1767 bc
160-180	,0798 b	,0552 ab	,0415 ab	,0563 ab	,1250 c	,0206 a	,1545 c	,1497 c
180-200	,1188 c	,0204 a	,1216 c	,0341 ab	,1549 c	,0427 ab	,0988 bc	,1592 c
200-220	,0995 b	,1000 b	,1256 bc	,0341 a	,1239 bc	,1777 c	,1348 bc	,1795 c
220-240	,1286 abc	,0880 ab	,1196 abc	,0804 a	,1548 bc	,1608 c	,1513 abc	,1239 abc
240-260	,1342 bc	,1367 bc	,0848 ab	,0341 a	,1509 c	,1502 c	,1346 bc	,1549 c
260-280	,1467 c	,1021 bc	,0540 ab	,0321 a	,1084 bc	,1555 c	,1230 c	,1254 c
280-300	,1379 b	,1083 b	,0942 ab	,0313 a	,1258 b	,1366 b	,1230 b	,1557 b
300-320	,2020 c	,1802 bc	,0924 a	,1182 ab	,1471 abc	,1635 abc	,1389 abc	,1438 abc
320-340	,1886 b	,1904 b	,2096 b	,0460 a	,1674 b	,1667 b	,0783 a	,1726 b
340-360	,1857 b	,1663 b	,1377 b	,0624 a	,1597 b	,1870 b	,0655 a	,2473 c
360-380	,1364 bc	,1877 c	,1283 bc	,0277 a	,1075 b	,1610 bc	,1151 b	,2565 d
380-400	,0804 abc	,1706 e	,1012 bcd	,1330 cde	,0466 ab	,1403 de	,0387 a	,2483 f
0-400 cm	**0,2557 c**	**0,2629 c**	**0,2339 b**	**0,1860 a**	**0,2569 c**	**0,2544 c**	**0,2641 c**	**0,3000 d**

Table 14. Content of $C_{residue}$ by layers up to 400 cm according to the fertilization variants

The percent of $C_{organic}$ in comparison to C_{total} of soil varied within a wide range: from 43.11 % in the variant with $N_{180}P_{60}K_{60}$ to 71.51% in $N_{18}P_0K_0$. With the exception of systematic introduction of $N_{180}P_{60}K_{60}$, all other fertilization variants involved in the study contributed to the higher percent of the total humus substances in C_{total} of soil. The results from the investigation on the changes of organic C of soil showed that the independent nitrogen fertilization, especially with annually applied high norms, had a strong negative effect on the mobility of the organic substance and caused serious decrease of the percent of carbon in the insoluble residue.

The carbon of humic and fulvic acids (HA and FA) also varied considerably depending on the mineral fertilization (Fig.5). Regardless of the lower values of C of HA and FA down the soil profile, the variations between the fertilization variants were significant to a maximum degree, averaged for 0-400 cm depth. They were not significant for C-HA in the 0-40 cm layer and for C-FA in the 200-220 cm and 240-260 cm layers. Highest differentiation in the values of C-Ha between the fertilization variants was found in the 80-100 cm and 360-380 cm layers, and of the values of C-FA – in the 360-380 cm layer. Averaged for the investigated depth of 400 cm, variant $N_{180}P_0K_0$ had highest content of C-HA, exceeding the check variant with 24.9 %. A considerable increase of C-FA according to the check was determined after systematic application of $N_{180}P_0K_0$ – with 60.1 %, of $N_{60}P_{180}K_0$ – with 72.3 %, and of $N_{180}P_{60}K_{60}$ – with 70.0 %.

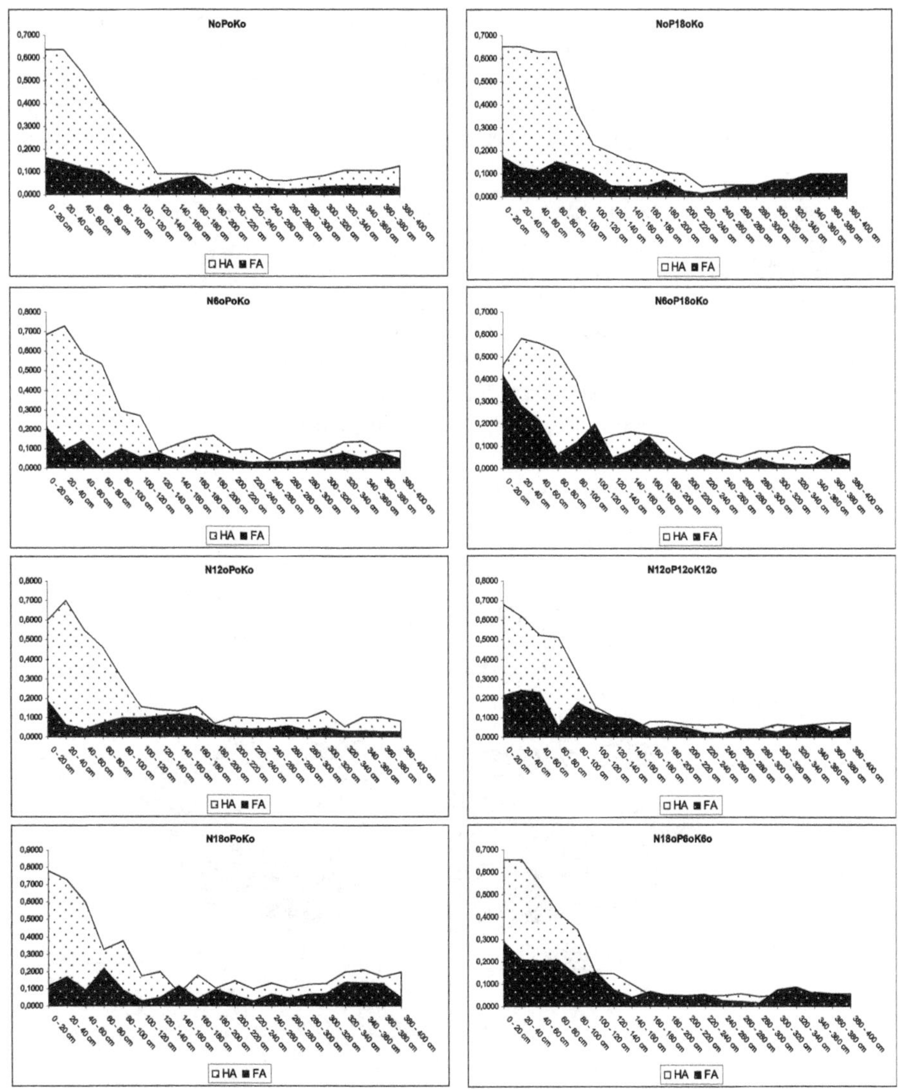

Fig. 5. Content of C_{HA} and C_{FA} by layers up to 400 cm according to the fertilization variants

The variants with independent introduction of all three nitrogen norms as well as of phosphorus had more C in HA in comparison to the check, as well as the variants with combinations of the main macro elements. The differentiation with regard to C content in FA was even better expressed. The check variant and the independent nitrogen fertilization with 60 and 120 kg N/ha had lowest amounts. In all other variants with combinations of the three macro elements, as well as in independent introduction of the highest nitrogen norm, the amount of FA increased. It reached maximum values after systematic application of $N_{60}P_{180}K_0$, similar to the fraction of aggressive FA. Averaged for this high depth profile, a tendency was observed towards higher HA content as a result from the long-term systematic mineral fertilization regardless of the norms and ratios of the main macro elements.

The changes in the content of HA and FA averaged for the 400 cm profile led to distinct differentiation between the fertilization variants with regard to the values of the ratio HA/FA (Fig.6). Variation was within a wide range: from 1.72 to 3.75. The long-term systematic nitrogen fertilization with 180 kg N/ha in combination with low fertilization norms of phosphorus and potassium determined the type of humus as fulvic-humic, averaged for the investigated depth 0-400 cm. In all other fertilization variants, regardless of the norms and ratios between the macro elements and in the check variant, the values of this ratio were above 2, which determined the type of humus as humic.

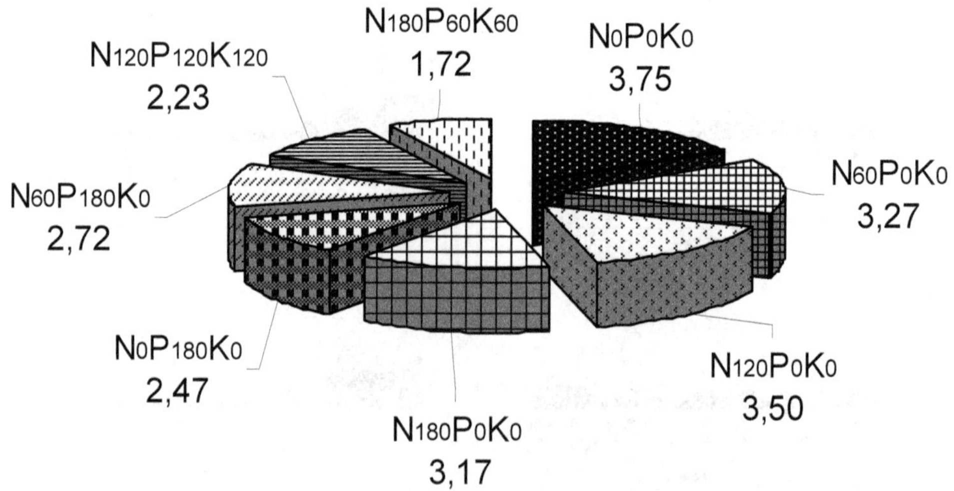

Fig. 6. Values of the ratio HA/FA average for the 0- 400 cm depth according to the fertilization variants

The similarities found between the separate fertilization variants with regard to the amount of HA linked to calcium were the reason for the lower rate of differentiation in their values. Systematic fertilization with $N_{180}P_{60}K_{60}$ and $N_{120}P_{120}K_{120}$ led to lower amount of HA-Ca, below the level of the check and all other investigated fertilization variants (Fig. 7). There was a well expressed tendency towards increase of carbon in HA-Ca as a result from the independent phosphorus and nitrogen fertilization, regardless of the nitrogen norm. Averaged for the 400 cm depth, the independent nitrogen fertilization with 180 kg/ha contributed most to the higher carbon in HA-Ca. The greater amounts of C_{HA} were due to this higher content of C in the HA fraction linked to calcium.

As a result from the long-term mineral fertilization averaged for the investigated depth, the degree of humification of the organic substance (OS) varied according to the type of the fertilization variant. The differentiation in the values of this index was extremely high regardless of the similarity and sameness in some of the variants. The 4 m profile had "very high" degree of humification of OS after systematic independent fertilization with 120 and 180 kg N/ha. Only at systematic application of NPK=3:1:1 the rate of humification averaged for the investigated profile was lowest (25.57 %) and according to Orlov and Grishina (1981) can be considered "moderate". In the other variants the values were 30-40 %, which determined humification as "high". It should be noted that in the check variant and in the variant with independent nitrogen fertilization with 60 kg/ha, the humification rate was at the upper limit tending towards "very high", while in the independent phosphorus fertilization and in the variants with $N_{60}P_{180}K_0$ and $N_{120}P_{120}K_{120}$ the values were closer to the lower limit.

Averaged for the tested variants of long-term fertilization, the slightly leached chernozem soil in the trial field can be referred to the low humic type according to the classification of Orlov and Grishina. C_{total} was highest in the 0-20 cm layer (1.62 %) and gradually decreased down the 4-m profile. In the last investigated layer (380-400 cm) its content was 0.26 %, but the 260-280 cm layer had lowest content.

Along the entire investigated profile, organic C was represented by humic acids which exceeded fulvic acids. The amount of HA was highest in the 0-20 cm layer and gradually decreased down the profile and at 380-300 cm it was 0.0919 %. Similar to organic carbon (total humic substances, THS), the amount of HA also slightly increased at depth below 300 cm. The above tendencies remained the same with regard to the changes in FA down the profile. According to the classification of Orlov and Grishina (1981), the slightly leached chernozem soil in the trial field possessed high to very high content of HA, expressed as percent from the organic C in soil. The values of this ratio were more than 80 % at depth from 20 to 80 cm. They gradually decreased down the profile but nevertheless remained within 60 - 80 %. The greater part of HA along the entire 400 cm profile was linked to calcium. Down the profile their amount gradually decreased and in the 220-300 cm zone the lowest concentrations were detected. At the 4th meter their concentration slightly increased. At depth 40-100 cm the amount of HA-Ca was very high (>80 % of HA). At all other depths the percent of HA-Ca/HA was 60-80 % and can be considered high. The amount of C in HA, free and linked to R_2O_3, was lower than the amount of the HA linked to Ca and also slightly decreased down the profile, the lowest values being registered in the 320-340 cm layer especially at $N_{180}P_{60}K_{60}$.

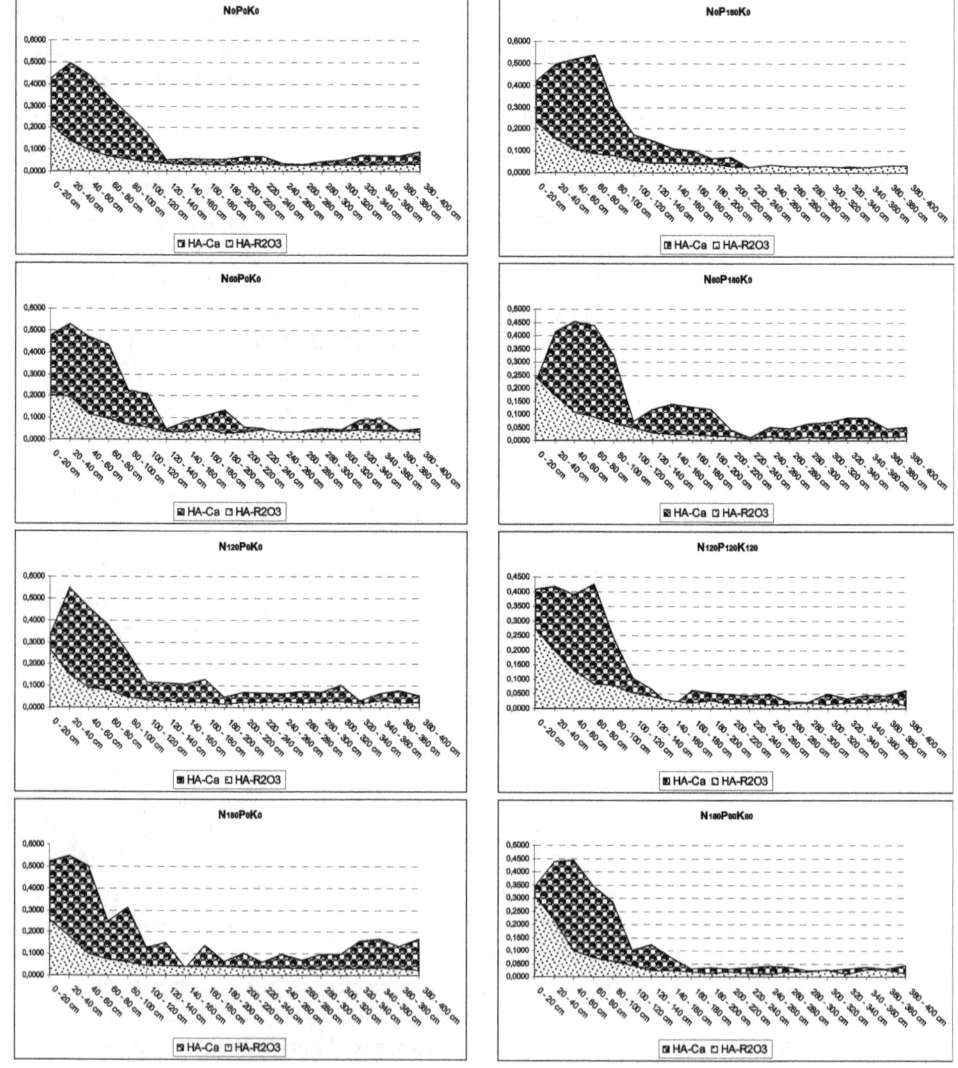

Fig. 7. The amount of C in HA, linked to Ca and free and linked to R_2O_3 by layers up to 400 cm according to the fertilization variants

2.3.3 Soil organic matter reserves in depth up to 60 cm

The systematic introduction of macro elements at different norms and ratios during a period of 40 years of cultivation of the trial field lead to formation of different reserves of total carbon in soil at depth up to 60 cm with well expressed differentiation (Fig 8). The long-term 2-field agricultural use of the trial field without mineral fertilization was characterized with lowest reserves of total C. The independent nitrogen fertilization with increasing norms caused their increase according to the check variant with 12.6 %.This increase, however, was lower than the increase registered in all other variants. Highest reserve in absolute values at the moment of taking samples was found in the variants with 40-year fertilization with $N_0P_{180}K_0$ and $N_{60}P_{180}K_0$. The main reason for this fact is that besides the variation in the content of total C, respectively humus, variation in the values of the other component was found when determining reserves - volume density of soil. According to Yankov (2007, Personal Communication), highest values of volume density averaged for the 0-60 cm layer were demonstrated by the variant with systematic introduction of phosphorus (180 kg/ha) - 1.43 g/m³, and lowest mean values - by the variant with $N_{180}P_{60}K_{60}$ (1,22 g/m³). Over 36 % of the total carbon reserves in soil at depth up to 60 cm were concentrated in the 20-40 cm layer, followed by the layer lying beneath (Table 12). Regardless of the low differentiation in the content of total C down the soil profile up to the 60th cm, the differentiation of the layers according to their reserves was very well expressed. Humus reserves in soil were highest in the 20-40 cm layer. The layer 10-20 cm have a negative C balance according to check variant. The maximum increase according to the control in 0-10 cm and 10-20 cm layer was established in the variants with $N_{180}P_0K_0$ and $N_0P_{180}K_0$ (136,3 and 135,6 %, respectively for 0-10 cm layer and 103,2 % and 105,3 % for 10-20 cm layer).

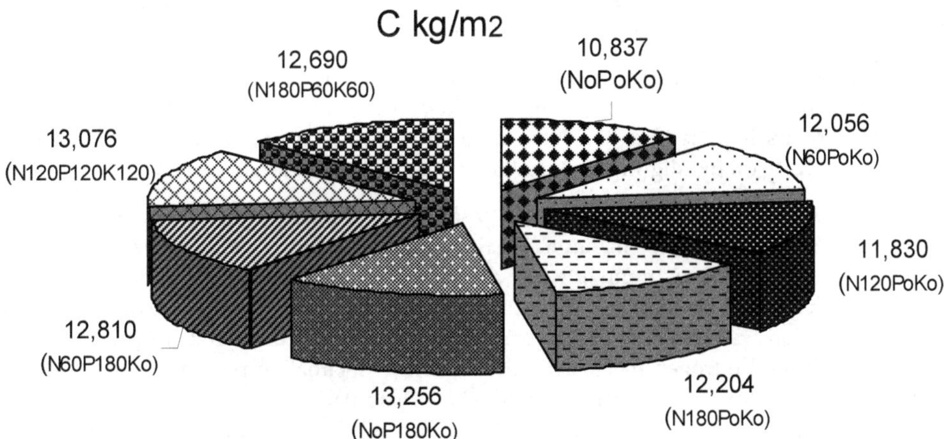

Fig. 8. Total Carbon reserves for layer 0-60 cm, C kg/m²

Combination between macroelements affected positively the humus reserves in soil. The variant with balanced fertilization $N_{120}P_{120}K_{120}$ contributed enrichement of soil carbon reserves in 40-60 cm layer with 37,1% according to the same layer in check variant. This tendency was established also for long-term fertilization with $N_{180}P_{60}K_{60}$.

Soil depth, cm	$N_0P_0K_0$	$N_{60}P_0K_0$	$N_{120}P_0K_0$	$N_{180}P_0K_0$	$N_0P_{180}K_0$	$N_{60}P_{180}K_0$	$N_{120}P_{120}K_{120}$	$N_{180}P_{60}K_{60}$
0 - 10	1,659	1,762	1,955	2,262	2,250	2,162	2,069	1,801
10 - 20	2,323	2,104	2,301	2,397	2,446	2,080	2,127	2,240
20 - 40	3,694	4,233	4,593	4,384	4,754	4,891	4,548	4,645
40 - 60	3,161	3,957	2,981	3,161	3,807	3,677	4,333	4,005

Table 15. Reserves by depth up to 60 cm according to fertilization, C - kg/m_2

3. Conclusion

The systematic mineral fertilization for a period of 40 years with different norms and at different ratios between nitrogen, phosphorus and potassium had high effect on the agrochemical condition of slightly leached chernozem (Haplic Chernozems) down the soil profile.

The soil acidity forms, averaged for the investigated depth of the 0-400 cm profile, were significantly affected by the type of fertilizer combination. The depth was the factor with decisive effect in all forms of soil acidity. Influence of mineral fertilization was higher on exchangeable Al^{3+}, Ca^{2+} and the $\sum Ca+Mg$, and significantly lower - on the values of residual hydrolytic acidity and the rate of alkali saturation. The amount of exchangeable Mg^{2+} had a clear tendency toward increasing down the soil profile

Independent long-term mineral fertilization with 180 kg N/ha and with $N_{180}P_{60}K_{60}$ caused the occurrence of exchangeable Al^{3+} in the soil absorption complex in the surface layers 0-10 and 10-20 cm. It was not present further down the profile.

The variant with independent nitrogen fertilization with 180 kg/ha has also the lowest $\sum Ca+Mg$ and the lowest value of sorption capacity. Intensive mineral fertilization with $N_{180}P_{60}K_{60}$ caused decreasing of degree of saturation with bases.

The value of the pH, sorption capacity, acidity on strongly acid positions and degree of saturation with bases showed a tendency of decreasing at the end of 40th year of trail beginning comparing the end of 30th year. In the same way independently of fertilizer variant the acidity on the slightly acid positions is strongly increased.

The mineralization ability down the soil profile was affected to a maximum degree of significance by the mineral fertilization and the incubation. Depth had decisive effect on the value of the index. The maximum effect of this factor was registered after 28-day incubation - 92.3 %. The role of mineral fertilization on nitrogen mineralization according to the incubation period was significantly less expressed – 9.2 %, 3.2 5 and 4.0 %, respectively. The amount of nitrified nitrogen increased with the longer incubation periods with 52.7 % (28 days) and with 118.8 % (56 days), respectively, in comparison to 14-day incubation. The long-term mineral fertilization with $N_{180}P_{60}K_{60}$ had the highest values of mineralization ability for all three incubation periods. The established strong effect of systematic mineral fertilization regardless of the norm and ratios of nitrogen, phosphorus and potassium on the mineralization ability of soil in comparison to the check variant was highest at depths 0-100 cm and 300 – 400 cm.

Systematic mineral fertilization carried out for 40 years in two-field crop rotation (wheat – maize) affected the content of C_{total} deep down the profile of the slightly leached chernozem soil. Systematic use of $N_{180}P_{60}K_{60}$ contributed most for the increase of its content at average depth 0-400 cm. Fertilizer variants $N_{180}P_{60}K_{60}$ and $N_{120}P_{120}K_{120}$ led to lower amount of HA-Ca, below the level of the check and all other investigated fertilization variants. There was a well expressed tendency towards increase of carbon in HA-Ca as a result from the independent phosphorus and nitrogen fertilization, regardless of the nitrogen norm.

The ratio C_{HA}/C_{FA} putted under average for the 400 cm profile to distinct differentiation between the fertilization variants.Variation was within a wide range: from 1.72 to 3.75. The long-term systematic nitrogen fertilization with 180 kg N/ha in combination with low fertilization norms of phosphorus and potassium determined the type of humus as fulvic-humic, averaged for the investigated depth 0-400 cm. In all other fertilization variants, regardless of the norms and ratios between the macro elements and in the check variant, the values of this ratio were above 2, which determined the type of humus as humic. Independent nitrogen fertilization, especially with annually applied high norms, had a strong negative effect on the mobility of the organic substance and caused serious decrease of the percent of carbon in the insoluble residue.

As a result from the systematic mineral fertilization in the 20-40 cm layer, higher reserves were formed by the layers lying above and below. Triple NPK combinations ($N_{120}P_{120}K_{120}$ and $N_{180}P_{60}K_{60}$) enriched organic mater reserves in 40-60 cm layer.

These results showed that regardless of the intensive agricultural activities and changes of some agrochemical characteristics, slightly leached chernozem soil (Haplic chernozems) in Sough Dobrudzha region in Bulgaria preserved its main genetic characteristics at the lower depths of the soil profile.

The effect of long-term fertilizer treatments on detail nutrient balances, technological quality of crops, concentration of available forms of macro elements and trace elements in soil and plant biomass dynamics and many other aspects of this experiment await detailed analysis.

4. Acknowledgment

The author is deeply indebted to Prof Maria Petrova for the initiated of this long term trail in 1967. The author thank to Dobrudzha Agricultural Institute and staff of Agrochemistry lab for correct support.

5. References

FAO, (2006). World Reference Base of Soil Resources. Rome, Italy.

Filcheva E., Tsadilas, C. (2002). Influence of Cliniptilolite and Compost on Soil Properties. *Commun. Of Soil Sci. and Plant Analysis, 33, 3-4, 595-607*

Ganev,S.& Arsova, A. (1980). Methods for determining weak acid cation exchange of soil. *Soil science and agrochemisty,* Vol XV, No 3,pp. 22-32

Goulding K.W.T., Poulton P.R., Webster C.P., Howe M.T. (2000). Nitrate leaching from Broadbalk wheat experiment, Rothamsted, UK, as influenced by fertilizer and manure inputs and the weather. *Soil Use Manage, 16, 244-250*

Iliev I. (2000). Mechanized soil sampler. *BG 385 Y1. Patient of useful model*

Iliev I., Nankova M. (1994). Another type motor-driven portable soil sampler. *ESNA XXIV th Annual Meeting,* September 12-16, 1994, Varna, BULGARIA: 140-147

Khan S.A., Mulvaney R.L., Ellsworth T.R., Boast S.W. (2007). The myth of nitrogen fertilization for soil carbon sequestration. *J.Environ.Qual. 36, 1821-1832*

Kononova M.M., Belchikova, N.P. (1961). Rapid methods of determining the humus composition of mineral soil. *Sov. Soil.Sci.* No10, pp.75-87

Koteva V. (2010) Effects of 45-years mineral fertilization on mobile potassium condition of the *Pellic Vertisols.* 45th *Croatian and 5th International Symposium of Agriculture,* pp. 787-791.

Kunzova,E., Hejcman M. (2009). Yield development of winter wheat over 50 years of FYM, N,P and K fertilizer application on black earth soil in the Czech Republic. *Field Crops Reseach,* No 111, pp.226-234

Nankova Margarita (2005). Personal Communication

Nankova M., Djendova R., Penchev E., Kirchev H., Yankov P. (2005). Effect of some intensive factors in agriculture on the ecological status of slightly leached chernozems. *Proceedings National Conference with International Participation "Management, Use and Protection of Soil Resources", 15-19 May 2005, Sofia:* pp.155-159.

Nankova M., Tonev T., Stereva L. (1994). Humus Fraction Composition of the Slightly Leached Chernozem Depending on Duration of Fertilization and Rotation Type, II. Influence of the Rotation Type. *Soil Science and Strategy for Sustainable Agriculture, Proceeding of the Fifth National Conference with International Participation, 10-13 May, Sofia, BULGARIA*

Nankova Margarita (2010). Long-term mineral fertilization and its effect on humus conditio of the Haplic Chernozems in Dobroudja. *"ADVANCES IN NATURAL ORGANIC MATTER AND HUMIC SUBSTANCES RESEARCH 2008-2010", XV Meeting of the International Humic Substances Sosiety, Puerto de la Cruz, Tenerife, Canary Island, 27 June-2 July 2010: 419-423*

Panayotova G. (2005). Influence of 35-year long mineral fertilization over the agro-chemical characteristics of two-way soil profile of *Pellic Vertisols. Soil Scence, agrochemistry and ecology,* vol. XL, pp. 66-71

Sims, G.K.(2006). Nitrogen starvation promotes biodegradation of N-heterocyclic compounds in the soil. *Soil Biol.Biochem.,* No 38, pp. 2478-2480. SPSS 16.0 2007

Takahaski S., Anwar, M. 2007. Wheat grain yield, phosphorus uptake and soil phosphorus fraction after 23 years of annual fertilizer application to an Andosol. *Field Crop Research, 101, 160-171*

Yankov P. (2007) Personal Communication

6

Spatial Patterns of Water and Nitrogen Response Within Corn Production Fields

Jerry L. Hatfield
National Laboratory for Agriculture and the Environment
USA

1. Introduction

Agricultures role on environmental quality has been debated for many decades and although there has been advances in our understanding of the linkage between agricultural management and environmental quality, there is still much we don't understand about the combination of field-scale and watershed scale management changes (Hatfield et al., 2009). There is increasing interest in developing solutions to environmental quality problems originating from agricultural management; however, the challenge remains on how we integrate the pieces of a very complex puzzle together to achieve solutions which transcend space and time scale. Within the Midwestern United States, the reoccurrence of the hypoxic zone within the Gulf of Mexico and the large increase in the size of the hypoxic zone after the 1993 floods in the Midwest focused attention on the role of agriculture in nonpoint source pollution. Burkart and James (1999) evaluated the nitrogen (N) balance for the Mississippi River Basin and concluded that mineralization of soil organic matter and application of commercial fertilizer was two primary contributors to N load. Jaynes et al. (1999) after examination of a small watershed (5400 ha) in central Iowa found that nitrate-N losses averaged 20 kg ha^{-1} for this watershed and reached a level in excess of 40 kg ha^{-1} during 1993. Hatfield et al. (1998) found that drainage from Walnut Creek was the primary transport pathway for nitrate and the annual loads were related to precipitation differences among years. Hatfield et al. (1999) found that for Walnut Creek watershed that drainage through the subsurface drain lines accounted for approximately half of the annual precipitation with evapotranspiration accounting for the other half. This movement of water through the soil profile and the solubility of nitrate in water produce large amounts of nitrate-N loss (Jaynes et al, 1999). These results suggested a strong link between precipitation, crop water use patterns, and nitrate losses. Occurrence of the hypoxic zone has prompted an increased level of debate about the need to reduce N inputs into agricultural systems. Opponents of this conclusion argue that over the past 20 years the input of N fertilizer has not increased, soil organic matter levels haven't changed, and crop production levels have increased suggesting that the efficiency of the agronomic production system has increased and any change in nitrate loss would be difficult to achieve. Hatfield et al. (2009) performed an analysis of the temporal changes in the nitrate-N concentrations in the Raccoon River watershed in central Iowa and observed the changes in nitrate-N concentrations since the 1970's were dominated more by changes in land use practices than on nutrient management. This change in land use affected the soil water balance and the patterns of water movement within fields and throughout the season rather than application rates as the primary transport mechanism.

Crop yield response to N has been the focus of agronomic studies and has been primarily conducted on small plots under relatively controlled conditions. However, the advent of application systems to differentially apply N across fields has opened up new possibilities for quantifying N response at a scale which producers would have confidence of the information being applicable to their operations. Understanding field scale responses to N management practices is critical to the development of precision agriculture tools and methods for determining the patterns to change application rates within a field. One method that has been applied to the precision application of N has been to use leaf color or reflectance as a measure of N status. Detection of N stress in crop plants using leaf color has created opportunities for field-scale evaluation of N response. Some of the methods that have been employed are the Soil-Plant Analyses Development (SPAD) chlorophyll meter, color photography, or canopy reflectance factors to assess N variation across corn fields (Schepers et al. 1992, 1996; Blackmer et al. 1993, 1994, 1996a, 1996b; Blackmer and Schepers, 1996). These methods have been based on comparisons of an adequately-fertilized strip in the same field with a strip with an altered N rate. This approach eliminates the requirements for prior knowledge of the relationship between nutrient concentration and crop reflectance.

There have continued to be advances in the use of remote sensing methods to estimate N status in crops. Lee et al. (2008) used a single waveband at 0.735 µm to quantify N status in rice (*Oryza sativa* L.). In their method they used the first derivative of the reflectance at 0.735 µm as difference of the reflectance at 0.74 - 0.73 µm and dividing by ten. They found this to be as accurate as other indices including the normalized difference vegetative index (NDVI) expressed as $(R_{NIR} - R_{RED})/(R_{NIR} - R_{RED})$ where R_{NIR} reflectance in the near-infrared waveband (0.78 - 0.79 µm) and R_{RED} the reflectance in the 0.61 - 0.68 µm waveband. They also compared a simple ratio vegetative index expressed as R_{NIR}/R_{RED}. They used these wavelengths to create maps of canopy N status across fields at the panicle formation stage. They showed the variation in canopy N status but didn't evaluate the spatial patterns within fields. There continues to be advancements in the use of remote sensing to detect N status in crops, Chen et al. (2010) developed a double-peak canopy N index to predict plant N concentration in both corn (*Zea mays* L.) and wheat (*Triticum aestivum* L.) and found this worked well for N status; however, did not extend this approach to assess field variation. This is an extension of an earlier method proposed by Haboudane et al. (2002) in which they proposed leaf chlorophyll indices could be used to predict leaf chlorophyll content for application to precision agriculture. One of the lingering questions that remain is the degree of variation present in N response across a corn production field.

Spatial variation of crop yield across fields has prompted a series of questions about the role of nutrient management. Jaynes and Colvin (1997) showed that yield variation within central Iowa fields was related to precipitation differences among years. However, in their study there was not a measure of crop water use within the field with the implied assumption that precipitation was uniform across the field. Hatfield and Prueger (2001) and Hatfield et al. (2007) found a large amount of spatial variation in water use across fields related to soil types and N management. These results have prompted a series of studies developed to further elucidate these interactions. Studies conducted on water use and N application has been done within a single soil type or management zone. There is a lack of information on the interactions among soil, crop water use, and N management that would help develop an understanding of better N management in cropping systems typical of the Midwest United States. The increasing concern about the role of agricultural practices on water quality have prompted us to ask a series of questions about the interactions between crop water use, N use, yield, and soils. The objectives of our studies have been to quantify the interactions among N application, crop water use, yield, and soils for central Iowa production fields on the Clarion-Nicollet-Webster soil association and to determine the impact of these interactions on N management and potential offsite impacts from drained agricultural lands.

Spatial analysis of field scale changes have recently been evaluated by Inman et al. (2008). They determined the relationship between early season NDVI, soil color-based management zones, and relative corn yields. Values of NDVI were collected at the eight-leaf growth stage and regression models explained between 25 to 82% of the variability in relative yield where relative yield was defined as the ratio of the observed yield to the maximum yield for the field as defined by Brouder et al. (2000). Inman et al. (2008) found coupling NDVI with the color-based soil zones didn't increase the ability to explain yield within the field. A complimentary study conducted by Massey et al. (2008) on ten years of site-specific data for corn, soybean (*Glycine max* (L.) Merr.), and grain sorghum (*Sorghum bicolor* L.) for a 36.5 ha field in central Missouri with claypan soils to quantify temporal changes in crop yield response. They developed profitability maps for the field and found that large areas of the field had negative profit due to areas of the field in which there was significant topsoil erosion. Brock et al. (2005) showed that high yielding management zones in a corn-soybean rotation were associated with poorly drained level soils while low yielding zones were associated with eroded or more sloping soils. Sadler et al. (2000a and 2000b) conducted a field scale study on drought stressed corn and found that although there was a relationship between soil map units and grain yield these relationships did not explain the reason for the yield variation. They found that variation among sites within soils was significant and suggested that improved understanding of yield variation would require more attention to within season observations of crop water stress. These studies have shown that there is a large amount of variation present within fields and that our understanding of the reasons for these patterns of variation would improve the knowledge base for precision agriculture applications.

There have been several studies that have related remote sensing indices to crop yield and detection of N status in plant leaves; however, there has been little research on the spatial patterns within the field that could lead to improved understanding of the changes that occur within the growing season. There is a lack of understanding of the interactions between the spatial patterns of crop water use and N response. This study was conducted to evaluate the spatial and temporal patterns of crop water use and couple these observations with observations collected from N strips within fields as part of a N evaluation study. The objective of this study was to evaluate the spatial patterns within different fields observed by remote sensing and yield maps collected at harvest with yield monitors to determine the information content contained in spatial analyses of agricultural fields.

2. Methodological approach

2.1 Water-nitrogen interaction studies

Studies have been conducted in the Clarion-Nicollet-Webster Soil Association in central Iowa within the Walnut Creek watershed. This 5400 ha watershed was described in Hatfield et al. (1999). Production sized fields have been used as experimental units for these studies because of the need to quantify the effects of different N rates on crop yield and water use across soil types. These production fields ranged in size from 32 to 96 ha each with a similar experimental design. The experimental design was a replicated plot design with strips of different N rates being a treatment and each treatment replicated at least three times across a field. Nitrogen rates were applied in 1997 and 1998 using a starter application at planting of 50 kg ha^{-1} only with the second treatment having the N starter rate and the sidedress rate determined by the Late Spring Nitrate Test (LSNT). The third treatment was the starter plus a sidedress rate to represent a non-limiting N rate of an additional 150 kg ha^{-1}. This experimental procedure was described by Jaynes et al. (2001). In 1999 and 2000, the N application procedure was modified to further refine N application rates based on the use of the leaf chlorophyll measurements and soil test based on the results obtained from the 1997 and 1998 experiments. The rates applied were 50, 100, and 150 kg ha^{-1} to different soils, planting rates, and plant population densities (75,000 and 85,000 plants ha^{-1}). Nitrogen rates were applied uniformly across each field using liquid UAN (32% N) for these studies. These applications were applied with production scale equipment to mimic producer operations.

Soil N concentrations were measured prior to spring operations, after planting, and at the end of the growing season after harvest to a soil depth of 1.5 m using a 5 cm core. Cores were subdivided into depth increments to estimate the N availability throughout the root zone. Sample position was recorded with a GPS unit to ensure accurate location of each subsequent sample. Nominal plot size for plant measurements and yield determinations was 15 x 15 m. Yields were measured on 5 m length of row for five subsamples within each plot in which no plants had been removed or measured during the growing season. Plots were replicated three times within each treatment.

Crop water use rate was measured from planting to harvest with an energy balance method (Bowen ratio in 1997 and 1998, and eddy correlation in 1999 and 2000). These units consisted of a net radiometer positioned at 3 m above the canopy, soil heat flux at 10 cm (within the row, middle of the row, and adjacent to the row), wind speed, air temperature, water vapor pressure (positioned at 0.5 and 1.5 m above the canopy), three dimensional sonic anemometer and krypton hygrometer (positioned at 1 m above the canopy), and an infrared thermometer (15° fov) positioned at 45° from nadir in a south-facing direction. All measurements were recorded at 10-second intervals and either 15 minute or 30 minute averages stored in the data acquisition unit. Data were screened to ensure proper data quality and then converted to equivalent water depth and summed for the growing season to determine crop water use. Within each soil-N combination we had a single energy balance station and this measurement technique provides a sample representative of a 50 m^2 area. These instruments were positioned to provide a measurement of the representative area around the plant measurements. Missing data for a given station were treated by using the relationships among instruments within a soil type across N management practices to determine the relationships among variables. These relationships were then used to fit in any missing data using an approach similar to the method described by Hernandez-Ramirez (2010). Generally the length of any missing data was less than 3 hours. The amount of missing data for these experiments was less than 3% of the total data record.

Crop transpiration rates were estimated from an energy balance model that determined the soil water evaporation rate based on the leaf area index of the crop and previous precipitation amounts following the approach described by Ritchie and Burnett (1971). Soil water evaporation rates were estimated from a surface energy balance model based on crop residue cover amounts and the energy balance. Precipitation for these studies was available from a tipping bucket raingauge located at a meteorological station within 1 km of the field sites.

Crop growth and development were measured in a variety of ways. In 1997 measurements were made of yield at harvest. Beginning in 1998 and 1999, a more intensive plant regime of weekly plant measurements consisting of leaf area, phenological stage, number of leaves, dry weight, and plant height were made on 10 plants from each plot and each plot was replicated three times. In 2000, the frequency was decreased to four destructive plant samplings to represent the 6-leaf stage, 12-leaf stage, tasseling, and mid-grain fill. Leaf chlorophyll measurements were made on 30 plants in each plot with a leaf chlorophyll meter at two times per week commencing with the 6 leaf stage and continuing through late grain fill. The upper leaf was measured at the mid-leaf position until the tassel appeared then the leaf immediately above the ear position was measured. Leaf carbon and N contents were determined on dry, ground samples from same stages as the 2000 plant samples were collected. For 1999 and 2000 experiments, stalk sugar content was measured with a refractive method using sap collected from freshly cut stalks. This was done on ten plants from each plot. Grain quality parameters of protein, oil, and starch were measured on subsamples of grain collected from the hand-harvest samples. Field yields were measured with yield-monitors mounted on the producers combine. These data were registered with a GPS unit to obtain field locations.

Data analyses for these studies are based on crop yield, total seasonal water use, and N application. Water use and N use efficiency was determined by the ratio of crop yield to either seasonal crop transpiration or N application rates. Intensive plant sampling data are not described in this report but were used to understand the dynamics of plant response to changes in within season N management decisions. Likewise, the leaf chlorophyll and stalk sugar content data were used to guide decisions in the 1999 and 2000 experiments. These data sets represent a complete analysis of crop-soil-water-N interactions.

2.2 Spatial nitrogen studies

Experiments were conducted in 12 different fields located in central Iowa from 2000 to 2002 (Table 1). These fields varied in size from 15 to 130 ha. Each of the field experiments was similar with strips arranged in the field with different N rates and in three of the study fields, variable planting rates were also used as a treatment variable. The strips were a minimum of 50 m wide covering at least 60 rows of corn. The arrangement of the N treatments was randomized across the field and strips were treated as replicates. Nitrogen was applied as UAN (Urea and Ammonium Nitrate) as a preplant treatment in all fields and incorporated into the soil. Rates of N application were determined by using soil test records from the producer's fields to determine the typical rate applied and then using 0.5 and 1.5 times that rate in the other strips. All production practices were conducted by the producer during the course of the experiment. In one experiment, planting rates were evaluated and randomly assigned as blocks across the field with N rates as subplots within a block. Planting rates were also randomized across the field. In each field the same corn hybrid was planted; however, hybrids varied among fields and years. For each field there were overlays of soil type, elevation collected from Real Time Kinetic GPS equipment to create 1 m contours, and N application rate. Soils within fields were a mixture of Canisteo (Fine-loamy, mixed (calcareous), mesic Typic Haplaquolls), Clarion (Fine-loamy, mixed, mesic Typic Hapludolls), Nicollett (Fine-loamy, mixed, mesic Aquic Hapludolls), Okoboji (Fine, montmorillonitic, mesic Cumulic Haplaquolls), and Webster (Fine-loamy, mixed, mesic Typic Haplaquolls) soils. An example of the field layout for this experiment is shown in Fig. 1.

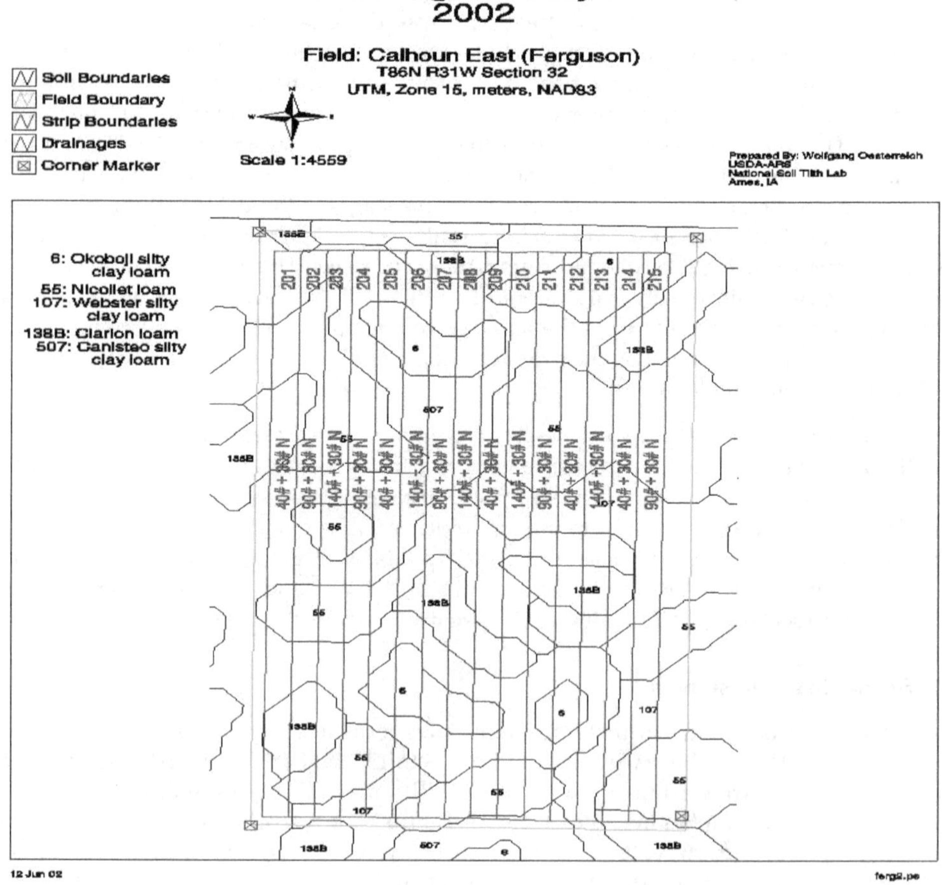

Fig. 1. Experimental layout for field-scale N management study conducted across central Iowa from 1999 to 2002.

Year/Field	Nitrogen Rate (kg/ha)	Agronomic Variables
2000/Carroll 1	70, 116, 162, 210	72,000 plants/ha
2000/Carroll 2	70, 116, 162, 210	72,000 plants/ha
2000/Sac	70, 116, 162, 210	72,000 plants/ha
2000/Shelby	70, 116, 162, 210	72,000 plants/ha
2000/Story 1	52, 100, 145	72,000 plants/ha
2000/Story 2	52, 100, 145, 191	72,000 plants/ha
2001/Story 1	133, 180	75,000 plants/ha
2001/Story 2	51, 102, 144	75,000 plants/ha
2000/Story 3	57, 126, 173	75,000 plants/ha
2002/Calhoun East	78, 134, 190	79,000 plants/ha
2002/Dallas South	78, 134, 190	79,000 plants/ha
2002/Coon Rapids	67,134,201	57,000, 69,200, 81,500 plants/ha

Table 1. Nitrogen rates and associated agronomic parameters for the field experiments from 2000 to 2002.

Soil map units were extracted from the soil survey data and detailed topographic data were collected with GPS equipment for each field. Soil nutrient content data were collected from each field for the upper 1 m of the profile for a minimum of 30 locations within each field prior to the growing season. Yield data were obtained from each field using yield monitors on the producers combine and these were calibrated prior to harvest. Yields were corrected to 15.5% grain moisture content.

Remote sensing observations for each field were obtained from an aircraft based hyperspectral unit with 35 wavebands (Table 2). Details on the wavebands and radiometric resolution are provided in Table 2 for each waveband. Data were obtained on clear days at four times during the growing season in May, July, August, and September. These times were selected to obtain data after planting and before there was sufficient growth to affect the soil background (May), maximum vegetative growth (July), mid-grain fill (August), and near physiological maturity (September). The pixel sizes were nominally 2 x 2 m across the study sites. These data were collected with GPS signals to provide a location of each pixel and these were georeferenced to the yield data and other parameters collected from each field with a positional accuracy of 5 m. All data were georeferenced and all data were aggregated into 8 x 8 m pixels for further analysis. This size was selected because of the width of harvest equipment for the yield monitor data.

Band Number	Center Wavelength (µm)	Band Width (µm)	Band Number	Center Wavelength (µm)	Band Width (µm)	Band Numbe r	Center Wavelength (µm)	Band Width (µm)
1	0.426	0.003	13	0.681	0.006	25	0.797	0.006
2	0.445	0.006	14	0.689	0.006	26	0.805	0.006
3	0.506	0.006	15	0.698	0.006	27	0.819	0.005
4	0.520	0.006	16	0.707	0.006	28	0.826	0.005
5	0.530	0.006	17	0.715	0.006	29	0.833	0.005
6	0.540	0.006	18	0.724	0.006	30	0.851	0.005
7	0.549	0.006	19	0.734	0.006	31	0.860	0.006
8	0.561	0.006	20	0.743	0.006	32	0.869	0.006
9	0.580	0.006	21	0.755	0.006	33	0.880	0.006
10	0.599	0.006	22	0.766	0.006	34	0.890	0.006
11	0.620	0.006	23	0.774	0.006	35	0.899	0.006
12	0.639	0.006	24	0.785	0.006			

Table 2. Wavebands and bandwidth for the AISA hyperspectral data collected over the field sites in 2000 to 2002.

Data were analyzed on the individual N strips within the field and across the field. Vegetative indices were computed from the reflectance data obtained from the aircraft data and correlated to strip yield and field level yield observations. Correlation and regression analysis were conducted between N rate and corn yield for each individual field and across all fields for each year of the study. T-tests among means were conducted on differences between soils within a field strip while analysis of variance was conducted on N rates across a field with an interaction term based on the soil type by N rate comparison. These analyses were made using ANOVA and GLM models in SAS (SAS 2009). Spatial analysis was conducted for each field using GS+ version 5.1 to determine the spatial relationships among yield and the vegetative indices across different fields. Autocorrelation values were computed across the field using the field location points as coordinates to compute the range and sill for yield and the red/green index.

3. Results and discussion

3.1 Spatial variation in soil water use

Results from the studies where crop water use was quantified across the field revealed major differences among soil types. An example is shown in Fig. 2 for crop water use patterns for a Clarion and Webster soil with two different N rates. There was only a single meteorological unit in each soil type; however, error analysis for daily measurements are less than 10% of the daily total water use so the differences shown among soil types and N rates in Fig. 2 are significantly different at the end of the growing season. Crop water use patterns began to deviate shortly after crop establishment and continued throughout the growing season with the most noticeable difference occurring after Day of Year (DOY) 200 at the beginning of the grain-filling period (Fig. 2). The differences between the Clarion and Webster soil are due to their organic matter content with the Clarion soil having organic matter content between 1-2% and the Webster soil between 4-5%. This leads to a difference in water holding capacity of nearly 100 mm in the upper 1 m for these two soil profiles. By extension, this creates a difference in the amount of soil water available to the growing crop during the season. In this comparison, Clarion soil with the lower amount of N applied showed the larger amount of crop water use during the season and ultimately showed the larger grain yield at the end of the growing season compared to high rates of N application. In typical growing seasons in central Iowa, there is adequate soil water available for plant growth early in the growing season and it is not until the onset of the reproductive period in which crop water use rates increase and precipitation amounts begin to diminish and to meet atmospheric demand there is a reliance on the amount of water stored within the soil profile.

A recent study by Hatfield et al. (2007) revealed large differences in the daily and seasonal amounts of crop water use across corn and soybean fields in central Iowa. They found the primary reasons for these differences were due spatial variation in precipitation and spatial variation in soils across various fields. These results confirm the observations collected from multiple soils within the same field. The observations collected from several different studies across multiple years reconfirm the observations shown in Fig. 2.

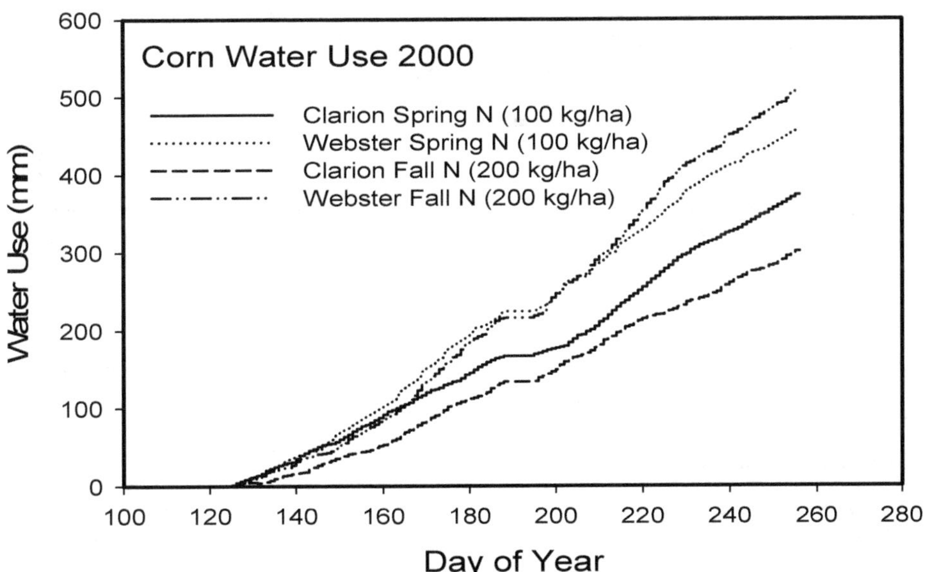

Fig. 2. Crop water use patterns during the 2000 growing season for corn on two soil types with two nitrogen application rates.

An extension beyond the crop water use patterns is the development of an assessment of water use efficiency. Water use efficiency is expressed as the amount of grain yield relative to the seasonal total of transpiration. In this case, transpiration was determined by removing the soil water evaporation component from the total crop water use amounts. In this analysis, water use efficiency was calculated for the 150 kg ha^{-1} N application rate. What is striking in this figure is the observation that we have a large number of points which are below the line which can be interpreted as using water but not producing a large yield. Some of these are related to low N rates; however, others have large amounts of N (Fig. 3). The primary reason for this type of relationship is due to the seasonal pattern of crop water use (Fig. 2) which causes water deficit conditions in the grain-filling period. Water is the primary variable causing variation in corn yields more than N application rates. These observations prompted the extension of the N studies to be conducted across a production fields in central Iowa.

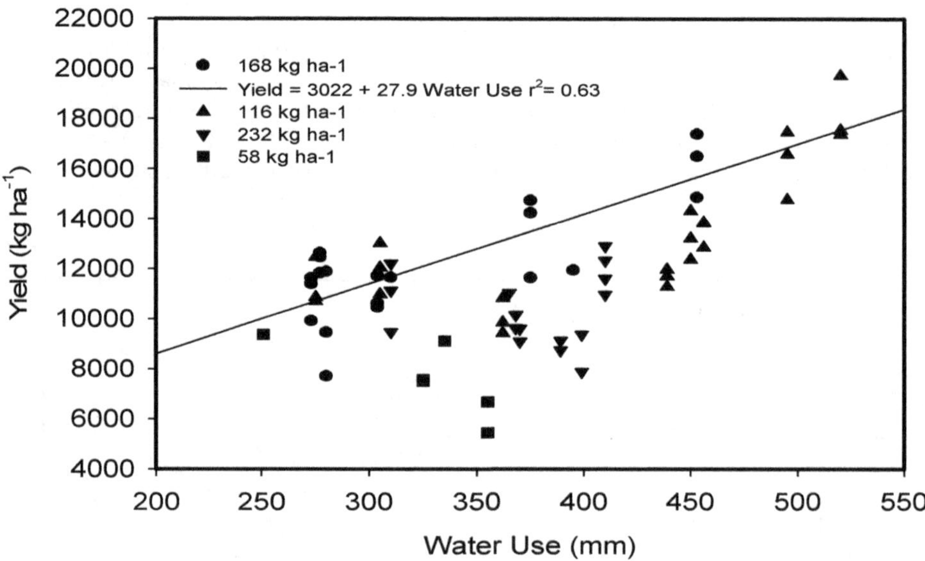

Fig. 3. Water use efficiency for corn for multiple years with different nitrogen application rates based on observation of seasonal totals of transpiration and grain yield.

3.2 Field scale response to nitrogen

Corn yield response to N applications varied across fields and years (Fig. 4). These data were separated by field and year to remove the potential confounding effects of weather variation among years. Analysis of the relationship between N rate and yield revealed that most fields showed no response to N (Table 3). The Sac Field 1 showed a negative response to N rate while the other fields showed a positive response to N (Table 3). The different response to N poses a problem in developing a general set of guidelines for N management and raises questions about why these fields differ in their yield response to N rate. However, this observation provides an opportunity for application of precision agricultural tools if the underlying mechanism for the difference can be identified and is consistent among growing seasons to allow for application of this information for decision-making.

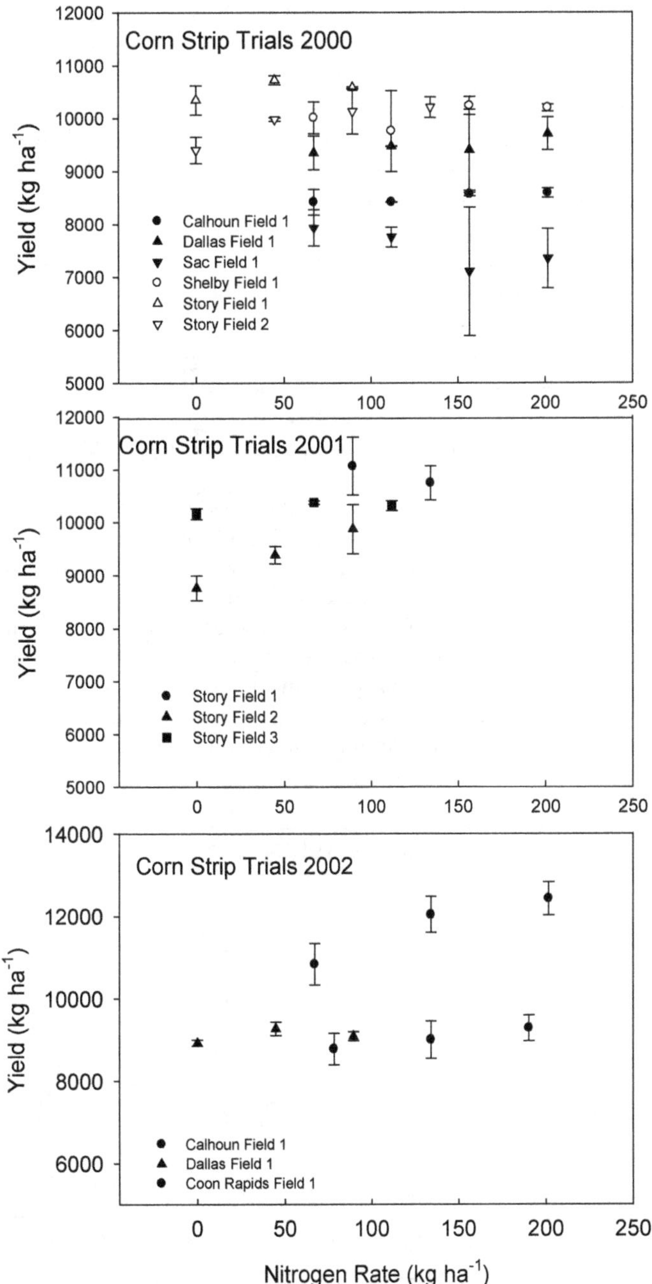

Fig. 4. Corn yield response to applied nitrogen in different fields in central Iowa from 2000 to 2002.

Year/Field	Significance	Slope
2000/Carroll 1	ns	
2000/Carroll 2	ns	
2000/Sac	*	-
2000/Shelby	ns	
2000/Story 1	ns	
2000/Story 2	**	+
2001/Story 1	ns	
2001/Story 2	***	+
2000/Story 3	**	+
2002/Calhoun East	*	+
2002/Dallas South	ns	
2002/Coon Rapids	**	+

ns- Not significant, *-p<0.1, **-p<0.05, ***-p<0.001

Table 3. Analysis of the effect of nitrogen rates on corn yields from the fields in central Iowa.

There was a range of soils within these fields and across the study sites. Variation among strips within a field for a given N rate was not significant for all fields when evaluated with a simple analysis of variance (ANOVA) using like strips as replicates. Understanding that N response is not consistent among the different fields creates questions about reasons for N responses observed among fields. Development of processes for the application of N within a field that takes advantage of information about corn response to N would greatly enhance the efficiency of N use. To explore this question and the lack of a consistent response, further evaluations were conducted on the data set. An additional data set on N response across a single field obtained by Hatfield and Prueger (2001) showed that N response was related to soil water holding capacity (Fig. 5).

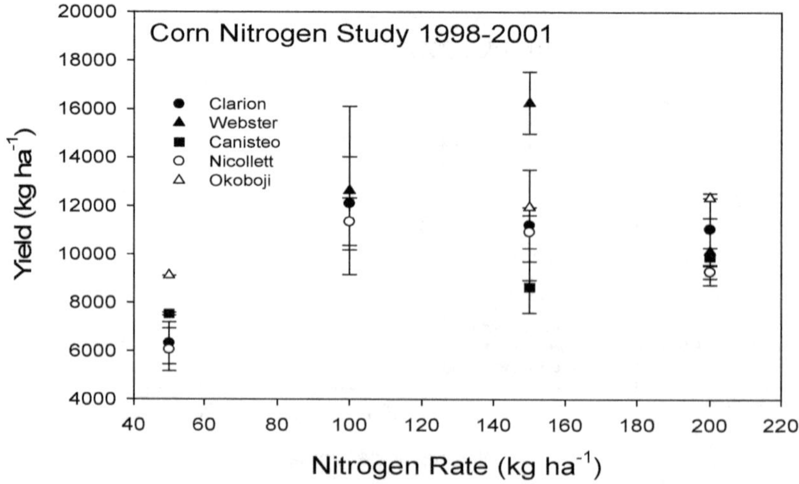

Fig. 5. Corn yield response to applied nitrogen across five different soils for a field in central Iowa for 1998 to 2001.

All soils showed an increase in yield above the 56 kg ha^{-1} rate but only the Webster soil showed a positive relationship to increasing N above this rate (Fig. 5). These results show that soils with a higher water holding capacity (Webster) had a different response to N compared to soils with lower water holding capacity (Clarion, Canisteo). Even though the studies were conducted at a different location, the soils were similar to those in these N studies so these results help explain the results shown in Fig. 4. The lack of response to N across the majority of the fields was related to the distribution of soils within the field. The Sac Field, in 2000, had a decrease in yield with increasing N was caused by the large amount of highly eroded Clarion soils within the field resulting in a limitation on the water holding capacity in the soil profile causing yields to be severely limited by water. These results are similar to those reported by Massey et al. (2008) in which the eroded soils had the lower yields and in this study lower yields were associated with soils having lower water holding capacity. Similar findings were reported by Sadler et al. (2000b) in which they suggested that understanding of site-specific yield maps would be enhanced by observations of water stress within the field. Their observations and those from this study suggest that spatial yield patterns in response to N management are dictated by soil types within the field and the interaction with soil water availability.

In order to understand the spatial patterns of yield within a strip, a stepwise approach was taken to evaluate these patterns. The first step was to evaluate the frequency distribution of yields for the different N rates as shown for the Coon Rapids field (Fig. 6). These frequency distributions are similar to other fields in this study. In all fields, the lower N rate had a lower mean yield but a similar range of minimum and maximum yields compared to the other N rates; however, the distribution showed a wider dispersion and more variation. As the N rate increased the variation pattern showed reduced dispersion with a higher frequency of values near the mean value (134 kg ha^{-1} compared to 200 kg ha^{-1}). In all fields we observed that the higher the N rate the less variation in frequency distribution with a similar shape of the yield distribution and the range of maximum and minimum values. The distribution of the yields based on the percentiles showed the 134 and 200 kg ha^{-1} rates were the same. All three rates showed a skewed distribution toward the lower yields.

To further evaluate the spatial patterns within fields, yields were summarized by each soil type within each field for N rates. Spatial variation of yields within fields was significant in their relationship to soil variation within the field. Across all of the N rates there was a similar pattern with the higher yields in the Webster soils and the lower yields in the Clarion soils (Fig. 7). Yields in the Webster soils were larger than the yields in the Clarion soils at all N rates and showed less variation than those in the Clarion soils (Fig. 7). This type of analysis was completed for all of the different fields evaluated in this study and the results shown in Fig. 7 were consistent among all of the fields with the soils having a higher water holding capacity producing the larger yields compared to those soils with lower water holding capacity.

When the yields are aggregated to create a N response curve across fields, yield differences were significant using a simple T-test between these two soils. These two soils were chosen because they were the most dominant in all of the different fields measured in this study. Yield differences between the Clarion and Webster soil at the 134 and 190 kg ha^{-1} rates were over 1000 kg ha^{-1} and there was a decrease in the corn yield with the N rate of 190 kg ha^{-1} in

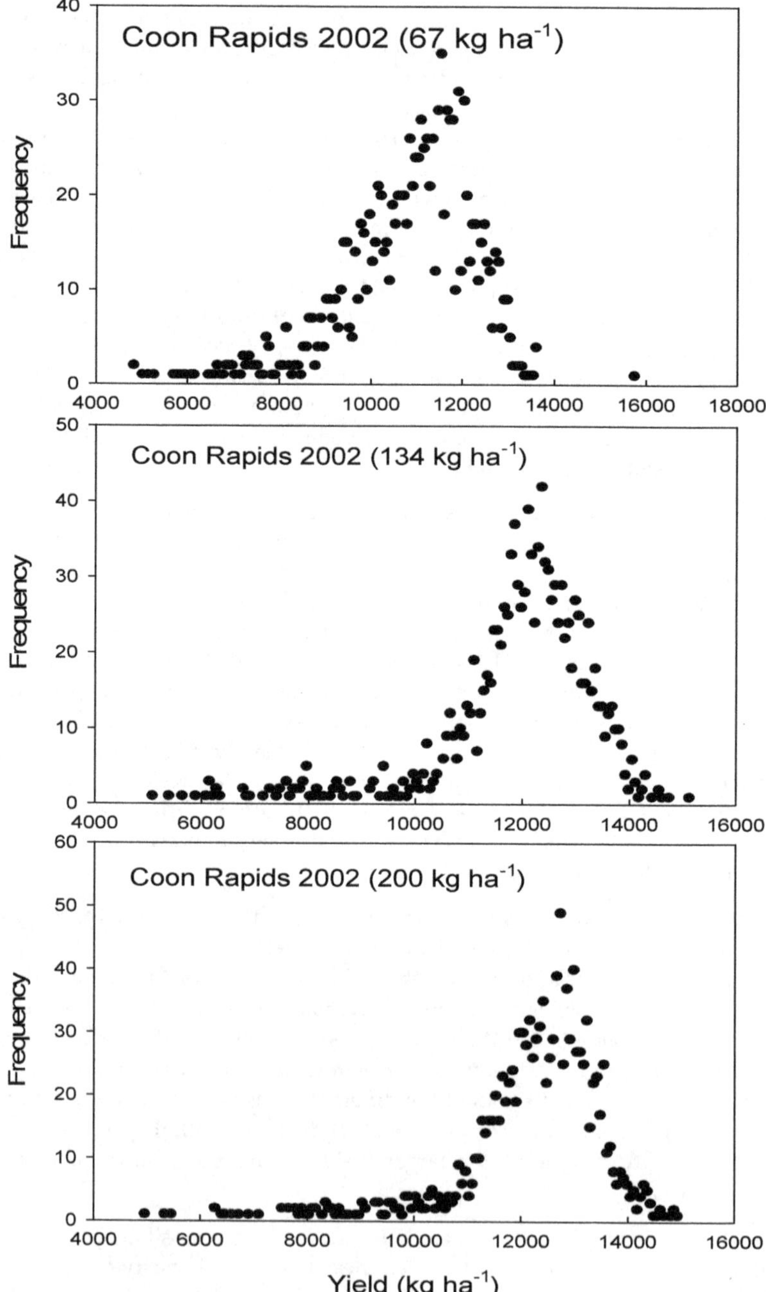

Fig. 6. Frequency distribution of corn yields at the 67, 134, and 200 kg N ha^{-1} rate for the Coon Rapids field in 2002.

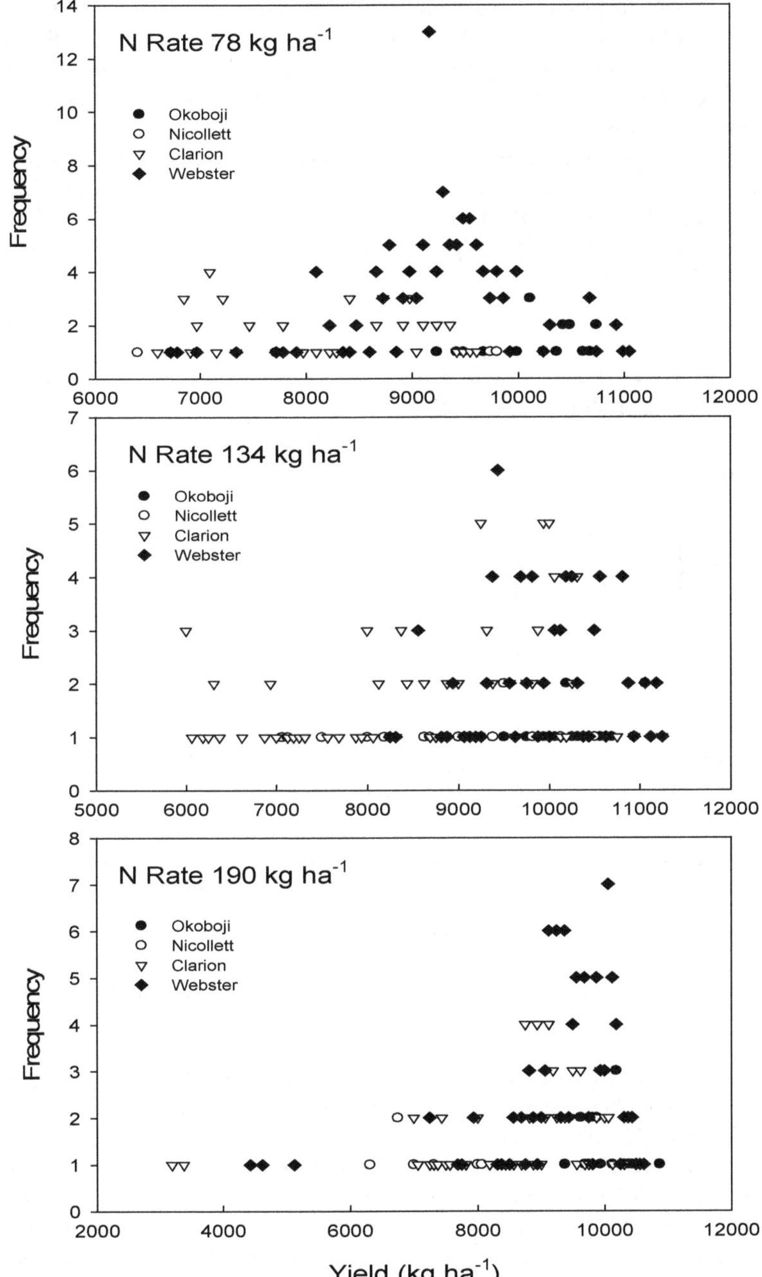

Fig. 7. Frequency distribution of corn yields for the 78, 134, and 190 kg N ha^{-1} rate for the four soil types within the field in the Calhoun East field in 2002.

the Clarion soil (Fig. 8). This was a similar response as to that observed in the Sac field in 2000 (Fig. 4). The N response curve shown in Fig. 5 suggests that the improved water holding capacity of the Webster soil allows for enhanced yield compared to the other soils because of the increased available soil water during grain-filling. Seasonal water use patterns between a Clarion and Webster soil were significant during the reproductive stage of growth because at this time of the year, crop water use was dependent upon stored soil water in the soil profile (Fig. 2). These water use patterns lead to differences in crop water stress which affects yield patterns as suggested by Sadler et al. (2000b). This was a consistent finding across all of the fields examined in this study in which the higher water holding capacity soils had a higher yield regardless of N rate. Observations from the 78 kg

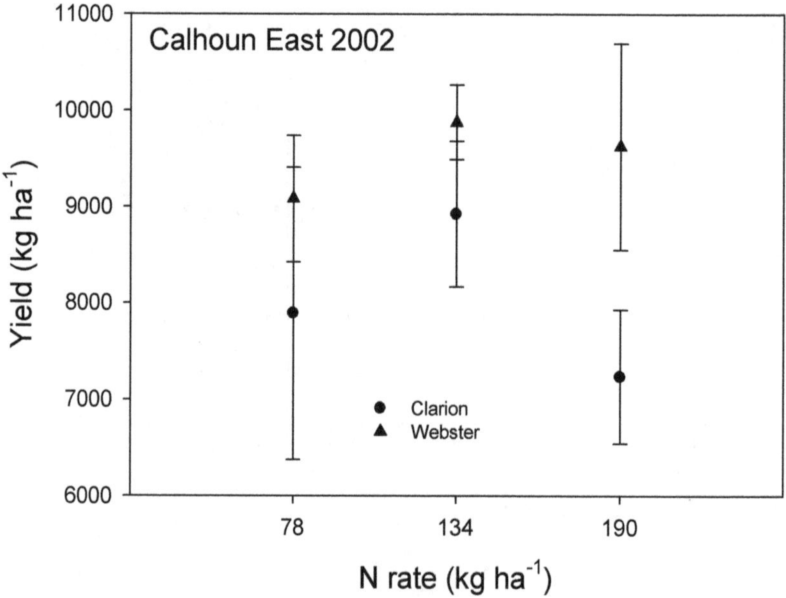

Fig. 8. Corn yield response to applied nitrogen for Clarion and Webster soils using three N rates in 2002.

N ha^{-1} rate showed that yields in the Clarion soil were distributed across the range of yields observed in the field. There were changes in the statistical moments for the yields in the different soils within N rates as shown in Table 3. Yields in the 134 and 190 kg N ha^{-1} rate were not significantly different for any of soils (Table 4). The yield distribution within the soil types reveals the effects of the soil water availability as a major factor in determining yield response to N rates (Fig. 7). Evaluation of the N response across fields will have to account for the water holding capacity of the soil and the precipitation during the growing season in order to interpret the results. Analysis of the yield distributions within fields segregated by soil type demonstrates the impact that available soil water has on determining the spatial pattern of corn yield.

Soil Type	N Rate	Mean	Std. Dev.	Skewness	Kurtosis
Clarion	78	129.9	14.8	-0.09	-1.46
	134	140.6	21.6	-0.59	-0.65
	190	140.8	20.7	-2.29	7.63
Nicollett	78	124.5	22.5	0.65	-1.72
	134	143.2	15.9	-0.37	-0.49
	190	114.7	10.6	0.41	-1.21
Webster	78	147.0	12.3	-0.50	1.62
	134	158.8	11.5	-0.21	-0.48
	190	149.2	16.9	-2.55	8.80
Okoboji	78	162.2	7.4	-7.7	-0.42
	134	165.4	6.9	0.08	0.35
	190	160.5	5.9	0.42	-0.03

Table 4. Mean, standard deviation, skewness, and kurtosis for corn yields within each soil type for different N rates within the Calhoun East field in 2002.

3.3 Seasonal patterns in fields

Harvested yield represents one point in the season which is the result of all of the interacting factors during the season. One question is whether the factors that affect yield patterns at harvest persist throughout the growing season or are there changes which occur and are detectable only in grain yield. Application of techniques related to improved management decisions require that observations within a field be able to detect a plant response that is ultimately related to crop yield as part of the decision making process. Sadler et al. (2000b) suggested that yield patterns could be explained by following the patterns of crop stress during the season. These data sets contain a sequence of measurements during the growing season that may be related to crop yield. The hyperspectral remote sensing data allowed for several indices to be calculated; however, one of the relationships we examined was the red (0.681μm) /green (0.561μm) ratio. This ratio was selected because of the strong relationship to crop biomass and crop yield. There was an inverse relationship between yield and the August red/green index for the fields (Fig. 9). Although there was a large variation about the regression line, this index showed a significant relationship with yield compared to other vegetative indices. Seasonal patterns of different vegetative indices provide insights into the spatial patterns of vegetative response during the course of the growing season. In this study we evaluated the NIR (0.819 μm) /red (0.681 μm) ratio and found that this vegetative index was not consistently related to yield across all of the fields while the red/green relationship showed a more consistent relationship across all of the fields. There are large varieties of vegetative indices that can be computed from the wavebands shown in Table 2; however, the consistency of this index across the fields in this study was one of the primary reasons for its use in these analyses. Hatfield et al. (2008) reviewed the different indices derived from remote sensing signals and their relationship to various agronomic variables and there are a variety of different indices which can be applied to these fields and in this study the red/green index provided a useful method of assessing response across fields.

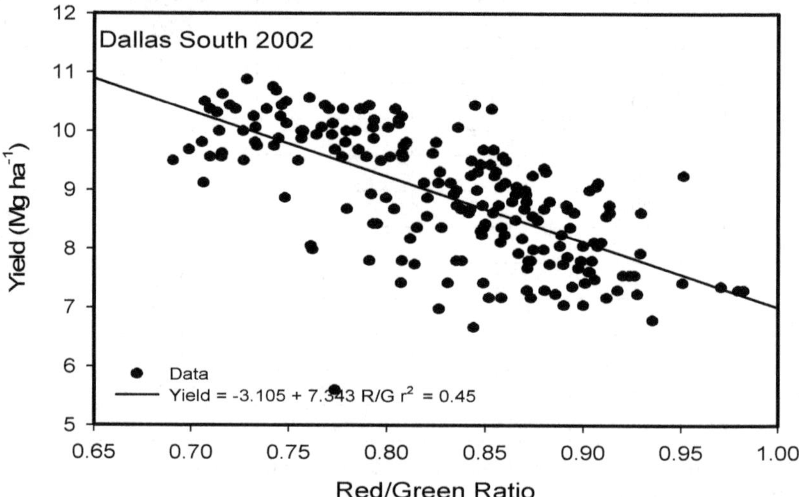

Fig. 9. Relationship between corn yield and red/green ratio observed in August with an aircraft scanner for the Dallas South field in 2002.

A more detailed examination of the red/green ratio was conducted with the observations collected four times during the growing season. Observations throughout the season represent unique characteristics of the growing season, May observations represent the soil background, July represents maximum vegetative cover, August the point of mid-grain fill, and September the time near physiological maturity (Fig. 10). At each of these times the frequency distribution of the red/green ratio was computed for each N rate within the field. There is a seasonal trend in the frequency distributions with a decrease in the variation found in the distribution from the May to July or August observations and then an increase in variation for the September observations (Fig. 10). Variation in the red/green ratios early in the season was related to the soil variation within each N rate. The variation in the July and August observations was small for all three N rates. Observations of the water use patterns among soils within a field showed little difference at this time of the growing season because there was adequate soil water in all soils to meet crop demands. Later in the growing season the crop water demand exceeds the precipitation and crop water use is dependent upon stored soil water and variation among soils becomes evident and the variation in the red/green ratio is similar to the bare soil distribution (Fig. 10). There was no significant difference among the 67, 134, or 200 kg N ha^{-1} rates for the frequency distributions of the red/green ratio (Fig. 10). The frequency patterns of the red/green ratios within N rates follow the yield patterns. Spatial patterns of reflectance reveal the seasonal dynamics of the interactions of soil types with N rates. These same patterns of red/green reflectance throughout the season were the same across all of the fields within this study. There is a consistent pattern in terms of a decreasing variation as the crop develops until mid-grain fill and then variation increases during the later grain-fill stages. The only difference among fields was whether the early grain-fill observations began to reveal spatial variation because of the lack of soil water in the profile and limited precipitation to meet the crop water demands. In fields with adequate soil water during grain-fill the variation is less pronounced.

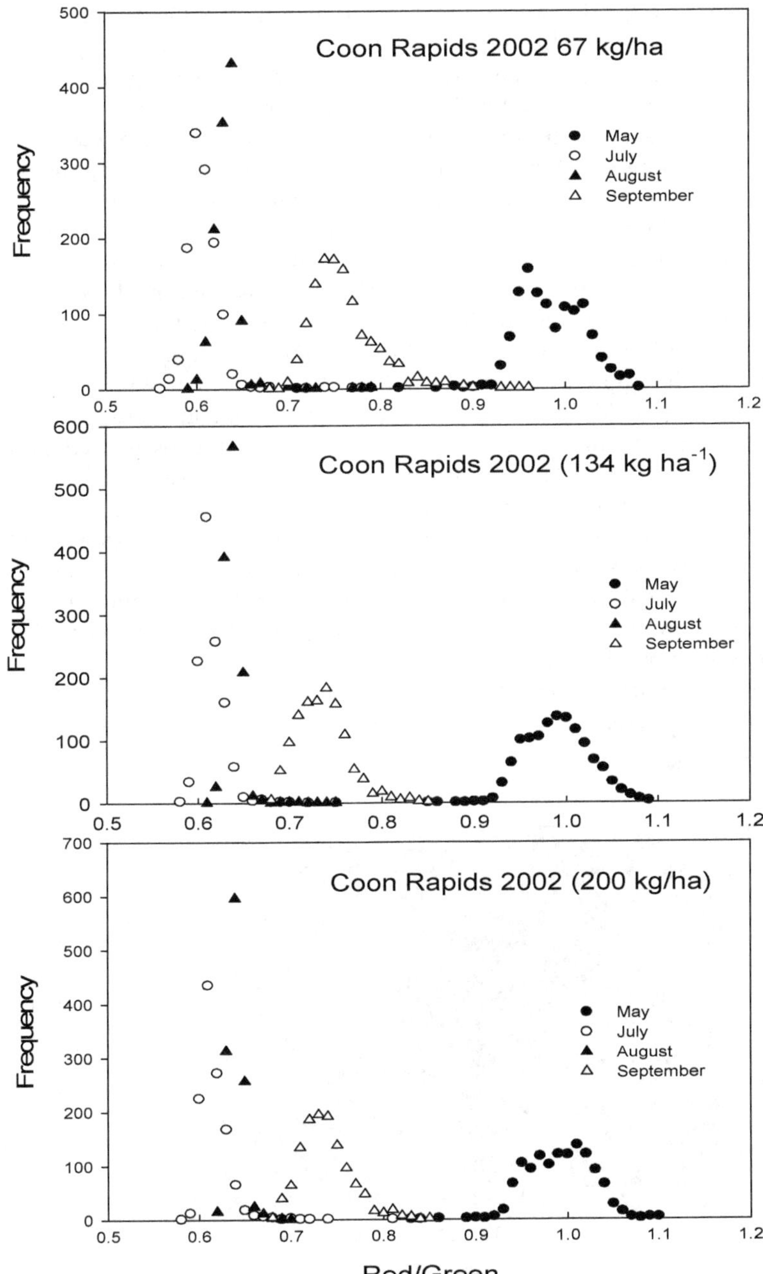

Fig. 10. Frequency distribution of red/green ratio for the four observation times during the 2002 growing season for the Coon Rapids field with 67, 134, and 200 kg N ha^{-1} rate.

Spatial analysis of the red/green index for the May and August periods during the 2002 growing season showed the effect of the soil differences for the May image with the differences from west to east that were related to the distribution of soil types within the field (Fig. 11). The presence of waterway was very evident in this kriged map of the field. The range of the samples was 20 m indicating there were detectable differences over relatively short distances within the field. In other fields the range was considerably longer and on the order of 80 to 100m. Spatial analysis was able to reveal the patterns of the soil types within the fields. This is in contrast with August image in which there is little variation across the field in the red/green ratio (Fig. 12). There is one spot with poor plant growth that was detectable in the field. In this analysis there was no stable range in the data because there were no significant spatial patterns detected within the field. These interpolated maps confirm the analysis conducted within each strip that showed the July and August periods have the least variation in vegetative indices across the field because the crop growth is uniform (Fig. 12). The growth of the crop reduces the variation within the field and there is no detectable variation caused by the N rates within this field. Spatial analysis of the September red/green ratios showed the variation had reoccurred within the field (Fig. 13). This temporal pattern was common across all of the fields in which the variation in the red/green ratio decreased in the July and August observations and there was no correlation of these ratios with soil types within the field. The reason for this pattern is that during this phase of crop growth the water use rate is small and with the soil profile completely recharged at the beginning of the growing season there is more than sufficient soil water along with the precipitation to produce a uniform growth across the field. During the grain-fill period when crop water use rates are larger and precipitation is more infrequent then soil water availability from the soil profile becomes a critical factor and influences the red/green ratio because of the effect on leaf senescence.

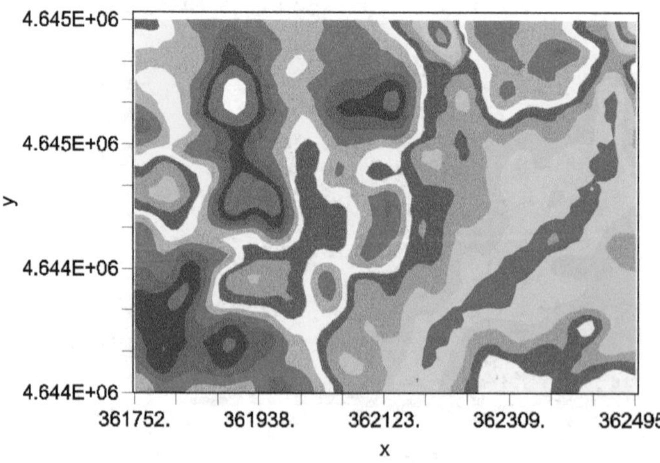

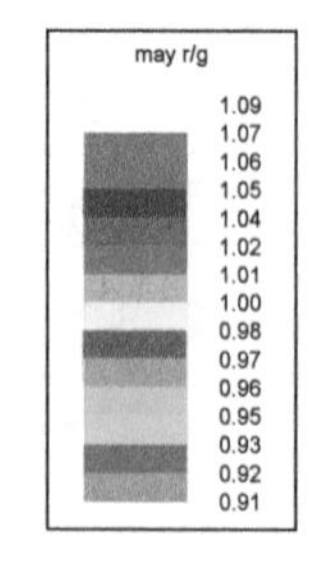

Fig. 11. Spatial map of the red/green ratio for the Coon Rapids field in 2002 for the May image.

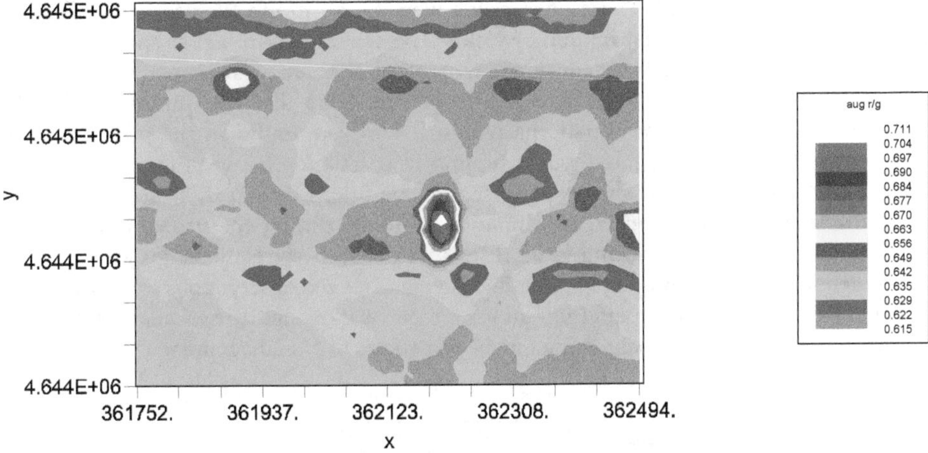

Fig. 12. Spatial map of the red/green ratio for the Coon Rapids field in 2002 for the August image.

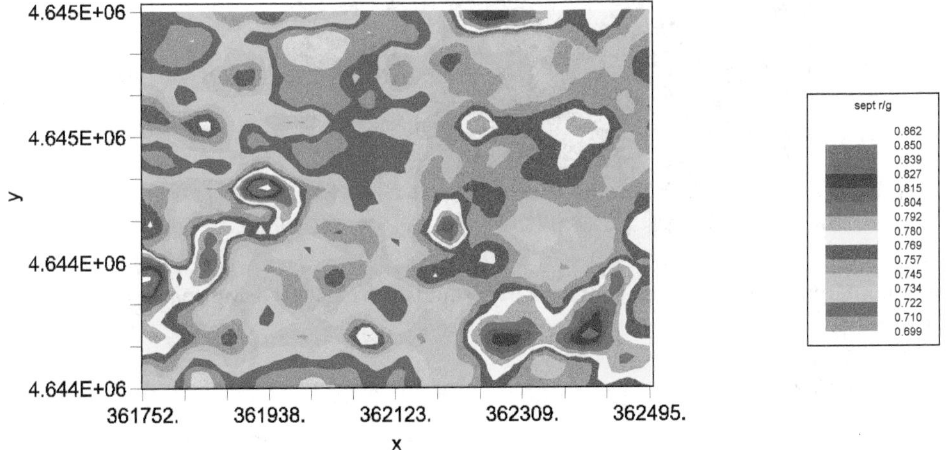

Fig. 13. Spatial map of the red/green ratio for the Coon Rapids field in 2002 for the September image.

Spatial variation maps for the Coon Rapids field in 2002 showed the yield variation patterns that indicated evidence of a relationship to N rates at the higher N rate of 200 kg ha^{-1}. The May spatial analysis image had a significant correlation with the yield of 0.65. These strips had higher yields in parts of the field but were not consistent across the entire strip and there was a difference between the west and east end of the field (Fig. 14). The September red/green patterns were significantly correlated with the May red/green values with a value of 0.70 and are indicative of the role of soil variation and the effect on soil water dominating the effect of N. The higher yields were found in the Webster soils as shown earlier with the frequency distribution of yields. The spatial map of yields show these high yields but the high yields are not consistent with N rates. There was the reemergence of the presence of the waterway within the field in the yield map that was present in the May spatial map of the field. These patterns within fields were found in the other fields we examined in this study and there was a significant correlation between the yield and soil type across all fields of 0.58. The growing season conditions during these study years had normal or slightly below normal precipitation amounts during the growing season and above normal precipitation during grain fill would offset these relationships and allow for more potential benefit of applied N. This finding confirms the observations from Jaynes and Colvin (1997) in which the spatial variation of yield was related to seasonal precipitation and extends their results to include the interacting effects of N management.

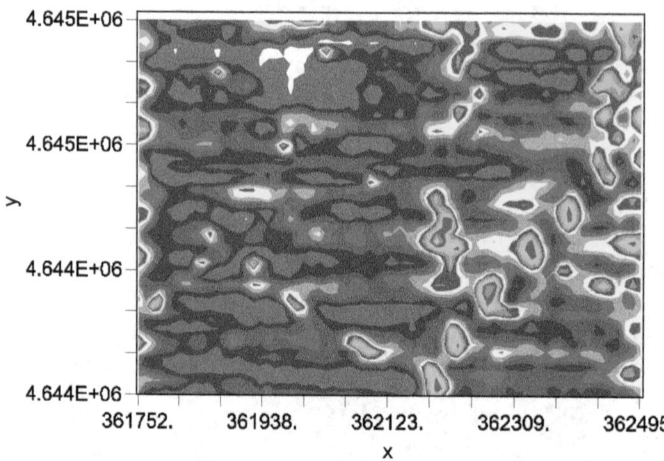

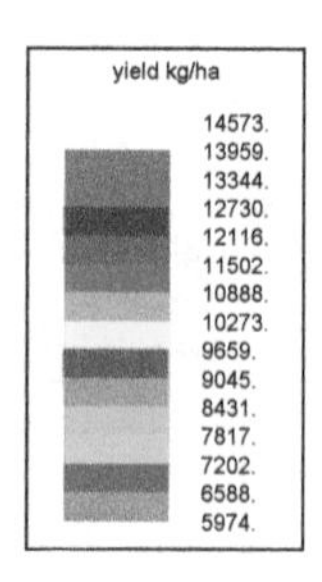

Fig. 14. Interpolated yield map from the Coon Rapids field in 2002 using spatial analysis software.

4. Conclusions

Nitrogen response across agricultural fields is more complex than observing a consistent response across a change in management practices. Observations among fields has shown that when multiple soils are encountered within a production field there are spatial patterns in both water use and N impacts on crop yield. There have been few studies which have coupled water and N dynamics across corn production fields. It has been assumed that water patterns operate separately from N management practices; however, the spatial patterns within a field show there is a temporal and spatial pattern determined by the combination of the precipitation patterns during the season, the soil water holding capacity, and the crop growth (crop water use) patterns. Observations of N impacts on corn yield across production scale fields revealed that yield responses were dependent upon the soil type and within a rate strip there were a range of yields and when further dissected into the spatial patterns, these spatial patterns were related to the water use patterns and soil water holding capacity. The observations from this study revealed that N impacts on crop yield were directly related to soil water holding capacity and to improve N response an improvement in soil water availability during grain-filling would be necessary.

Observations of the changes in the spatial patterns during the growing season have shown that there is complex interaction between the patterns of soil within the field and the final pattern of corn yield as a function of the patterns of soil water use and N management inputs. Agriculture will benefit from an enhanced understanding of the interactions of soil water use and N management and how these interact across a production field. The combination of remote sensing along with yield maps offers an enhanced method to evaluate field scale responses to both weather and management which will benefit production efficiency. These efforts will lead toward improved production efficiency and enhance the capability of agricultural systems to become more efficient in terms of water and N use.

5. Acknowledgements

The support of the Risk Management Agency and especially Virginia Guzman and Dave Fulk are greatly acknowledged and this research is under the agreement 07-IA-0831-0210. This effort would not be possible without the capable support of Brooks Engelhardt, Wolf Oesterreich, and Bert Swalla in their efforts to collect and process the data from the field experiments and the interactions with Galen Hart for his encouragement and insights. Likewise, the support of the producers Don Ferguson, Mike Hermanson, Nels Leo, Kriss Lightener, Dale Pennington, and David Schroeder who willingly let us use their fields for these studies and the efforts of the Farnhamville Cooperative of Farnhamville, IA (Jeff True, Gabe Tar) for helping identify the producers. The support from CALMIT at the University of Nebraska (Rich Perk and Don Rundquist) to obtain the hyperspectral data is greatly appreciated.

6. References

Blackmer, T.M., Schepers, J.S. (1996). Aerial photography to detect nitrogen stress in corn. *J. Plant Physiol.* 148:440-444.

Blackmer, T.M., Schepers, J.S., Varvel, G.E. (1994). Light reflectance compared with other nitrogen stress measurements in corn leaves *Agron. J.* 86:934-938.

Blackmer, T.M., Schepers, J.S., Varvel, G.E., Meyer, G.E. (1996a). Analysis of aerial photography for nitrogen stress within corn fields. *Agron. J.* 88:729-733.

Blackmer, T.M., Schepers, J.S., Varvel, G.E., Walter-Shea, E.A. (1996b). Nitrogen deficiency detection using reflected shortwave radiation from irrigated corn canopies. *Agron. J.* 88:1-5.

Blackmer, T.M., Schepers, J.S., Vigil, M.F. (1993). Chlorophyll meter readings in corn as affected by plant spacing. *Commun. Soil Sci. Plant Anal.* 24:2507-2516.

Brock, A., Brouder, S.M., Blumhoff, G., Hoffman, B.S. (2005). Defining yield-based management zones for corn-soybean rotations. *Agron. J.* 97:1115-1128.

Brouder, S.M., Mengel, D.B., Hoffman, B.S. (2000). Diagnostic efficiency of the black layer stalk nitrate and grain nitrogen tests for maize. *Agron. J.* 92:1236-1247.

Burkart, M.R., James. D.E. (1999). Agricultural nitrogen contributions to hypoxia in the Gulf of Mexico. *J. Environ. Qual.* 28:850-859.

Chen, P., Haboudane, D., Trembaly, N., Wang, J., Vigneault, P., Li, B. (2010). New spectral indicator assessing the efficiency of crop nitrogen treatment in corn and wheat. *Remote Sens. Environ.* 114:1987-1997.

Haboudane, D., Miller, J.R., Trembaly, N., Zarco-Tejada, P.J., Dextraze, L. (2002). Integrated narrow-band vegetation indices for prediction of crop chlorophyll content for application to precision agriculture. *Remote Sens. Environ.* 81:416-426.

Hatfield, J.L., Prueger, J.H. (2001). Increasing nitrogen use efficiency of corn in Midwestern cropping systems. Proceedings of the 2nd International Nitrogen Conference on Science and Policy. *TheScientificWorld* (2001) 1(S2):682-690.

Hatfield, J.L., Jaynes, D.B., Burkart, M.R., Cambardella, C.A., Moorman, T.B., Prueger, J.H., Smith, M.A. (1999). Water quality in Walnut Creek watershed: Setting and farming practices. *J. Environ. Qual.* 28:11-24.

Hatfield, J.L., Prueger, J.H., Kustas, W.P. (2007). Spatial and temporal variation of energy and carbon dioxide fluxes in corn and soybean fields in central Iowa. *Agron. J.* 99:285-296.

Hatfield, J.L., Gitelson, A.A., Schepers, J.S., and Walthall, C.L. (2008). Application of Spectral Remote Sensing for Agronomic Decisions. *Agron. J.* 100:S-117-S-131.

Hatfield, J.L., McMullen, L.D., Jones, C.W. (2009). Nitrate-nitrogen patterns in the Raccoon River Basin as related to agricultural practices. *J. Soil and Water Conserv.* 64:190-199.

Hernandez-Ramirez, G., Hatfield, J.L., Prueger, J.H., and Sauer, T.J. (2010). Energy balance and turbulent flux partitioning in a corn-soybean rotation in the Midwestern U.S. *Theor. Appl. Climatol.* 100:79-92.

Jaynes, D.B., Colvin. T.S. (1997). Spatiotemporal variability of corn and soybean yield. *Agron. J.* 89:30-37.

Jaynes, D.B., Hatfield, J.L., Meek, D.W. (1999). Water Quality in Walnut Creek Watershed: Herbicides and Nitrate in Surface Waters. *J. Environ. Qual.* 28:45-59.

Jaynes, D.B, Colvin, T.S., Karlen, D.L., Cambardella, C.A., Meek, D.W. (2001). Nitrate loss in subsurface drainage as affected by nitrogen fertilizer rate. *J. Environ. Qual.* 30:1305-1314.

Inman, D., Khosla, R., Reich, R., Westfall, D.G. (2008). Normalized difference vegetation index and soil color-based management zones in irrigated maize. *Agron. J.* 100:60-66.

Lee, Y., Yang, C., Chang, K., Shen, Y. (2008). A simple spectral index using reflectance of 735 nm to assess nitrogen status of rice canopy. *Agron. J.* 100:205-212.

Massey, R.E., Myers, D.B., Kitchen, N.R., Sudduth, K.A. (2008). Profitability maps as an input for site-specific management decision making. *Agron. J.* 100:52-59.

Ritchie, J.T., Burnett. E. (1971). Dryland evaporative flux in a subhumid climate: II. Plant influences. *Agron. J.* 63:56-62.

Sadler, E.J., Bauer, P.J., Busscher, W.J. (2000a). Site-specific analysis of a droughted corn crop: I. Growth and grain yield. *Agron. J.* 92:395-402.

Sadler, E.J., Bauer, P.J., Busscher, W.J., Miller, J.A. (2000b). Site-specific analysis of a droughted corn crop: II. Water use and stress. *Agron. J.* 92:403-410.

SAS Institute (2009) User's guide. Statistics. Version 9, SAS Institute, Cary, North Carolina.

Schepers, J.S., Blackmer, T.M., Wilhelm, W.W., Resende, M. (1996). Transmittance and reflectance measurements of corn leaves from plants with different nitrogen and water supply. *J. Plant Physiol.* 148:523-529.

Schepers, J.S., Francis, D.D., Vigil, M., Below, F.E. (1992). Comparison of corn leaf nitrogen concentration and chlorophyll meter readings. *Commun. Soil Sci. Plant Anal.* 23:2173-2187.

7

Effect of Mixed Amino Acids on Crop Growth

Xing-Quan Liu[1] and Kyu-Seung Lee[2*]
[1]*School of Agriculture and Food Science,*
Zhejiang Agriculture and Forestry University, Hangzhou,
[2]*Department of Bio-Environmental Chemistry,*
College of Agriculture and Life Sciences, Chungnam National Univeristy, Taejon,
[1]*P.R. China*
[2]*Korea*

1. Introduction

1.1 Nitrogen uptake and assimilation

Among the mineral nutrient elements, nitrogen is a kind of macronutrient. Most plant species are able to absorb and assimilate nitrate (NO_3^-), ammonium (NH_4^+), urea and amino acids as nitrogen sources. Most soils do not have sufficient N in available form to support desired production levels. Therefore, addition of N from fertilizer is typically needed to maximize crop yields. Many kinds of N fertilizers are used which contain varying forms of N such as NO_3^-–N, NH_4^+–N and urea. However, NO_3^- form of nitrogen is the predominant form of N absorbed by plants, regardless of the source of applied N (Breteler and Luczak, 1982). This preference is due to several autotrophic soil bacteria, which rapidly oxidize NH_4^+ to NO_2^-, and then to NO_3^- in warm, well-aerated soils. Even though NO_3^- is the most available form of N to plants, it can be more readily lost from the root zone because it is very mobile and easy to leach. This economically and environmentally undesirable process perpetuates a large amount of the uncertainty associated with N fertilizer management (Pessarakli, 2002).

In the soil solution, nitrate is carried towards the root by bulk flow and is absorbed into the epidermal and cortical symplasm. Within the root symplasm, nitrate has four fates: (1) reduced to nitrite by the cytoplasmic enzyme nitrate reductase; (2) efflux back across the plasma membrane to the apoplasm; (3) influx and stored in the vacuole; or (4) transported to the xylem for long–distance translocation to the leaves (Andrews, 1986; Ashley et al., 1975; Black et al., 2002; Cooper and Charkson, 1989). Translocated from the xylem, nitrate enters the leaf apoplasm to reach leaf mesophyll cells, where nitrate is again absorbed and either reduced to nitrite or stored in the vacuole.

Nitrate translocated from the roots through the xylem is absorbed by a mesophyll cell via one of the nitrate–proton symporters into the cytoplasm, reduced to nitrite by nitrate reductase (NR) in the cytoplasm, and then reduced to ammonium by nitrite reductase (NiR)

* Corresponding author

in the chloroplast, which is then incorporated into amino acids by the glutamine synthetase–glutamine– 2–oxoglutarate amidotransferase (GS/GOGAT) enzyme system, giving rise to glutamine (Gln) and ultimately other amino acids and their metabolites (Fig. 1; Taiz and Zeiger, 2002). Therefore, NR, NiR and GS constitute the first three enzymes of the nitrate assimilatory pathway. The NR activity is the limiting step of NO_3^-–N conversion to amino acid synthesis (Campbell, 1999). In most plant species only a proportion of the absorbed nitrate is assimilated in the root, the remainder being transported upwards through the xylem for assimilation in the shoot where it is reduced and incorporated into amino acids (Forde, 2000).

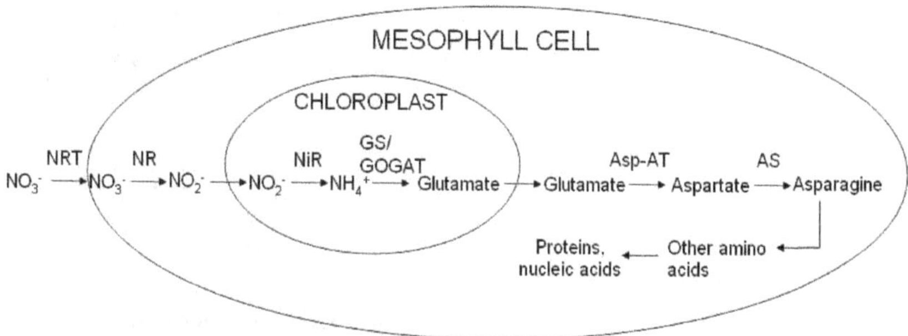

Fig. 1. The main process of nitrate assimilation.

1.2 Availability of amino acids

Traditional models of nutrient cycling assume that organic N matter must be decomposed by soil microorganisms to release inorganic N, before that N becomes available for plant uptake. But, there are growing evidences that plant can absorb organic N directly. Earlier studies of nutrient absorption demonstrated that higher plants could take up amino acids (Virtanen and Linkola, 1946). More recent studies of amino acid absorption have further focused on the characteristics of the carrier systems and other mechanistic aspects of the uptake process and a wide array of amino acid transporters has been identified in several different plants species (Frommer et al., 1993; Montamat et al., 1999; Neelam et al., 1999).

In the moist tundra of the arctic, inorganic N supplied to plants by mineralization is not sufficient to meet their requirement of N due to low temperatures and anoxic soils. But these soils have large stocks of water–extractable free amino acids (Atkin, 1996). The studies of nitrogen cycling in artic tundra have indicated that some non–mycorrhizal plant species, such as *Eriophorum vaginatum*, could absorb amino acids rapidly, accounting for at least 60% of total the nitrogen absorbed (Chapin et al., 1993). Ectomycorrhizal species have higher amino acid uptake than non–mycorrhizal species (Kielland, 1994). Amino acid uptake was the general ability found widely in plants from boreal forest (Näsholm et al., 1998; Persson and Näsholm, 2001).

1.3 Amino acids and nitrate uptake

Plants can store high levels of nitrate, or they can translocate it from tissue to tissue without deleterious effect. However, hazardous effects may occur when livestocks and humans consume plant material with rich nitrate, they may suffer from methemoglobinemia or carcinoma by converting nitrate to nitrite of nitrosamines. Some countries limit the nitrate content in plant material sold for human consumption.

Several authors reported that free amino acids could down regulate nitrate uptake and nitrate content in plant. It was found that exogenously supplied amino acids and amides could decrease the uptake of nitrate by soybean (Muller and Touraine, 1992); wheat (Rodgers and Barneix, 1993); maize (Ivashikina and Sokolov, 1997; Padgett and Leonard, 1993, 1996; Sivasankar et al., 1997); barley (Aslam et al., 2001) (Table 1). Plants appear to have multiple mechanisms for regulating nitrate uptake in addition to amino acids or N-status (Padgett and Leonard, 1993).

Work by Breteler and Arnozis (1985) determined that pretreatment of dwarf bean roots with many different individual amino acids inhibited nitrate uptake to varying degrees dependent upon prior exposure of the plants to nitrogen and the specific amino acid treatment. No significant effect of amino acids on nitrate transport was detected when both NO_3^- and amino acids were present in the bathing solution, and no correlation emerged between inhibition of nitrate uptake and inhibition of nitrate reductase relative to specific amino acids. A more detailed study, presented by Muller and Touraine (1992), demonstrated inhibition of uptake by 50% or greater by alanine, glutamine, asparagines, arginine, β-alanine and serine when soybean seedlings were pretreated for 18 h prior to exposure to NO_3^-. The mechanisms of inhibition by arginine and alanine appeared to differ, however. Arginine stopped NO_3^- uptake immediately upon introduction to the uptake solution, kinetically similar to NH_4^+ inhibition. The authors suggested that this may be the result of a non-metabolic response such as alteration of membrane potentials. Inhibition by alanine was slower to develop, suggesting a metabolic component to the regulation rather than a physical or chemical interference.

Plant materials	Amino acid treatment	NO_3^- supplied	NO_3^- uptake	Remarks
	(mM)		(%)	
Soybean	10	0.5	5−85	14 amino acids
	100	0.5	40−120	(Muller, 1992)
Wheat	1.0	0.3	50−105	3 days N starvation
	1.0	0.3	89−106	Non−starvation (Rodger, 1993)
Maize root	15	5.0	84	(Padgett, 1993)
Barley root	1.0	0.1	40−50	(Aslam, 2001)
	1.0	10	70	

Table 1. Effect of amino acid on NO_3^- uptake in several plants

The N status of the plants could also affect the inhibitory effect of amino acids on nitrate uptake. Rodger and Barneix (1993) had supplied amino acids exogenously to N starved or non−starved wheat seedlings. Exogenously supplied amino acids and amides had no effect on the wheat seedlings under well nourishment. However, some of the amino acids and amides supplied seedlings starved of N for 3 days inhibited up to 50% of the nitrate uptake rate.

Aslam et al. (2001) had conducted study on differential effect of amino acids (Glu, Asp, Gln and Asn) on nitrate uptake and reduction systems in barley roots. Similar results were observed i.e. 50-60% inhibition in the NO_3^- uptake when the roots were supplied with 0.1 mM NO_3^-. However, no inhibition occurred at 10 mM NO_3^-. In contrast, Kim (2002) had conducted study on effect of mixed amino acids on nitrate uptake in rice, pea, cucumber and red pepper. The result showed that the effect of mixed amino acids (MAA) on nitrate uptake in nutrient solution was unaffected in low MAA concentration and accelerated in high MAA concentration. The results indicated that external MAA could regulate nitrate uptake.

1.4 Amino acids and enzyme regulation

Nitrate reductase (NR) is a substrate inducible enzyme involved in the nitrate assimilation in higher plant, and the enzyme occupying a control point in the pathway of nitrate assimilation. Activity of the NR fluctuates widely in response to many environmental or physiological factors, such as the presence of NH_4^+ or amino acids in the growth medium. In studies of the possible regulation of NR activity by amino acids in higher plants, the results have often been conflicting. For example, Radin (1975, 1977) had shown that the reduction of nitrate to nitrite in cotton roots is inhibited by specific amino acids. On the other hand, Oaks (1977) had found using an *in vitro* assay those amino acids results in enhanced levels of NR and also cause only minor inhibitions in both intact and excised corn roots. Aslam et al. (2001) reported that the amino acids partially inhibited (35%) the induction of nitrate reductase activity (NRA) in barley roots supplied with 0.1 mM NO_3^-, but no inhibition occurred at 10 mM NO_3^-. He has concluded that the inhibition of induction of NRA by the amino acids is a result of the lack of substrate availability due to inhibition of the NO_3^- uptake system at low NO_3^- supply. It has been suggested that glutamate inhibited NRA in roots, but not in shoots (Ivashikina and Sokolov, 1997). This inhibition seems be dependent on plant materials, age of plants, growth conditions, nitrate concentration, amino acid kinds, amino acids concentration and other factors.

Effect of amino acids on the regulation of NR gene expression has been studied at the molecular level. Deng et al. (1991) reported that the addition of 5 mM glutamine to the nutrient solution of tobacco plants grown in 1 mM NO_3^- resulted in a pronounced inhibition of NR mRNA accumulation in the roots. Vincentz et al. (1993) showed, under low light conditions (limiting photo synthetic conditions), the supply of glutamine or glutamate led to a drop in the level of NR mRNA, while glutamine and glutamate were less efficient at decreasing NiR mRNA than NR mRNA levels. Li et al. (1995) also demonstrated that 5 mM glutamine added together with NO_3^- resulted in reduced levels of NR mRNA in both root and shoot of maize. Sivasankar et al. (1997) observed that Gln and asparagine (Asn) inhibited the induction of NR activity (NRA) in corn roots at an external supply of 250 μM and 5 mM NO_3^-. They concluded that inhibition was not the result of altered NO_3^- uptake, and tissue nitrate accumulation was reduced at 250 μM external nitrate in the presence of 1mM Asn, but not at 5mM Asn.

In the studies of the possible regulation of NR activity by multiple amino acids in higher plants, the conclusions are again contradictory. The inhibition on NR activity by glycine, asparagines, and glutamine could be partially or wholly prevented by the presence of other amino acids during the induction (Radin, 1977). However when glutamine and asparagine were included along with the "corn amino acid mixture", the inhibition on the induction of NR in corn roots was more severe (Oaks et al., 1977). Chen and Gao (2002) have applied different mixture of glycine, isoleucine and proline replacing nitrate of solution partially (20%) to Chinese cabbage and lettuce in hydroponics. Amino acids enhanced the NR activity in Chinese cabbage, while it decreased in lettuce.

1.5 Influence on yield and N assimilation

L–tryptophan, considered as a physiological precursor of auxins in higher plants, was applied to soil to evaluate its influence on yield of several crops. Kucharski and Nowak (1994) found that L–tryptophan did not affect the yield of above ground part and roots of field bean. On the other hand, positive effects on corn and cabbage growth were reported (Sarwar and Frankenberger, 1994; Chen et al., 1997).

Amino acids were used to partially replace NO_3^- or foliar spray in many plants. In most case, the application of amino acids led to decreased nitrate content and increased total nitrogen content in lettuce, Chinese cabbage, onion, pakchoi or other leafy crops (Gunes et al., 1994, 1996; Chen and Gao, 2002; Wang et al., 2004). Some authors suggested that plants probably preferred amino acids as sources of reduced nitrogen, and nitrate uptake was inhibited by amino acids. In fact, there was little evidence or data to support the conclusions. It has not been distinguished that increased total nitrogen came of nitrate or amino acids.

1.6 Objectives

Regulation of induction of the NO_3^- uptake and reduction systems by nitrogen metabolites has been attributed to feed–back inhibition (Pal'ove–Balang, 2002). It was found that nitrate uptake rate follows a biphasic relationship with external nitrate concentration, suggesting the existence of at least two different uptake systems (Cerezo et al., 2000). At high external nitrate concentration (> 0.5 mM), a low affinity transport system (LATS), which shows linear kinetics, contributes significantly to the uptake rate and appears to be constitutively expressed and essentially unregulated. At low external concentrations (< 0.5 mM), two high affinity transport systems (HATS) operate, one of these being constitutive whereas the other is induced by nitrate. The HATS for nitrate uptake is sensitive to metabolic inhibitors and appears to be an active transport system (Daniel–Vedele et al., 1998).

Although the regulatory effect of amino acids on nitrate uptake and NR has been examined extensively, its effect on GS has not been examined in detail. Otherwise, a lot of amino acids were investigated about their regulation on nitrate uptake and assimilation, but very little information has been reported about effect of mixed amino acids (MAA).

In fact, there are two possible reasons for the increase of total N content in the plants: preference for amino acids as sources of reduced nitrogen and regulation of amino acids on inorganic nitrogen uptake and assimilation.

The solution experiments were carried out to investigate the regulation of the induction of NO_3^- uptake, NRA, NiRA and GSA in radish and red pepper by applying mixed amino acids (MAA) under the conventional fertilization. These two plants were selected because radish is NO_3^- preferred crop and red pepper is NH_4^+ preferred crop. The amino acids used in this experiment were alanine (Ala), β–alanine (β–Ala), aspartic acid (Asp), asparagine (Asn), glutamic acid (Glu), glutamine (Gln) and glycine (Gly). These amino acids were selected for the reasons include: (1) their structural role in proteins, (2) significant effect on NO_3^- uptake which was found in many works, and (3) considerable amounts in plant phloem and xylem (Caputo and Barneix, 1997; Lohaus et al., 1997; Peeters and Van Laere, 1994; Winter et al., 1992).

In the frame of the studies on the effect of the mixed amino acids (MAA) on nitrate uptake and assimilation, the pot experiments were focused on the role of MAA in process of NO_3^- uptake and assimilation. In order to distinguish the origin of N in radish, ^{15}N labeled nitrate was used.

In order to develop an approach for more efficient N fertilizer use and to prevent environmental pollution due to nitrate leaching, the aim of the study presented here, is to investigate the effect of amino acid fertilizer (AAF) on nitrate removal in high nitrate soils.

2. Hydroponic experiment of radish

2.1 Materials and method

Seeds of Ilsan radish (*Raphanus sativus*) soaked for 6 h allowed to germinate on paper towels were soaked in water in the dark. After 5 days the seedlings were transferred to 50 mL plastic tubes containing 10 mL inorganic nutrient solution. The nutrient solution was renewed every day. The composition of the inorganic nutrient solution is given in Table 2. Iron (Fe-EDTA), boron (H_3BO_3), manganese ($MnCl_2{\cdot}4H_2O$), zinc ($ZnSO_4{\cdot}7H_2O$), copper ($CuSO_4{\cdot}5H_2O$) and molybdenum ($H_2MoO_4{\cdot}H_2O$) were supplied to all treatments at rates of 40, 460, 90, 7.7, 3.2 and 0.1 μM, respectively. Seedlings were grown in a growth chamber maintained at 25°C, 70–80% relative humidity, with a 14 h light/10 h dark cycle and a light intensity of 300 μmol $m^{-2}s^{-1}$.

Chemicals	K^+	NO_3^-	Ca^{2+}	$H_2PO_4^-$	Mg^{2+}	SO_4^{2-}
KNO_3	1.25	1.25				
$Ca(NO_3)_2$		2.50	1.25			
KH_2PO_4	0.25			0.25		
$MgSO_4$					0.50	0.50
Total	1.50	3.75	1.25	0.25	0.50	0.50

Table 2. The main compositions of the nutrient solution for hydroponic experiment (mM)

The mixed amino acids (MAA) solution contained 7 equal concentrations of amino acids were as follows: alanine (Ala), β–alanine (β–Ala), aspartic acid (Asp), asparagine (Asn), glutamic acid (Glu), glutamine (Gln) and glycine (Gly). After 10 days, radish seedlings were placed in 10 ml inorganic nutrient solution containing 5.0 mM NO_3^- and 0, 0.3 or 3.0 mM MAA, as indicated in Table 3. The pH of the nutrient solutions were maintained between 6.0–6.1 by adding 1.0 M KOH appropriately. The nutrient solutions were renewed at 4, 8 and 16 h, respectively. The choice of the levels of MAA and renewed time of the nutrient solutions were according to the study of Kim (2002)

Treatments	K^+	NO_3^-	Ala	β-Ala	Asp	Asn	Glu	Gln	Gly
A0	5.25	5.0	—	—	—	—	—	—	—
A1	6.78	5.0	0.3	0.3	0.3	0.3	0.3	0.3	0.3
A2	13.10	5.0	3.0	3.0	3.0	3.0	3.0	3.0	3.0

Table 3. The compositions of the treatment solutions for radish in hydroponic experiment (mM)

Plants were harvested 24 h after treatment and separated into roots and shoots for enzymes assay and N content analysis. Net NO_3^- uptake rates were determined by amount of NO_3^- disappeared from the initially treated solution.

2.2 Results and discussion

2.2.1 Effect on NO_3^- uptake

The MAA treatments showed different effect on nitrate uptake depending on the concentrations (Fig. 2). The NO_3^- uptake in treatment A1 was similar to that of A0 after 8 h exposure to NO_3^-. However, exposure for longer hours (16 or 24 h) to 0.3 mM MAA inhibited the NO_3^- uptake by 38% compared with A0. In contrast, the highest NO_3^- uptake was found in treatment A2 that showed 305% higher than A0.

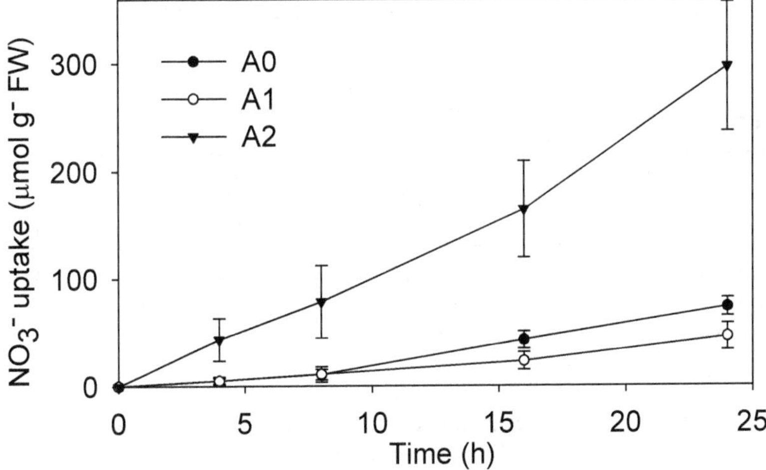

Fig. 2. Effect of mixed amino acids on the nitrate uptake in radish supplied with 5.0 mM NO_3^-. Values are means ± SD (n=5).

Several authors reported that free amino acids could down regulate NO_3^- uptake. It was found that exogenously supplied amino acids and amides could decrease the uptake of NO_3^- by soybean (Muller and Touraine, 1992); wheat (Rodgers and Barneix, 1993); maize (Ivashikian and Sokolov, 1997; Padgett and Leonard, 1996; Sivasankar et al., 1997); barley (Aslam et al., 2001). In this experiment, the effectiveness of the MAA treatments on NO_3^- uptake was similar to above references at low MAA treatment rate (0.3 mM MAA, Fig. 2). However, contrary result was found at high MAA treatment rate (3.0 mM MAA, Fig. 2), in which NO_3^- uptake was 4-fold higher than the control. This result was similar to rice, pea, cucumber and red pepper, which were treated with 5.0 mM MAA (Kim, 2002).

The effect on nitrate uptake seems to respond to kinds and concentration of amino acids. Muller and Touraine (1992) had examined the effect of 14 different amino acids on nitrate uptake in soybean seedlings supplied with 0.5 mM nitrate. After 10 mM single amino acid pretreatment, about half of the tested amino acids had a substantial inhibitory effect on nitrate uptake, mainly Ala, Glu (almost 100% inhibition), Asn and Arg (about 80%), and Asp, βAla, Scr, and Gln (from 70% to 48%). However, when supplied at 100 mM amino acid to the tip-cut cotyledons, only eight of fourteen amino acids had inhibitory effect, and four amino acids had enhanced nitrate uptake.

2.2.2 Effect on NO_3^- and NO_2^- accumulation

The application of MAA increased the NO_3^- concentrations both in shoots and in roots regardless of application rates (Table 4), resulting in the highest concentration in A1 and the lowest concentration in A0. The high concentration of NO_3^- in A2 was attributed to the high NR activity (Fig. 3). Although A1 treatment showed the lowest uptake of NO_3^- (Fig. 2), the highest concentration of NO_3^- was found by the reason of that low NR activity in A1 (Fig. 3) led to a blocking of the reduction of NO_3^- to NO_2^-. With respect to the NO_2^- values (Table 4), in our experiments, the highest NO_2^- concentrations in both shoots and roots were found in the A2. In shoots, the lowest NO_2^- concentration was found in A1 and the lowest in A0 in roots.

Treatments	NO_3^- (μmol g^{-1})		NO_2^- (nmol g^{-1})	
	Shoot	Root	Shoot	Root
A0	62.47 ± 4.06 a	16.30 ±1.88 b	6.76 ± 0.62 b	11.43 ± 1.67 c
A1	67.73 ± 7.49 a	22.99 ±2.23 a	3.77 ± 0.34 c	17.14 ± 2.10 b
A2	63.37 ± 3.58 a	17.41 ±1.92 b	29.70 ± 2.78 a	30.39 ± 4.13 a

Data are means ± SD (n=5). Analysis of variance (ANOVA) was employed followed by Duncan's new multi range test. Values with similar superscripts are not significantly different ($P>0.05$).

Table 4. Effect of mixed amino acids on NO_3^- and NO_2^- concentration in fresh weight of radish at 24 h after treatment

Although many authors agree that amino acid can negatively regulate nitrate content in higher plants (Chen and Gao, 2002; Gunes et al., 1994, 1996; Wang et al., 2004), the results in the present experiment do not support this interpretation. Both in shoots and in roots, the MAA used in this study led to little increase of NO_3^- concentrations (Table 4). The contradiction may reside in treatment method and treatment period of amino acids. It was demonstrated in other studies that amino acid pretreatment decreased NO_3^- accumulation slightly, but Gln and Asn increased the NO_3^- concentration in barley roots when they were used together with nitrate (Aslam et al., 2001). The reason of difference between this experiment and others is that NO_3^- content of shoots includes portion of NO_3^- in xylem sap. Concentrations of NO_3^- in xylem sap can be quite high, especially in plants that transport most of the NO_3^- taken up to the shoot for reduction (e.g., maize 10.5 mM, Oaks, 1986; barley 27 to 34 mM, Lews et al., 1982).

As interim product of NO_3^- assimilation procedure, the concentration of NO_2^- depended on the reduction rate of nitrate and nitrite. The highest concentration of NO_2^- found in A2 (Table 4) was due to high NR activity (Fig. 3), and the lowest concentration of NO_2^- in shoots in A1 (Table 4) was due to low NR activity (Fig. 3) too. However, low NiR activity (Fig. 4) led to a blocking of the reduction of NO_2^- to NH_4^+ in roots in A1, so that concentration of NO_2^- showed higher than A0 (Table 4).

2.2.3 Effect on NRA, NiRA and GSA

For NO_3^- assimilation, NO_3^- is reduced to NO_2^- by catalysis of NR. In this experiment, low concentration and high concentration of MAA treatments led to different effects on NR activity (Fig. 3). Both in the shoots and in the roots NR activities were inhibited slightly in A1. Significant increases of NR activities were found in A2 treatment, with 75% in shoots and 340% in roots respectively, relative to A0.

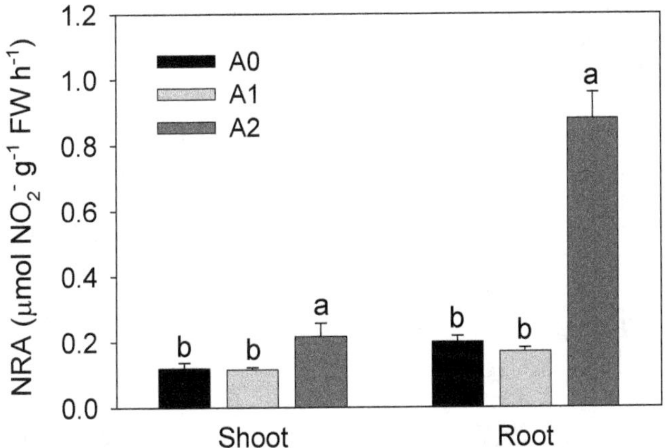

Fig. 3. Effect of mixed amino acids on nitrate reductase activity in radish at 24 h after treatment. Values are means ± SD (n=5).

There are contradictory results for the possible regulation of NR activity by amino acids for higher plants. For example, Radin (1975, 1977) has shown that the reduction of NO_3^- to NO_2^- in cotton roots is inhibited by specific amino acids. On the other hand, Oaks et al. (1979) have found that amino acids inhibited minor levels of NR in both intact and excised corn roots using an *in vitro* assay. Aslam *et al.*, (2001) reported that the amino acids partially inhibited the increase of NR activity in barley roots where most NO_3^- uptake was facilitated via high affinity transport system (HATS) but had little effect where low affinity transport system (LATS) is operative. It has been suggested that glutamate inhibited NR activity in roots, but no inhibition in shoots (Ivashikian and Sokolov, 1997). Sivasankar et al. (1997) observed that Gln and asparagine (Asn) inhibited the induction of NR activity in corn roots at both 250 μM and 5 mM of external NO_3^- supply. They concluded that inhibition was not the result of altered NO_3^- uptake, and tissue nitrate accumulation was reduced at 250 μM external nitrate in the presence of 1 mM Asn, but not at 5 mM Asn.

In the studies of the possible regulation of NR activity by multiple amino acids in higher plants, the conclusions are also contradictory. The inhibition on NR activity by glycine, asparagines, and glutamine could be partially or wholly prevented by the presence of other amino acids during the induction (Radin, 1977). However when glutamine and asparagines were included along with the "corn amino acid mixture", the inhibition on the induction of NR in corn roots was more severe (Oaks et al., 1979). Chen and Gao (2002) have applied different mixture of glycine, isoleucine and proline to Chinese cabbage and lettuce in hydroponic experiment. They found the amino acids treatment enhanced NR activity in Chinese cabbage, while decreasing it slightly in lettuce.

In this experiment, at 5.0 mM NO_3^- which is facilitated by LATS, the presence of 0.3 mM MAA partially inhibited NR activity, as observed in other works, whereas the 3.0 mM MAA increased the NR activity more than 4 times (Fig. 3). In addition, the very high NO_2^- content was found in A2 (Table 4). These results suggest that high concentration MAA can increase NO_3^- uptake by enhancing NR activity in radish, especially in roots.

The next step in NO_3^- assimilation is the conversion of NO_2^- to NH_4^+ by the action of NiR. Both enzymes, NR and NiR, are induced by the same factors, and therefore the response of NiR to the MAA treatments resembled that of NR in roots, but was a little different with that of the NR in shoots (Fig. 3 and Fig. 4). NiR activities in shoots and roots in A1 were inhibited by 17% and 52% respectively in relation to A0. In A2, NiR activity was inhibited by 15% in shoots and enhanced 8 times in roots. In the present study, the decrease of NiR activity in shoots in A2 might be attributed to the low concentration of amino acids in shoots, too. The increase of NiR activity in roots in A2 was due to the same reason with NR.

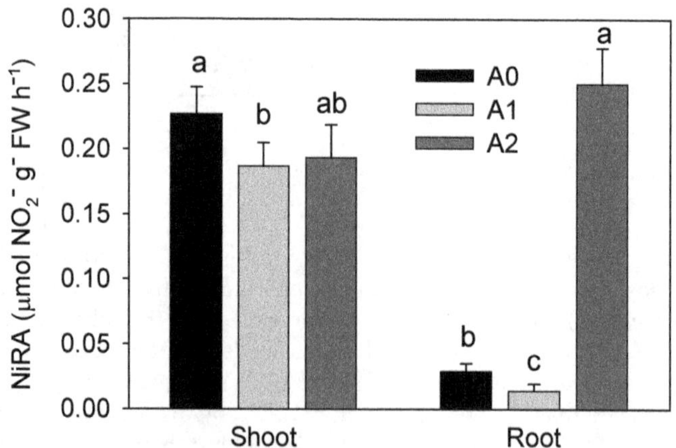

Fig. 4. Effect of mixed amino acids on nitrite reductase activity in radish at 24 h after treatment. Values are means ± SD (n=5).

The principal NH_4^+ pathway is the glutamine synthetase (GS)/glutamate synthase (GOGAT) cycle. The behavior of GS activities in shoots was not affected by MAA treatments (Fig. 5). However differences were found in roots between treatments, showing 22% inhibition in A1 and 17% increase in A2 in relation to A0.

The NH_4^+ originating in the plant from NO_3^- reduction is incorporated into an organic form primarily by the enzyme GS. In the present experiment, GS activity was inhibited by 0.3 mM MAA treatment in radish roots, whereas 3.0 mM of MAA treatment enhanced the activity (Fig. 5). It is also striking that effect of MAA on NO_3^- assimilation in the roots was higher than in the shoots, presumably NO_3^- was more available and the MAA content was higher in the roots.

The results of the present experiment clearly indicated that NO_3^- uptake and NO_3^- assimilation was regulated by MAA in radish, especially at high concentration of MAA treatment. In conclusion, the application of high MAA rates (principally A2) could be the direct cause of increased activities of the three enzymes (NR, NiR and GS) of the NO_3^- assimilatory pathway and the NO_3^- uptake was enhanced when supplied with LATS range of NO_3^-.

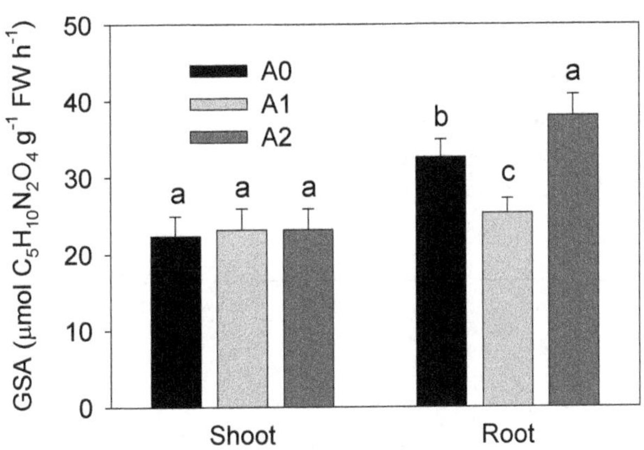

Fig. 5. Effect of mixed amino acids on glutamine synthetase activity in radish at 24 h after treatment. Values are means ± SD (n=5).

3. Hydroponic experiment of red pepper

3.1 Materials and methods

Seeds of Chongok red pepper (*Capsicum annuum*) were sown in February 2005. The seedlings were grown in individual pots filled with commercialized artificial soil in an experimental greenhouse for 35 days and then transferred to 50 mL plastic tubes containing 20 mL inorganic nutrient solution. The nutrient solution was renewed every day. The composition of the inorganic nutrient solution and the cultural condition were the same with hydroponic experiment of radish.

The mixed amino acids (MAA) solution was the same with that used in hydroponic experiment of radish which contained 7 equal concentrations of amino acids. At 7 days after transferring, red pepper seedlings were placed in inorganic nutrient solution containing 1.0 mM NO_3^- and 0, 0.3 or 3.0 mM MAA, as indicated in Table 5. The pH of the nutrient solutions were maintained between 6.0-6.1 by adding 1.0 M KOH appropriately. The nutrient solutions were renewed at 4, 8, and 16 h, respectively.

Treatments	K^+	NO_3^-	Ala	β-Ala	Asp	Asn	Glu	Gln	Gly
A0	10.25	10.0	–	–	–	–	–	–	–
A1	11.78	10.0	0.3	0.3	0.3	0.3	0.3	0.3	0.3
A2	18.10	10.0	3.0	3.0	3.0	3.0	3.0	3.0	3.0

Table 5. The compositions of the treatment solutions for red pepper in hydroponic experiment (mM)

Plants were harvested 24 h after treatment and separated into roots and leaves for enzymes assay and N content analysis. Net NO_3^- uptake rates were determined by amount of NO_3^- disappeared from the initially treated solution.

3.2 Results and discussion

3.2.1 Effect on NO_3^- uptake

The MAA treatments showed different effect on nitrate uptake depending on the concentrations (Fig. 6). Application of MAA at both 0.3 mM and 3.0 mM concentrations increased NO_3^- uptake in red pepper ($P < 0.001$) and the highest NO_3^- uptake was found in treatment A2 showing 7 fold increases over A0.

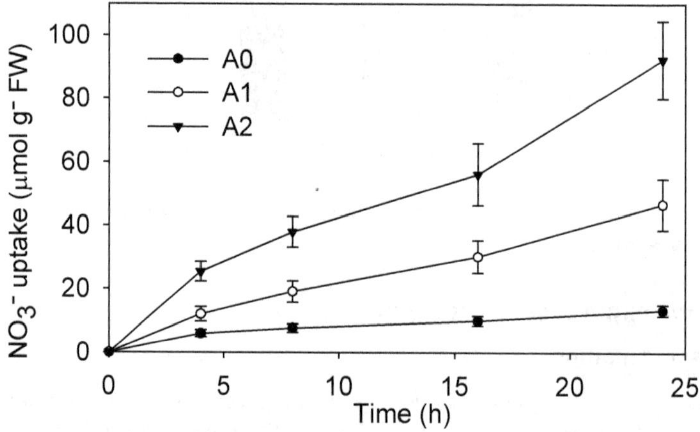

Fig. 6. Effect of mixed amino acids on the nitrate uptake in red pepper supplied with 10.0 mM NO_3^-. Values are means ± SD (n=5).

3.2.2 Effect on NO_3^- and NO_2^- accumulation

The highest NO_3^- concentration both in the roots and leaves were found in treatment A0 (Table 6), with respect to the lowest NO_3^- content found in treatment A1 in the leaves ($P < 0.05$) and A2 in the roots ($P < 0.01$). With respect to the NO_2^- values (Table 6), in this experiment, the highest NO_2^- concentrations in roots were found in the A2 and the lowest in A0 ($P < 0.001$). In leaves, the lowest NO_2^- concentration was found in A2 and the lowest in A1 ($P > 0.05$).

Treatments	NO_3^-		NO_2^-	
	Leaf	Root	Leaf	Root
A0	9.12±0.58 a	6.20±0.23 a	0.036±0.003 b	0.596±0.032 c
A1	7.54±0.34 b	3.99±0.36 b	0.046±0.006 a	1.164±0.046 b
A2	8.31±0.43 ab	2.66±0.19 c	0.024±0.004 c	2.371±0.085 a

Values are means ± SD (n=5). Analysis of variance (ANOVA) was employed followed by Duncan's new multi range test. Values with similar superscripts are not significantly different (P>0.05).

Table 6. Effect of mixed amino acids on NO_3^- and NO_2^- concentration in fresh weight of red pepper at 24 h after treatment (μmol g^{-1})

3.2.3 Effect on NRA, NiRA and GSA

For NO_3^- assimilation, NO_3^- is reduced to NO_2^- by catalysis of NR. In this experiment, MAA treatments led to different effects on NR activity in leaves and in roots (Fig. 7). In the roots, treatment A1 and treatment A2 showed increases of 35% and 212% respectively in relation to A0 ($P < 0.01$). In contrast, NR activities were inhibited slightly in leaves by MAA treatments, showing 8.2% in A1 and 10.5% in A2, respectively ($P > 0.05$).

The response of NiR to the MAA treatments resembled that of NR in roots, but was different with that of the NR in leaves (Fig. 8). NiR activities in leaves and roots in A1 were increased by 18% and 60% respectively in relation to A0 (leaves: $P < 0.05$; roots: $P < 0.01$). In A2, NiR activities were the same with A0 in leaves and enhanced 138% in roots (leaves: $P > 0.05$; roots: $P < 0.01$).

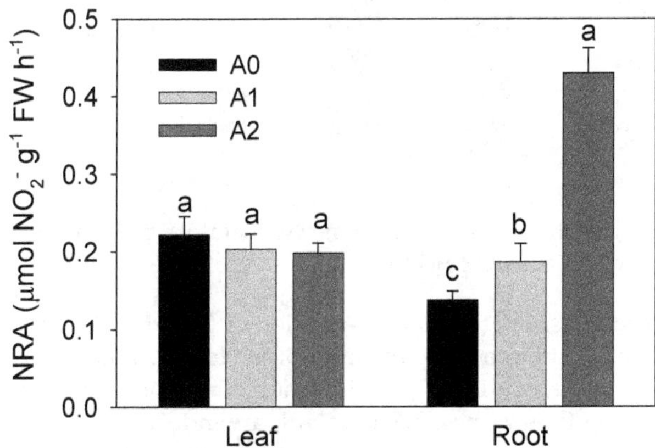

Fig. 7. Effect of mixed amino acids on nitrate reductase activity in red pepper at 24 h after treatment. Values are means ± SD (n=5).

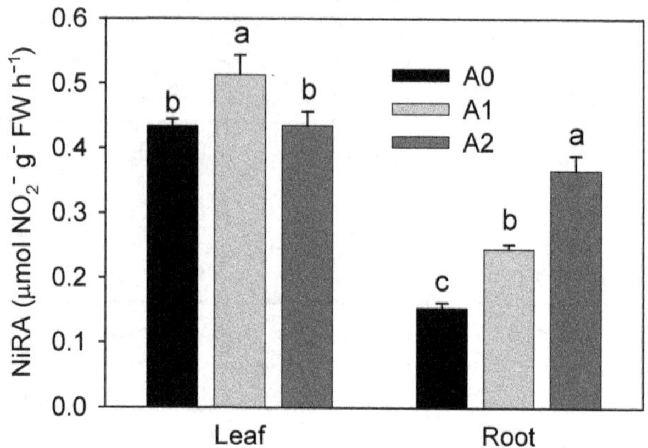

Fig. 8. Effect of mixed amino acids on nitrite reductase activity in red pepper at 24 h after treatment. Values are means ± SD (n=5).

The principal NH_4^+ pathway is the glutamine synthetase (GS)/glutamate synthase (GOGAT) cycle. The behavior of GS activities in leaves was increased by 16% in A1 but not affected in A2 (Fig. 9; $P > 0.05$). However, slight inhibitions were found in roots, showing 7% in A1 and 17% in A2 in relation to A0 ($P < 0.05$).

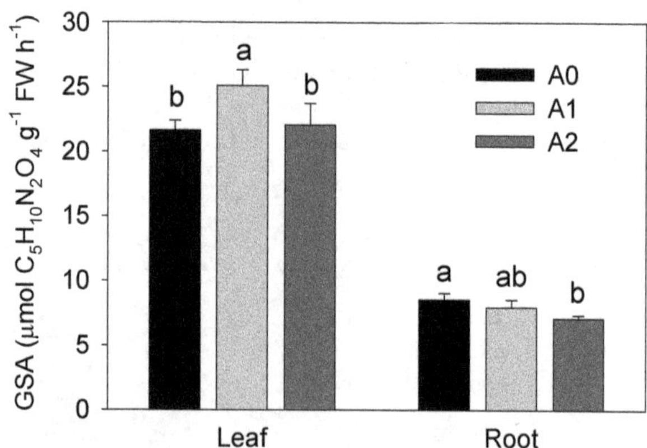

Fig. 9. Effect of mixed amino acids on glutamine synthetase activity in red pepper at 24 h after treatment. Values are means ± SD (n=5).

The first step in nitrate assimilation is the reduction of NO_3^- to NO_2^- by NR, the main and most limiting step, in addition to being the most prone to regulation (Sivasankar et al., 1997; Ruiz et al., 1999). The next step in NO_3^- assimilation is the conversion of the NO_2^- to NH_4^+ by the action of NiR. Both enzymes, NR and NiR, are induced by the same factors (Oaks, 1994). In our experiment, at 10 mM NO_3^- which is facilitated by LATS, the presence of MAA could increase the activities of NR and NiR in roots (Fig. 7 and Fig. 8). In addition, the very high NO_2^- content was found in MAA treatments in roots (Table 6). These results suggest that MAA can increase NO_3^- uptake by enhancing NR activity in roots of red pepper. It is also striking that effect of MAA on NO_3^- assimilation in the roots was higher than in the leaves, presumably NO_3^- was more available and the MAA content was higher in the roots.

Ammonium assimilation in higher plants was long thought to begin with the synthesis of glutamate by glutamate dehydrogenase (GDH). It is now believed that the major pathway of NH_4^+ assimilation is the GS-GOGAT pathway, and GDH generally acts in a deaminating direction (Milflin and Habash, 2001). However, a role in NH_4^+ detoxification would explain the increase in GDH expression under conditions that provoke high tissue NH_4^+ levels (Lancien et al., 2000).

Two possible effect ways of amino acids on N assimilation process had been suggested: direct effect on mRNA of NR (Deng et al., 1991; Li et al., 1995; Vincentz et al., 1993) and feed-back inhibition on NO_3^- reduction systems (King et al., 1993; Ivashikian and Sokolov, 1997; Sivasankar et al., 1997). The hypothesis is that these two effect ways can collectively influence N assimilation in higher plant. This might probably be the main reason for differential effects on NO_3^- uptake observed in different studies. In the present experiment, GS activity was inhibited slightly by MAA treatments in roots, whereas irregular results were obtained in leaves (Fig. 9).

3.2.4 Effect on amino acids and proteins accumulation

With respect to the main products of NO_3^- assimilation, amino acids and proteins (Table 7), the plants treated with MAA did not show increase in these compounds as being supposed apart from amino acids in roots ($P < 0.05$). In contrast, the concentration of proteins in the roots ($P < 0.05$) and leaves ($P > 0.05$) decreased with the MAA rate. Amino acids in leaves ($P > 0.05$) showed the same tendency too.

Amino acids are the building blocks for proteins and also the products of their hydrolysis (Barneix and Causin, 1996). In the present experiment, amino acids concentrations (Table 7) were higher in the roots than in leaves. This is normal since the N assimilation occurs primarily in the roots than in the leaves. In roots, proteins concentrations (Table 7) were decreased by MAA treatment due to the possibility that amino acids content had effect on protein breakdown.

Treatments	Amino acids		Proteins	
	Leaf	Root	Leaf	Root
A0	0.93±0.03 a	2.35±0.12 b	5.03±0.27 a	1.92±0.12 a
A1	0.78±0.06 b	2.81±0.16 a	4.53±0.18 b	1.90±0.10 a
A2	0.67±0.03 b	3.05±0.08 a	4.35±0.24 b	1.45±0.11 b

Data are means ± SD (n=5). Analysis of variance (ANOVA) was employed followed by Duncan's new multi range test. Values with similar superscripts are not significantly different (P>0.05).

Table 7. Effect of mixed amino acids on level of amino acids and proteins in fresh weight of red pepper at 24 h after treatment (mg g^{-1})

In conclusion, the results of the present experiment clearly indicated that NO_3^- uptake and NO_3^- assimilation were regulated by MAA in red pepper. The application of MAA rates could be the direct cause of increased activities of the enzymes (NR and NiR) of the NO_3^- assimilatory pathway and the NO_3^- uptake was enhanced when supplied with

LATS range of NO_3^-. In addition, NO_3^- uptake by red pepper in unit weight plant was less than that of radish due to the different preference on N form between these two plants.

4. Pot experiment of radish with high NO_3^- soil

4.1 Materials and methods

Commercialized artificial soil (pH, 5.2; EC, 1240 mS m^{-1}; NO_3^--N, 280 mg Kg^{-1}; available P_2O_5, 1020 mg Kg^{-1}) was mixed with ^{15}N labeled potassium nitrate (10 atom % ^{15}N) and incubated at room temperature for 14 days at 60% of their maximum water-holding capacity. Finally, the high nitrate soil (pH, 5.0; EC, 3230 mS m^{-1}; NO_3^--N, 1906 mg Kg^{-1}; available P_2O_5, 1060 mg Kg^{-1}) was obtained and used for this experiment. Seeds of radish were sown into 100 mL pots filled with the incubated soil and grown in a glasshouse.

The mixed amino acids (MAA) solution contained equal concentrations of amino acids viz., alanine (Ala), β-alanine (β-Ala), aspartic acid (Asp), asparagines (Asn), glutamic acid (Glu), glutamine (Gln) and glycine (Gly). From 17 or 24 days after sowing, seedlings of radish were sprayed with 0.2 or 0.5 mM MAA solution for 2 or 4 times, as indicated in Table 8. The pH of the MAA solutions was maintained between 6.0-6.1 by adding 1.0 M KOH appropriately.

Treatments	Composition of treated solutions (mM)								Applied time
	K^+	Ala	β-Ala	Asp	Asn	Glu	Gln	Gly	DAS
A0*	—	—	—	—	—	—	—	—	—
A1	0.78	0.2	0.2	0.2	0.2	0.2	0.2	0.2	17, 20, 24, 27
A2	2.10	0.5	0.5	0.5	0.5	0.5	0.5	0.5	17, 20, 24, 27
A3	0.78	0.2	0.2	0.2	0.2	0.2	0.2	0.2	24, 27

* Same amount of distilled water sprayed

Table 8. Composition of the treated solutions and application times for radish in pot experiment

Fresh leaves were collected at 28 days after sowing to determine the NO_3^- content and enzyme activities and at 30 days after sowing to determine the NO_3^-, amino acids and protein contents. Plant shoots were harvested at 30 days after sowing to determine crop yield and N assimilation. After harvest the soils were collected for chemical analysis.

4.2 Results and discussion

4.2.1 Effect of MAA on enzyme activities

Nitrate reductase is the first enzyme involved in the metabolic route of NO_3^- assimilation in higher plants. Significant differences were found in the NR activity between the treatments ($P < 0.01$) (Fig. 10). The highest activity was attained with A2, showing an increase of 30% compared with the activity attained with A0. Treatment A1 and A3 were less effective in increasing the activity of NR than A2, with increase of 21% and 7%, respectively.

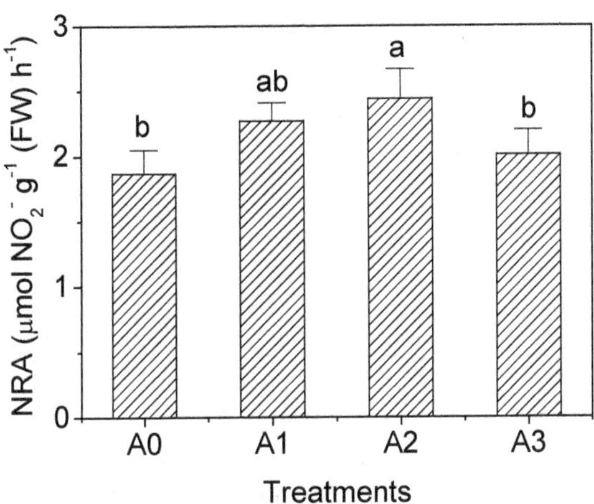

Fig. 10. Effect of mixed amino acids on nitrate reductase activity of radish leaves 28 day after sowing in pot experiment with high NO_3^- soil. Values are means ± SD (n=4).

The next step in NO_3^- assimilation is the conversion of the NO_2^- to NH_4^+ by the action of NiR. The MAA treatments showed different effects on NiR activity depending on the applied concentrations and times of MAA (Fig. 11). The highest activity of NiR was found in treatment A2, showing an increase of 7% compared with A0 ($P < 0.1$). However, the activity of NiR showed a decrease of 11% in A1 ($P < 0.05$).

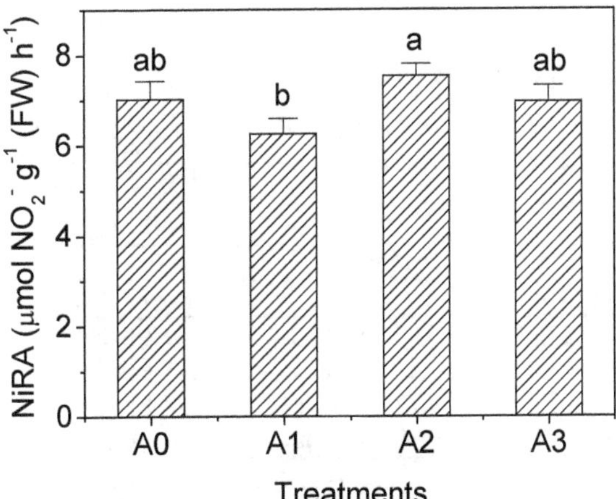

Fig. 11. Effect of mixed amino acids on nitrite reductase activity of radish leaves 28 day after sowing in pot experiment with high NO_3^- soil. Values are means ± SD (n=4).

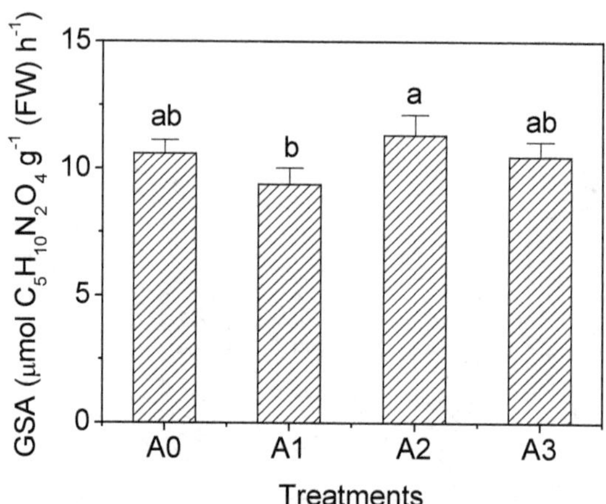

Fig. 12. Effect of mixed amino acids on glutamine synthetase activity of radish leaves 28 day after sowing in pot experiment with high NO_3^- soil. Values are means ± SD (n=4).

The response of GS to MAA treatments is showed in Fig. 12. The greatest activity was observed in treatment A2, with an increase of 7% over the reference treatment ($P > 0.1$). On the contrary, the least activity of GS was found in A1, with a 12% decrease compared with A0 ($P < 0.05$).

The results of activities of enzymes are similar to those of other research, which indicated that treatment of MAA (Section 3.1) and amino acid fertilizer (Section 3.7) could enhance activity of NR in radish when supplied with high rate NO_3^-. In the present experiment, the treatments of MAA led to different rates of increase in NR activity and also affect NiR and GS activities depending on applied rates. Higher activities of three enzymes were found in A2 for the reason that the positive effect on NR was stronger than the feed-back inhibition. However, decrease of NiR and GS was observed in A1 due to the feed-back inhibition on NO_3^- reduction systems which affected GS first.

4.2.2 Effect of MAA on N contents

The data in Table 9 showed that N contents of the plants were affected by using MAA. The NO_3^- content of radish was decreased by 24-38% by applying MAA ($P < 0.001$) compared with the reference treatment.

Treatments	Amino acids	Proteins	NO_3^-	Total N
	(mg g-1 FW)		(mg g-1 DW)	
A0	2.91 ± 0.10 a	9.05 ± 0.58 a	5.79 ± 0.59 a	44.9 ± 1.9 a
A1	2.93 ± 0.07 a	9.57 ± 0.46 a	3.85 ± 0.44 bc	38.4 ± 1.3 b
A2	3.03 ± 0.07 a	9.77 ± 0.54 a	3.57 ± 0.45 c	41.9 ± 1.8 ab
A3	2.99 ± 0.09 a	9.19 ± 0.69 a	4.38 ± 0.18 b	40.2 ± 1.7 ab

Data are means ± SD (n=4). Analysis of variance (ANOVA) was employed followed by Duncan's new multi range test. Values with similar superscripts are not significantly different (P>0.05).

Table 9. Effect of mixed amino acids on nitrogen contents of radish leaves 30 day after sowing in pot experiment with high NO_3^- soil

With respect to the main products of NO_3^- assimilation, amino acids and proteins (Table 9), the plants treated with MAA showed a little increase of these compounds ($P > 0.05$) and the highest contents were found in A2.

The total N content of the plants was affected significantly by using MAA ($P < 0.01$). Treatments of A1, A2 and A3 showed to decrease the total N content to 14%, 7% and 10% compared with the control, respectively.

The result of NO_3^- content agrees with the interpretation that amino acid can negatively regulate nitrate content in higher plants (Chen and Gao, 2002; Gunes et al., 1994, 1996; Wang et al., 2004). In the present experiment, surged value of NO_3^- content was also found at 24 h after MAA treating (the data were not shown). This was probably due to the different response of individual plant to the complex mechanism of MAA in NO_3^- assimilation process in short period. However, 3 days after MAA application, regular result of NO_3^- content in shoots of radish was found.

The predominance of amino acids and proteins were attributed to high activities of main enzymes of NO_3^- assimilation and the direct uptake of amino acids from MAA.

The result of total N content was opposite from that of field experiment in which total N content was increased by applying amino acid fertilizer. These contradictory results were due to different stage of amino acids treatment. Possibly, young plants may lack a complete functional system for NO_3^- uptake and assimilation (Pessarakli, 2002). Wang et al. (2004) reported that application of amino acids in autumn could increase total N in pakchoi but no significant effect was observed when treated in summer.

4.2.3 Effect of MAA on radish yield and N utilization

The plant production in terms of fresh weight was found to be significantly higher ($P < 0.05$) in treatment A1 and A2, with increases of 13% and 12% compared with the control, respectively (Table 10). The response of production in dry weight to MAA treatments was more sensitive than that of fresh weight (Table 10), with significant influences in MAA application ($P < 0.01$). The highest yield in dry weight was found in A2, with an increase of 44% in relation to A0. The results of N utilization (Table 10) were similar to dry yield described above, again registering the highest value in A2, with an increase of 34% compared with A0 ($P < 0.01$). Furthermore, significant effects were also observed in A1 and A3, with increase of 27% and 13% respectively, relative to A0 ($P < 0.01$).

Treatments	Fresh weight (g/plant)	Dry weight (g/plant)	N utilization (mg/plant)
A0	13.32 ± 0.71 b	0.86 ± 0.10 b	37.60 ± 2.87 c
A1	14.99 ± 1.01 a	1.22 ± 0.13 a	47.60 ± 4.11 ab
A2	14.86 ± 0.57 a	1.23 ± 0.11 a	50.72 ± 2.53 a
A3	13.01 ± 0.71 b	1.06 ± 0.07 ab	42.43 ± 3.67 bc

Data are means ± SD (n=4). Analysis of variance (ANOVA) was employed followed by Duncan's new multi range test. Values with similar superscripts are not significantly different (P>0.05).

Table 10. Effect of mixed amino acids on radish yield and nitrogen utilization 30 day after sowing in pot experiment with high NO_3^- soil

For responses of growth, the application of MAA showed enhanced effects obviously. These results are in agreement with those observed by Chen et al. (1997), who reported that application of amino acids led to positive effects on Chinese cabbage growth. Among the treatments of MAA, the growth responses were increased by increasing the application rate of MAA. The increases of yield were due to the positive adjusting of MAA on growth of plants, thus contributing to the increases of N utilization (Table 10) even though the total N content was decreased in MAA treatments (Table 9).

4.2.4 Recovered fertilizer nitrogen

It has become evident that amino acids are a principal source of nitrogen for certain plants, such as mycorrihizal, heathland species (Read, 1993), non-mycorrihizal plants from arctic and alpine ecosystems (Chapin et al., 1993; Kielland, 1994) and boreal forest plants (Näsholm et al., 1998; Persson and Näsholm, 2001). These systems are similar in that N mineralization rates are heavily constrained by climate, and plant N demands cannot be met through the uptake of inorganic ions (Raab et al., 1999). Based on these researches, the amino acids were used to partially replace NO_3^- in hydroponic experiment or spray to leaves in many plants. In most case, the application of amino acids led to the decrease of nitrate content and total nitrogen content in lettuce, Chinese cabbage, onion, pakchoi or other leafy crops (Chen and Gao, 2002; Gunes et al., 1994, 1996; Wang et al., 2004). It had been suggested that plants probably preferred amino acids as sources of reduced nitrogen, and nitrate uptake was inhibited by amino acids.

In this study, the high NdfF was found in MAA treatments (Table 11), indicating that applied MAA did not act as a source of nitrogen for plants. On the contrary, plants had taken up more NO_3^--N from soil due to the regulation of MAA on NO_3^- uptake and assimilation. The results for the possible regulation of NO_3^- uptake and assimilation by amino acids for higher plants are contradictory. Many authors agreed that amino acids can down regulate the NO_3^- uptake and assimilation in higher plants (Aslam et al., 2001; Ivashikian and Sokolov, 1997; Oaks et al., 1979; Radin, 1975, 1977; Sivasankar et al., 1997). But Aslam et al. (2001) reported that inhibition did not occur when the concentration of NO_3^- in the external solutions had been increased to 10 mM. This result is consistent with the other research, which indicated that radish treated with mixed amino acids containing 5.0 mM NO_3^- in growth medium show significantly increased the NO_3^- uptake. In this experiment, the positive effect on NO_3^- uptake by applying MAA was due to very high NO_3^- content in soil (1906 mg Kg^{-1}).

Treatments	NdfF (%)	QNdfF (mg/plant)	NdfFRec (%)
A0	65.9 ± 1.5 b	24.8 ± 0.7 d	33.0 ± 1.3 d
A1	68.6 ± 2.2 ab	32.7 ± 1.1 b	43.6 ± 2.0 b
A2	71.6 ± 0.9 a	36.3 ± 0.8 a	48.4 ± 1.9 a
A3	67.2 ± 2.1 ab	28.5 ± 1.0 c	38.1 ± 2.1 a

NdfF; the percentage of N derived from fertilizer, QNdfF ; the quantity of N derived from fertilizer, NdfFRec ; the fertilizer-N recovery

Data are means ± SD (n=4). Analysis of variance (ANOVA) was employed followed by Duncan's new multi range test. Values with similar superscripts are not significantly different (P>0.05).

Table 11. Nitrogen derived from fertilizer in the radish shoots

4.2.5 Effect of MAA on chemical properties of soil

The chemical properties of soil at the end of experiment are showed in Table 12. The planting of radish affected these chemical properties of soil clearly. However, there were no differences in pH of soil among treatments planted with radish. On the other hand, either planting treatment or MAA treatment showed effect on nitrate in soil. Compared with the non planting treatment, the treatments of planting showed a decrease of 65~81% and 35~47% of nitrate and available P at 30 days after sowing, respectively. The different rates of decrease were due to the different growth rates led by MAA treatment.

Treatments	pH (1:5)	EC* ($mS\ m^{-1}$)	Available P_2O_5	NO_3^--N
			($mg\ Kg^{-1}$)	
NP	5.0	1985	599.7	1008.6
A0	5.7	981	390.6	339.4
A1	5.7	902	344.5	298.5
A2	5.7	674	316.5	192.0
A3	5.7	894	356.0	355.0

* The soil used in these experiments was commercialized artificial soil with lower soil density (about 0.4 $g\ cm^{-3}$) and higher water-holding capacity. Since determination of soil chemical properties is based on dry weight, the determined values of EC and NO_3^--N are quite high relative to ordinary soil. However, these are not very higher in soil solution.

Table 12. Chemical properties of soil at the end of pot experiment for radish with high NO_3^- soil

In conclusion, the results of the present experiment suggest that application of MAA can affect activities of three enzymes of N assimilation (NR, NiR and GS). However, the exact reason for this observation is unknown and further investigation is necessary. Furthermore, the application of MAA can enhance growth, N utilization, and concentrations of proteins and amino acids, and reduce the NO_3^- content in plant shoots. Considerable increase of N uptake from soil was indicated by the increased ^{15}N recovery by applying MAA compared with the control. These results suggest that the main role of MAA on nitrate uptake and assimilation might be relation with the regulation of NO_3^- uptake and assimilation, but not as sources of reduced nitrogen.

5. Pot experiment of radish with low NO_3^- soil

5.1 Materials and methods

Commercialized artificial soil (pH, 5.2; EC, 1240 $mS\ m^{-1}$; NO_3^--N, 280 $mg\ Kg^{-1}$; available P_2O_5, 1020 $mg\ Kg^{-1}$) was used for this experiment. Plant culture and MAA treatment were the same with that of pot experiment of radish with high NO_3^- soil. The sampling and analysis of plant and soil also were according to procedures adopted for radish with high NO_3^- soil.

5.2 Results and discussion

5.2.1 Effect of MAA on enzyme activities

Significant differences were found in the NR activity among the treatments ($P < 0.01$) (Fig. 13). The NR activity was inhibited by foliar application of MAA in this experiment, contrary to that of radish in which was planted in high nitrate soil. The lowest activity was attained with A2, showing a decrease of 28% compared with the activity attained in treatment A0. Treatment A1 and A3 were less effective in decreasing the activity of NR than A2, with decreases of 8% and 17%, respectively.

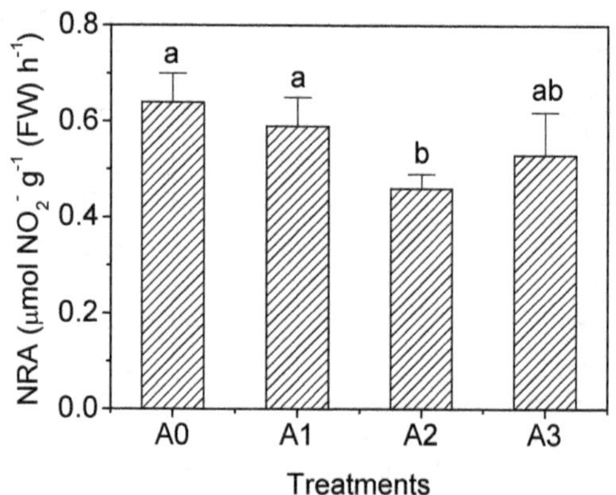

Fig. 13. Effect of mixed amino acids on nitrate reductase activity of radish leaves 28 day after sowing in pot experiment with low NO_3^- soil. Values are means ± SD (n=4).

The response of activity of NiR to the MAA application resembled that of NR (Fig. 14). The lowest activity of NiR was found in treatment A2, showing 40% decrease compared with the control treatment ($P < 0.001$). Treatment A1 and A3 also showed 23% and 32% decrease in relation to A0, respectively.

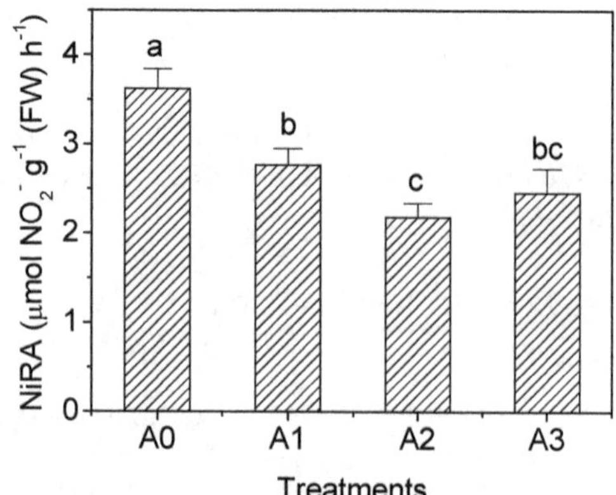

Fig. 14. Effect of mixed amino acids on nitrite reductase activity of radish leaves 28 day after sowing in pot experiment with low NO_3^- soil. Values are means ± SD (n=4).

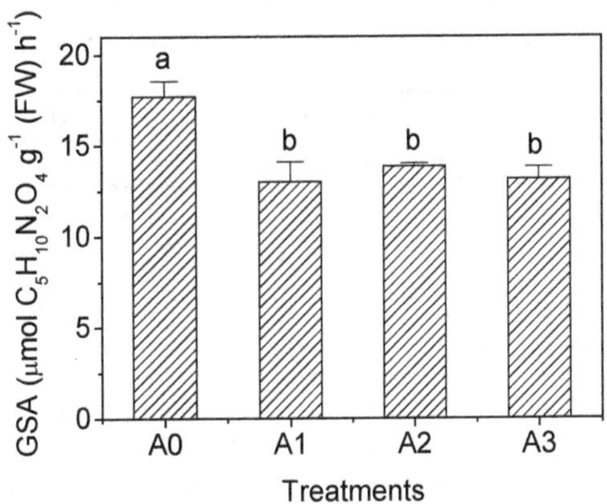

Fig. 15. Effect of mixed amino acids on glutamine synthetase activity of radish leaves 28 day after sowing in pot experiment with low NO_3^- soil. Values are means ± SD (n=4).

With respect to enzyme activity of GS (Fig. 15), the application of MAA led to significant decrease in the activity in leaves of radish in this experiment, the lowest activity being recorded in treatment A1, with a decrease of 27% in relation to the highest activity, found in the reference treatment A0 ($P > 0.01$). Treatment A2 and A3 also showed decreases of 22% and 26%, respectively.

Like some of the N transporters, NR is induced by its own substrate, NO_3^-, and this induction is fast, occurring within several minutes, and requires very low concentrations (< 10 μM) (Craw ford, 1995; Sueyoshi et al., 1995). NO_3^- is the primary factor, although other factors also influence the regulation of NO_3^- reduction and assimilation, including the end-products of assimilation such as amino acids. NiR and NR are similarly transcriptionally regulated for the reason of that NiR is strongly induced by the same factor, NO_3^-, probably to prevent the accumulation of toxic NO_2^- (Wang et al., 2000). The activities of NR and NiR were much lower than that of radish which was planted in high NO_3^- soil due to the poor NO_3^- in the soil used in the present experiment (Table 15).

In the present experiment, the activities of three enzymes decreased when treated with MAA. These results are in agreement with other researches which indicated that downstream N assimilation products such as amino acids can feed back to regulate NO_3^- uptake and reduction (Deng et al., 1991; Sivasanker et al., 1997; Vincentz et al., 1993). However, the effects of MAA on activities of enzymes are opposite to other experiments of ours, which indicated that treatment of MAA and amino acid fertilizer could enhance activity of NR in radish when supplied with high rate of NO_3^-. The contradictory results are due to the different NO_3^- levels of the soils.

5.2.2 Effect of MAA on N contents

The data in Table 13 showed that N contents of the plants were not affected significantly by using MAA. The highest concentrations of all N forms were observed in treatment A2 ($P > 0.05$). These results differed from radish which was planted in high NO_3^- soil. The different effects of MAA on N contents of radish in high NO_3^- soil and low NO_3^- soil are in agreement with the supposition that amino acids have different effect on NO_3^- uptake and assimilation.

Treatments	Amino acids	Proteins	NO_3^-	Total N
	(mg g^{-1} FW)		(mg g^{-1} DW)	
A0	1.89 ± 0.07 a	7.35 ± 0.40 a	1.30 ± 0.06 a	20.5 ± 1.5 a
A1	1.87 ± 0.11 a	7.60 ± 0.37 a	1.32 ± 0.16 a	20.3 ± 2.4 a
A2	1.77 ± 0.12 a	7.92 ± 0.46 a	1.35 ± 0.08 a	21.0 ± 0.7 a
A3	1.97 ± 0.12 a	7.06 ± 0.50 a	1.34 ± 0.08 a	20.0 ± 0.8 a

Data are means ± SD (n=4). Analysis of variance (ANOVA) was employed followed by Duncan's new multi range test. Values with similar superscripts are not significantly different (P>0.05).

Table 13. Effect of mixed amino acids on nitrogen contents of radish leaves 30 day after sowing in pot experiment with low NO_3^- soil

5.2.3 Effect of MAA on radish yield and N utilization

The plant production in fresh weight was found to be higher ($P < 0.01$) in treatment of A2, with an increase of 9% compared with the control treatment (Table 14). The response of production in dry weight to MAA treatments was not as sensitive as that in fresh weight (Table 14), only with slight influences. The results of N utilization (Table 14) were similar to dry yield, registering the highest value in A1, with an increase of 15% compared with A0 ($P < 0.01$).

Treatments	Fresh weight	Dry weight	N utilization
	(g/plant)		(mg/plant)
A0	8.38 ± 0.40 b	0.84 ± 0.05 a	17.06 ± 0.85 b
A1	8.57 ± 0.34 b	0.88 ± 0.09 a	19.71 ± 0.58 a
A2	9.83 ± 0.32 a	0.89 ± 0.05 a	18.52 ± 0.51 ab
A3	8.44 ± 0.26 b	0.85 ± 0.03 a	18.88 ± 0.42 a

Data are means ± SD (n=4). Analysis of variance (ANOVA) was employed followed by Duncan's new multi range test. Values with similar superscripts are not significantly different (P>0.05).

Table 14. Effect of mixed amino acids on radish yield and nitrogen utilization 30 day after sowing in pot experiment with low NO_3^- soil

5.2.4 Effect of MAA on chemical properties of soil

The chemical properties of soil at the end of experiment were showed in Table 15. The planting of radish affected these chemical properties of soil clearly. However, there were no differences in pH of soil among treatments planted with radish. On the other hand, either planting treatment or MAA treatment showed effect on soil nitrate reduction. Compared with the non planting treatment, the treatments of planting showed decrease of 86~88% for nitrate and decrease of 56~70% for available P at 30 days after sowing, respectively. The different rates of decrease were due to the different growth rates resulted from by MAA treatment. And the EC decreased accordingly.

Treatments	pH (1:5)	EC (mS m^{-1})	Available P_2O_5 (mg Kg^{-1})	NO_3^--N (mg Kg^{-1})
NP	5.4	1485	878	214.7
A0	5.7	482	384	68.2
A1	5.7	515	303	62.2
A2	5.7	349	264	70.3
A3	5.7	503	277	68.6

Table 15. Chemical properties of soil at the end of pot experiment for radish with low NO_3^- soil

The commercialized artificial soil used in this experiment was with lower soil density (about 0.4 g cm^{-3}) and higher water-holding capacity. Although the NO_3^- contents of 62.2~70.3 mg Kg^{-1} are not low in ordinary soil, available NO_3^- for plants is very poor in soil solution in this experiment. This might be the probable reason, that effects of MAA on N assimilation in the present experiment were different from that of radish which was planted in high NO_3^- soil. Whether in our experiments or in other researches, different effects of amino acids on NO_3^- reduction and assimilation were observed (Aslam et al., 2001).

In conclusion, the results of the present experiment suggest that application of MAA can decrease activities of three enzymes of N assimilation (NR, NiR and GS). However, except N utilization, the application of MAA did not have significant effects on growth, and concentrations of proteins, amino acids, total N and NO_3^- content in plant shoots. The difference in the results were found in both the present experiment and pot experiment which radish was planted in high NO_3^- soil may be due to different levels of NO_3^- content in soil solution. The hypothesis that effect of amino acids on NO_3^- uptake, reduction and assimilation depends on concentration of NO_3^- was justified.

6. Field experiment of radish

6.1 Materials and methods

The study was conducted in summer of 2005 at the experimental farm of the Chungnam National University, Daejeon, Korea. The average chemical properties of the soil of the field are described in Table 16. The fertilizer mixture was uniformly broadcasted onto the soil surface and incorporated before ridging. The seeds of radish were sown at the end of May 2005 and arranged in a completely randomized block design, with three replications. The plots were 5 m × 2 m consisting of 2 rows.

At 15 and 22 days after sowing, AAF was applied 2 times to plots by spraying to leaves after diluting 500, 1000 and 2000 times by water, respectively. The main chemical contents of the AAF and application quantities are shown in Table 17.

Soils	pH (1:5)	EC (mS m^{-1})	Organic matter (g Kg^{-1})	Available P_2O_5 (mg Kg^{-1})	Total N (g Kg^{-1})	NO_3^--N (mg Kg^{-1})
Before fertilization	6.0	122	15.6	170	0.81	80.2
After fertilization	6.0	191	15.8	279	0.87	191.2

Table 16. Chemical properties of soils used in field experiment of radish

Fresh leaves were collected at 23 days after sowing to determine the NO_3^-, amino acids and protein contents and enzyme activities. The plots were harvested at 35 days after sowing to determine crop yield and N assimilation. The topsoil samples (0–20 cm) were collected at 25 and 35 days after sowing for chemical analysis.

In order to compare the different AAF treatments for their N uptake, net N uptake was estimated by balancing N utilization and N input by applying AAF thus:

$$N_N = N_U - N_{AAF}. \tag{1}$$

where N_N is the net N uptake by plant; N_U is the total N utilization at harvest; N_{AAF} is N input by applying AAF.

It was assumed that N would have been either taken up by the plants or lost from the soil-plant system. In our experiment, leaching was the main way of N loss. Furthermore, N loss attributable to soil erosion and runoff was considered for our site with 2~5% slope. Since these losses may be influenced by protecting of the plants from the rain, the vegetation cover was observed at 25 and 35 days after sowing.

Classification (%)		Treatments				
		NP*	**A0**	**A1**	**A2**	**A3**
				(mg m^{-2})		
AAF application		–	–	750	1500	3000
Essential amino acid	2.22	–	–	16.7	33.3	66.6
Total amino acid	5.14	–	–	38.6	77.2	154.4
Total-N	3.80	–	–	28.5	57.0	114.0
Soluble P	3.12	–	–	23.4	46.8	93.6
Soluble K	4.97	–	–	37.3	74.6	149.2
Soluble B	0.13	–	–	0.98	1.95	3.90

* NP: No-planting

Table 17. Amino acid fertilizer applied to radish in the field experiment

6.2 Results and discussion

6.2.1 Effect of AAF on enzyme activities

Nitrate reductase is the first enzyme involved in the metabolic route of NO_3^- assimilation in higher plants. Significant differences were found in the NR activity between the treatments ($P < 0.01$) (Fig. 16). The highest activity was obtained with A1, showing an increase of 16% in relation to the activity obtained with A0. A2 was less effective in increasing the activity of NR than A1; whereas no increase of NRA occurred in A3, even treated with fourfold AAF than A1.

The next step in NO_3^- assimilation is the conversion of the NO_2^- to NH_4^+ by the action of NiR. The AAF treatments showed different effect on NiR activity depending on the applied rate of AAF (Fig. 17). The highest activity of NiR was found in treatment A1, showing an increase of 4% compared with A0 ($P < 0.05$). However, the activities of NiR were inhibited by 12 and 13% in A2 and A3, respectively.

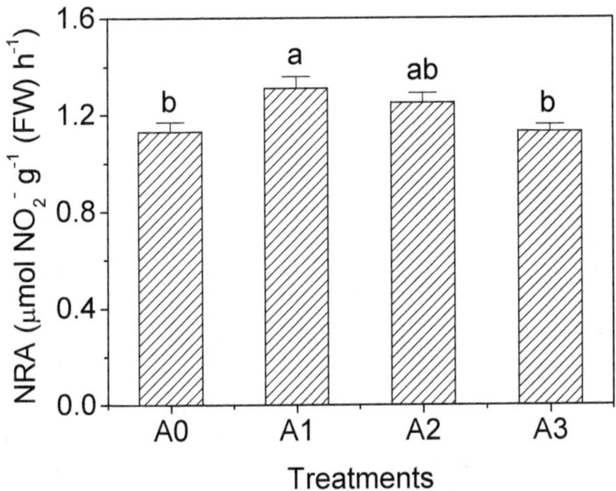

Fig. 16. Effect of amino acid fertilizer on nitrate reductase activity in leaves of radish 23 day after sowing. Values are means ± SD (n=3).

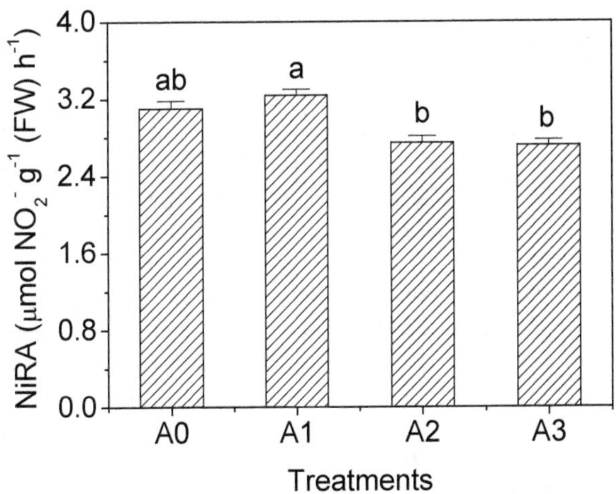

Fig. 17. Effect of amino acid fertilizer on nitrite reductase activity in leaves of radish 23 day after sowing. Values are means ± SD (n=3).

The reversible amination of 2-oxoglutarate to glutamic acid via GDH has long been considered as a major route of NH_4^+ assimilation (Srivastava and Singh, 1987). However the discovery of the enzyme GS-GOGAT system altered this point of view, and the incorporation of NH_4^+ to glutamine via GS and subsequently into glutamic acid by GOGAT is now widely accepted as the main route of NH_4^+ assimilation (Oaks, 1994). The response of GS (Fig. 18) to AAF treatments was similar to that of the NiR (Fig. 17). The greatest activity was reached in treatment A1, with an increase of 20% over the reference treatment ($P < 0.001$). On the contrary, the activity of GS was the lowest in A3, with a decline of 11% compared with A0 ($P < 0.05$).

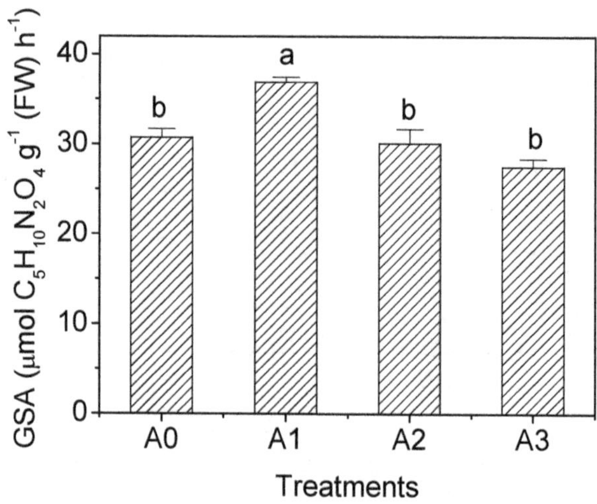

Fig. 18. Effect of amino acid fertilizer on glutamine synthetase activity in leaves of radish 23 day after sowing. Values are means ± SD (n=3).

The reduction of NO_3^- to NO_2^- by NR, is the main and most limiting step, in addition to being the most prone to regulation (Sivasankar et al., 1997; Ruiz et al., 1999). The synthesis of this enzyme is induced by nitrate (Oaks, 1994), but although its activity is known to be repressed by ambient ammonium, there are evidences that this enzyme can be regulated by certain amino acids. The results for the possible regulation of NR activity by amino acids for higher plants are contradictory. Many authors agree with that amino acids can inhibit the activity of NR in higher plants (Radin, 1975, 1977; Oaks et al., 1979; Ivashikian and Sokolov, 1997; Sivasankar et al., 1997; Aslam et al., 2001). But Aslam et al. (2001) reported that inhibition did not occur when the concentration of NO_3^- in the external solutions had been increased to 10 mM. This result is consistent with the other research, which indicates that radish treated with mixed amino acids containing 5.0 mM NO_3^- in growth medium showed significant increase of NR activity (Liu et al., 2005). The effect of amino acids on NR activity seems to be depended on plant materials, age of plants, growth conditions, nitrate concentration, kinds of amino acids, amino acids concentration and other factors. In this experiment, the positive effect on NR activity by applying AAF was due to high NO_3^- content in soil.

In the present experiment, the treatments of AAF led to different levels of increase of NR activity and inhibition on GS activity depending on applied rates. The high activities of three enzymes were found in A1 due to the positive effect of AAF on process of NO_3^- assimilation. However, inhibition on NiR and GS was observed in A2 and A3 for the reason that high rates of AAF application had high feed-back inhibition on NO_3^- reduction systems which affected GS first. This is probably the main reason why different effects on the enzymes were observed in this study.

6.2.2 Effect of AAF on biomass and utilization of N and P

The plant biomass production in fresh weight was found to be significantly higher ($P < 0.01$) in the AAF treatments (mean biomass in fresh weight of A1, A2 and A3 are 5.056, 4.738 and 4.653 Kg m^{-2}, respectively) compared with the control (mean biomass in fresh weight is 4.026 Kg m^{-2}) (Table 18). Among AAF treatments, the treatment with low concentration of AAF (A1) had a higher ($P > 0.05$) biomass production than the treatment with high concentration of AAF (A3). The response of biomass production in dry weight to AAF treatments resembled that in fresh weight (Table 28), with significant influence by applying AAF ($P < 0.01$). The highest biomass production in dry weight was found in A1, with an increase of 17% in relation to A0.

Treatments	Fresh weigh	Dry weight	N utilization	P utilization
A0	4026 ± 227 c	345.7 ± 14.2 c	9.33 ± 0.87 c	2.35 ± 0.09 b
A1	5056 ± 213 a	404.4 ± 11.6 a	14.48 ± 0.89 a	2.87 ± 0.11 a
A2	4738 ± 183 ab	394.8 ± 12.1 ab	13.04 ± 0.53 ab	2.68 ± 0.12 a
A3	4653 ± 189 b	382.0 ± 14.5 b	12.83 ± 0.67 b	2.64 ± 0.07 a

Values are means ± SD (n=3). Analysis of variance (ANOVA) was employed followed by Duncan's new multi range test. Values with similar superscripts are not significantly different (P>0.05)

Table 18. Effect of amino acid fertilizer on radish yield and utilization of nitrogen and phosphorus 35 day after sowing (g m^{-2})

The result of N utilization (Table 18) was similar to biomass production as described above, again registering the highest value in A1 (14.48 ± 0.89 g m^{-2}), with an increase of 55% compared with A0 (9.33 ± 0.87 g m^{-2}) ($P < 0.01$). Furthermore, significant effects were observed in A2 and A3 too, with increases of 40% and 37% respectively, in relation to A0 ($P < 0.01$). Even though P content was not influenced by the application of AAF (Table 20), P utilization increased in AAF treatments due to the increase of biomass production (Table 18).

The observed result of vegetation cover and calculated values of net N uptake are showed in Table 19. The treatments of AAF showed higher vegetation cover than the control. Besides the N input by applying of AAF, the treatments of AAF showed significant increase of 36~55%net N uptake compared with the control. Gunes et al. (1996) suggested that plants probably preferred amino acids as sources of reduced nitrogen, but they did not distinguish origin of the N contents in the plants. In our experiment, the increase of N uptake is about 200 times (Table 19) more than N supplied by applying AAF, indicating application of AAF could enhance the ability of uptake and assimilation of inorganic N by plants.

Treatments	Vegetation cover (%)		Net N uptake (g m^{-2})
	25 DAS	35 DAS	35 DAS
A0	63 ± 3 c	91 ± 2 b	9.33 ± 0.87 b
A1	85 ± 5 a	100 ± 0 a	14.45 ± 0.89 a
A2	79 ± 6 ab	100 ± 0 a	12.98 ± 0.53 a
A3	76 ± 3 b	100 ± 0 a	12.72 ± 0.67 a

Values are means ± SD (n=3). Analysis of variance (ANOVA) was employed followed by Duncan's new multi range test. Values with similar superscripts are not significantly different (P>0.05)

Table 19. Net nitrogen uptake and vegetation cover of radish

These results are in agreement with those observed by Chen et al. (1997), who reported that application of amino acids led to positive effects on cabbage growth. However, among the treatments of AAF, the growth responses were decreased by increasing the application rate of AAF. This may probably be related to the feed–back inhibition of high rate application of amino acids.

6.2.3 Effect of AAF on contents of N and P

The data in Table 20 showed that N contents of the plants were affected by using amino acid fertilizer. The NO_3^- content of radish was decreased by application of AAF ($P < 0.05$) compared with the reference treatment. Among the treatments, A1 gave the best result in reducing the nitrate to 1.16 mg g^{-1} (FW), with a decrease of 24% in relation to the highest NO_3^- content found in A0. This result agrees with the interpretation that amino acid can negatively regulate nitrate content in higher plants (Gunes et al., 1994, 1996; Chen and Gao, 2002; Wang et al., 2004). But this interpretation was not supported in all cases. It was observed that the mixed amino acids increased NO_3^- content slightly in radish when the plants growing in nutrient solution. The contradiction may reside in amino acids treatment method. It was demonstrated in other studies that amino acid pretreatment decreased NO_3^- accumulation slightly, but Gln and Asn led to NO_3^- concentration increase in barley roots when they were used together with nitrate (Aslam et al., 2001).

With respect to the main products of NO_3^- assimilation, amino acids and proteins (Table 20), the plants in treatment A1 gave the highest contents of these compounds ($P < 0.01$). In the A1 treatment, high activities of main enzymes of NO_3^- assimilation could explain the predominance of these nitrogenous compounds in radish. Under treatments of A2 and A3, the increases of amino acids and proteins derived from the direct uptake of amino acids from AAF.

Treatments	NO_3^-	Amino acids	Proteins	Total-N	Total-P
		(mg g^{-1} FW)		(mg g^{-1} DW)	
A0	1.53 ± 0.11 a	1.29 ± 0.02 b	1.28 ± 0.08 c	27.2 ± 1.6 c	6.8 ± 0.3 a
A1	1.16 ± 0.17 b	1.38 ± 0.02 a	1.98 ± 0.16 a	35.9 ± 1.6 a	7.1 ± 0.4 a
A2	1.32 ± 0.20 ab	1.33 ± 0.02 ab	1.74 ± 0.06 b	33.0 ± 0.9 ab	6.8 ± 0.4 a
A3	1.48 ± 0.08 a	1.32 ± 0.04 ab	1.70 ± 0.09 b	31.2 ± 1.2 bc	6.9 ± 0.5 a

Values are means ± SD (n=3). Analysis of variance (ANOVA) was employed followed by Duncan's new multi range test. Values with similar superscripts are not significantly different (P>0.05)

Table 20. Effect of amino acid fertilizer on contents of nitrogen and phosphorus in radish 23 day after sowing

The total N content of the plants was also affected significantly by the use of AAF ($P < 0.01$). Treatments of A1, A2 and A3 showed to increase the total N to 32%, 21% and 15% relative to the control, respectively. These increases were due to the positive adjusting of AAF on uptake and assimilation of N, and attributing to the increases of N utilization and net N uptake. The P content of radish was not affected significantly by the application of AAF (Table 20).

6.2.4 Effect of AAF on chemical properties of soil

The chemical properties of soil in middle growth period and at the end of experiment were showed in Table 21 and Table 22. The planting of radish affected total N of soil clearly, except at 35 days after sowing, with a fall of 10% compared with non planting treatment. However, there were no differences in total N of soil among treatments planted with radish. On the other hand, either planting treatment or AAF treatment showed effect on nitrate in soil.

Treatments	pH (1:5)	EC ($mS\ m^{-1}$)	Organic matter ($g\ Kg^{-1}$)	Available P_2O_5 ($mg\ Kg^{-1}$)	Total N ($g\ Kg^{-1}$)	NO_3^--N ($mg\ Kg^{-1}$)
NP	6.3	81	15.4	267	0.70	75.3
A0	6.3	57	15.2	297	0.67	52.5
A1	6.4	55	15.6	285	0.67	55.7
A2	6.5	65	15.1	310	0.67	58.9
A3	6.4	54	15.3	305	0.66	60.0

Table 21. Chemical properties of soil in the middle of growth period (25 day after sowing) for radish

In the soil of non planting, nitrate was decreased by leaching and runoff by rain. Compared with the non planting treatment, the treatments of planting showed 20~30% decrease at 25 days after sowing and 23~42% decrease at 35 days after sowing in the nitrate content of soil. Although with the lowest net N uptake, the lowest concentration of nitrate in soil was found in A0 treatment both at two sampling times. This was due to the fact that the vegetation covers of AAF treatments were higher than treatment of A0, and could effectively prevent nitrate of soil from leaching or runoff. The planting treatments showed lower values of EC than non planting treatment, but all were in the range of general soil. There were no significant differences among all treatments in pH and organic matter of soil. Moreover, very small differences were observed in available P due to different growth rate of the plants.

Treatments	pH (1:5)	EC ($mS\ m^{-1}$)	Organic matter ($g\ Kg^{-1}$)	Available P_2O_5 ($mg\ Kg^{-1}$)	Total N ($g\ Kg^{-1}$)	NO_3^--N ($mg\ Kg^{-1}$)
Before experiment	6.0	191	15.8	279	0.87	191.2
NP	6.4	99	15.9	308	0.70	94.3
A0	6.4	41	16.0	283	0.63	55.0
A1	6.4	42	15.8	263	0.63	65.9
A2	6.4	49	15.2	270	0.63	65.0
A3	6.4	45	15.3	272	0.63	73.2

Table 22. Chemical properties of soil at the end of field experiment (35 day after sowing) for radish

Treatments	NO_3^- removal (0~20cm) ($g\ m^{-2}$)	Removal rate by plant	Removal rate by leaching
		(%)	
NP	25.2	-	100.0
A0	35.4	26.3	73.7
A1	32.6	44.4	55.6
A2	32.8	39.7	60.3
A3	30.7	41.8	58.2

Table 23. Effect of amino acid fertilizer on nitrate removal from the soil

The data of NO_3^- removal are showed in Table 23. Even though the highest NO_3^- removal was found in treatment A0, the most removed NO_3^- was leached (73.7%) and would lead to pollution for groundwater. The application of AAF can enhance NO_3^- removal rate by planting, and avoid N losses through leaching and runoff due to increases of N utilization (Table 18) and vegetation cover (Table 19).

In conclusion, the results of the present experiment suggest that application of amino acid fertilizer can affect activities of three enzymes of N assimilation (NR, NiR and GS) and increase the growth and N assimilation in radish. However, the exact reason for this observation is not known and requires further investigation. The planting of radish proves very effective for nitrate removal in soil by its fast growth and very high biomass production (345.7~404.4 g DW m^{-2}) and N utilization (9.33~14.48 g m^{-2}) in short time (only 35 days in our experiment). Furthermore, the application of amino acid fertilizer can enhance biomass production, N utilization, and concentrations of proteins and amino acids, and it can reduce N losses through leaching and runoff.

7. Conclusions

By conducting these experiments, several findings were obtained: (1) increase of NO_3^- uptake by application of MAA, (2) different effect of MAA dictated by N status, (3) efficient NO_3^- removal by application of AAF, and (4) true role of MAA in process of NO_3^- uptake and assimilation.

Both for radish and red pepper, the application of MAA led to significant increase of NO_3^- uptake and activities of the three enzymes (NR, NiR and GS) of the NO_3^- assimilatory pathway in solution experiment. These results are different from other researches which inhibition was observed in most case. This difference was caused by two main reasons: (1) that effect of MAA was different to single amino acid, and (2) comparative high level NO_3^- was supplied in these experiments.

In pot experiments, responses to applied MAA were affected by plant species and NO_3^- level in soil. For radish, application of MAA led to increases of activities of three enzymes, growth, N utilization, and concentrations of proteins, and decrease of NO_3^- content in plant shoots, when the plants were planted in high NO_3^- soil. However, in the case that radish was planted in low NO_3^- soil, activities of the enzymes were decreased by using MAA, and growth, and concentrations of proteins, amino acids, total N and NO_3^- content were not affected. These phenomena indicate that the effect of MAA is dependent on NO_3^- level.

With respect to red pepper which was planted in high NO_3^- soil, foliar MAA sprays increased activities of the three enzymes, while reduced NO_3^- content, concentrations of proteins and amino acids, total N and N utilization. Partially different results were found in red pepper which was planted in low NO_3^- soil, including decreased activities of NiR and GS and increased of NO_3^- content in plant shoots by the application of MAA. The reason for these differences is the same to that of radish.

In field experiment of radish, the foliar sprays of AAF increased NO_3^- removal rate by planting, and avoid N losses through leaching and runoff due to increases of N utilization and vegetation cover. In addition, the application of AAF enhanced activities of three enzymes, biomass production, and concentrations of proteins and amino acids, reduced NO_3^- content in plant shoots. Similarly, for red pepper, the use of AAF led to increase of N utilization. However, decrease of total N content in red pepper plants was found in AAF treatments.

These results of ^{15}N labeled experiments and field experiments suggest that the main role of amino acids on nitrate uptake and assimilation might be relation with the regulation of NO_3^- uptake and assimilation, but not as sources of reduced nitrogen. In pot experiments, it was indicated that the N utilization of plants was depended on soil NO_3^- uptake which was regulated by application of MAA. In field experiment of radish, the increase of N utilization is about 200 times more than N supplied by applying AAF, indicating application of AAF could enhance the ability of uptake and assimilation of inorganic N by plants.

Finally, the effect of amino acids on NO_3^- uptake and assimilation was also influenced by stage of plant growth. For leaf radish, response of enzymes activity and yield was not affected by the stage of growth, while N accumulation (total N content) was more sensitive to applied amino acids in vegetative stage than that of young stage. With regard to red pepper, effects of amino acids on enzymes activity and N content in different growth stage were quite similar, while growth (dry biomass) showed to be increased significantly in vegetative stage.

A better understanding of effect of amino acid on process of NO_3^- uptake and assimilation will undoubtedly help in developing an approach to improve the management of fertilizer nitrogen and to prevent N loss through leaching or runoff. In the further study, more detailed researches should be carried out to investigate the precise manner by which MAA influences NO_3^- uptake and assimilation. The researches will focus on the effect of MAA on NR gene expression and relation between GDH and GS.

8. References

Andrews, M. 1986. The partitioning of nitrate assimilation between root and shoot of higher plants. Plant Cell Environ. 9, 511–519.

Ashley, D. A., W. A. Jackson, and R. Volk. 1975. Nitrate uptake and assimilation by wheat seedlings during initial exposure to nitrate. Plant Physiol. 55, 1102–1106.

Aslam, M, R. L. Travis, and D. W. Rains. 1996. Evidence for substrate induction of a nitrate efflux system in barley roots. Plant Physiol. 112, 1167–1175.

Aslam, M., R. L. Travis, and D. W. Rains. 2001. Differential effect of amino acids on nitrate uptake and reduction systems in barley roots. Plant Sci. 160, 219–228.

Aslam, M., R. L. Travis, and R. C. Huffaker. 1992. Comparative kinetics and reciprocal inhibition of nitrate and nitrite uptake in roots of uninduced and induced barley (*Hordeum vulgare* L.) seedlings. Plant Physiol. 99, 1124–1133.

Atkin, O. K. 1996. Reassessing the nitrogen relations of Arctic plants. A mini-review. Plant Cell Environ. 19, 695–704.

Barneix, A. J., and H. F. Causin. 1996. The central role of amino acids on nitrogen utilization and plant growth. J. Plant Physiol. 149, 358-362.

Barneix, A. J., D. M. James, E. F. Watson, and E. J. Hewitt. 1984. Some effects of nitrate abundance and starvation on metabolism and accumulation of nitrogen in barley (*Hordeum vulgare* L. cv Sonja). Planta 162, 469–476.

Black, B. L., L. H. Fuchigami, and G. D. Coleman. 2002. Partitioning of nitrate assimilation among leaves, stems and roots of poplar. Tree Physiol. 22, 717–724.

Botrel, A., and W. M. Kaiser. 1997. Nitrate reductase activation state in barley roots in relation to the energy and carbohydrate status. Planta 201, 496–501.

Bradford, M. M. 1976. A rapid and sensitive method for the quantification of microgram quantities of protein utilizing the principle of protein-dye binding. Anal. Biochem. 72, 248–254.

Breteler, H., and W. Luczak. 1982. Utilization of nitrite and nitrate by dwarf bean. Planta 156, 226–232.

Callaci, J. J., and J. J. Smarrelli. 1991. Regulation of the inducible nitrate reductase isoform from soybean. Biochim. Biophys. Acta 1088, 127-130

Campbell, W. H. 1999. Nitrate reductase structure, function and regulation: bridging the gap between Biochemistry and Physiology. Annu. Rev. Plant Physiol. Mol. Biol. 50, 277–303.

Caputo, C., and A. J. Barneix. 1997. Export of amino acids to the phloem in relation to N supply in wheat. Physiol. Plant 101, 853–860.

Cataldo, D. A., M. Haroon, L. E. Schrader, and V. L. Young. 1975. Rapid colorimetric determination of nitrate in plant tissue by nitration of salicylic acid. Comm. Soil Sci. Plant Anal. 6, 71–80.

Cawse, P. A. 1967. The determination of nitrate in soil solutions by ultraviolet spectrophotometry. Analyst 92, 311–315.

Cerezo, M., V. Flors, F. Legaz, and P. García-Agustín. 2000. Characterization of the low affinity transport system for NO_3^- uptake by *Citrus* roots. Plant Sci. 160, 95–104.

Chapin, F. S., L Moilainen, and K. Kielland. 1993. Preferential use of organic nitrogen by a non - mycorrhizal arctic sedge. Nature 361, 150–153.

Chen, G., and X .Gao. 2002. Effect of partial replacement of nitrate by amino acid and urea on nitrate content of nonheading Chinese cabbage and lettuce in hydroponics (Chinese). Sci. Agr. Sinica 35, 187–191.

Chen, Z., J. Huang, J. He, and K. Cai. 1997. Influence of L-tryptophan applied to soil on yield and nutrient uptake of cabbage (Chinese). Acta Ped. Sinica 34, 200–205.

Cooper, H. D., and D. T. Clarkson. 1989. Cycling of amino-nitrogen and other nutrients between shoots and roots in cereals –A possible mechanism integrating shoot and root regulation of nutrient uptake. J. Exp. Bot. 40, 753–762.

Cramer, M. D., O. W. Nagel, S. H. Lips, and H. Lambers. 1995. Reduction, assimilation and transport of N in wild type and gibberellin-deficient tomato plants. Physiol. Plant 95, 347–354.

Crawford, N. M., and A. D. M. Glass. 1998. Molecular and physiological aspects of nitrate uptake in plants. Trends Plant Sci. 3, 389-395.

Crété, P., M. Caboche, and C. Meyer. 1997. Nitrite reductase expression is regulated at the post-transcriptional level by the nitrogen source in *Nicotiana plumbaginifolia* and *Arabidopsis thaliana*. Plant J. 11, 625-634.

Criddle, R. S., M. R. Ward, and R. C. Huffaker. 1988. Nitrogen uptake by wheat seedlings, interactive effect of four nitrogen sources: NO_3^-, NO_2^-, NH_4^+, and urea. Plant Physiol. 86, 166-175.

Daniel-Vedele, F., S. Filleur, and M. Caboch. 1998. Nitrate transport: a key step in nitrate assimilation. Curr. Opin. Plant Biol. 1, 235-239.

Deng, M. D., T. Moureaux, I. Cherel, J. P. Boutin, and M. Caboche. 1991. Effects of nitrogen metabolites on the regulation and circadian expression of tobacco nitrate reductase. Plant Physiol. Biochem. 29, 139-247.

Fedorova, E., J. S. Greenwood, and A. Oaks. 1994. *In-situ* localization of nitrate reductase in maize roots. Planta 194, 279-286.

Fischer, W. N., B. André, D. Rentsch, S. Krolkiewicz, M. Tegeder, K. Breitkreuz, and W. B. Frommer. 1998. Amino acid transport in plants. Trends Plant Sci. 3, 188-195.

Ford, B. G. 2000. Nitrate transporters in plants: structure, function and regulation. Biochim. Biophys. Acta 1465, 219-235.

Forde, B. G., and D. T. Clarkson. 1999. Nitrate and ammonium nutrition of plants: Physiological and molecular perspectives. *In* Advances in Botanical Res. 30, 1-90.

Galván, A., A. Quesada, and E. Fernández. 1996. Nitrate and nitrite are transported by different specific transport systems and by a bispecific transporter in *Chlamydomonas reinhardtii*. J. Biol. Chem. 271, 2088-2092.

Gazzarrini, S, L. Lejay, A. Gojon, O. Ninnemann, W. Frommer, and N. Wirén. 1999. Three functional transporters for constitutive, diurnally regulated, and starvation-induced uptake of ammonium into Arabidopsis roots. Plant Cell 11, 937-947.

Gebhardt, C., J. E. Oliver, B. G. Forde, R. Saarelainen, and B. J. Miflin. 1986. Primary structure and differential expression of glutamine synthetase genes in nodules, roots and leaves of Phaseolus vulgaris. EMBO J. 5, 1429-1435.

Gerendás, J., and B. Sattelmacher. 1999. Influence of Ni supply on growth and nitrogen metabolism of *Brassica napus* L. grown with NH_4NO_3 or urea as N source. Ann. Bot. 83, 65-71.

Glaab, J., and W. M. Kaiser. 1993. Rapid modulation of nitrate reductase in pea roots. Planta 191, 173-179.

Glass A. D. M., J. Shaff, and L. Kochian,. 1992. Studies of the uptake of nitrate in barley. IV. Electrophysiology. Plant Physiol. 99, 456-463.

Granato, T. C., and C. D. Raper Jr. 1989. Proliferation of maize (*Zea maize* L.) roots in response to localized supply of nitrate. J. Exp. Bot. 40, 263-275.

Gunes, A., A. Inal, and M. Aktas. 1996. Reducing nitrate content of NFT grown winter onion plants (*Allium cepa L.*) by partial replacement of NO_3 with amino acid in nutrient solution. Sci. Hortic. 65, 203-208.

Gunes, A., W. H. K. Post, E. A. Kirkby, and M. Akas. 1994. Influence of partial replacement of nitrate by amino acid nitrogen or urea in the nutrient medium on nitrate accumulation in NFT grown winter lettuce. J. Plant Nutr. 17, 1929-1938.

Haynes, R., and K. M. Goh. 1978. Ammonium and nitrate nutrition of plants. Biol. Rev. 53, 465-510.

Hirose, N., and T. Yamaya. 1999. Okadaic acid mimics nitrogen-stimulated transcription of the NADH-glutamate synthase gene in rice cell cultures. Plant Physiol. 121, 805-812.

Hirose, N., T. Hayakawa, and T. Yamaya. 1997. Inducible accumulation of mRNA for NADH-dependent glutamate synthase in rice roots in response to ammonium ions. Plant Cell Physiol. 38, 1295-1297.

Hugh, A., L. Henry, and R. L. Jefferies. 2003. Plant amino acid uptake, soluble N turnover and microbial N capture in soils of a grazed Arctic salt marsh. J. Ecol. 91, 627-636.

Imsande, J., and B. Touraine. 1994. N demand and the regulation of nitrate uptake. Plant Physiol. 105, 3-7.

Ivashikian, N. V., and O. A. Sokolov. 1997. Regulation of nitrate, nitrite, ammonium and glutate. Plant Sci. 123, 29-37.

Jensen, E. S. 1996. Rhizodeposition of N by pea and barley and its effect on soil N dynamics. Soil Biol. Biochem. 28, 65-71.

Jonasson, S., and G. R. Shaver. 1999. Within-stand nutrient cycling in arctic and boreal wetlands. Ecology, 80, 2139-2150.

Jones, D. L. 1999. Amino acid biodegradation and its potential effects on organic nitrogen capture by plants. Soil Biol. Biochem. 31, 613-622.

Jones, D. L., and A. Hodge. 1999. Biodegradation kinetics and sorption reactions of three differently charged amino acids in soil and their effects on plant organic nitrogen availability. Soil Biol. Biochem. 31, 1331-1342.

Kaiser, J. J., and O. A. H. Lewis. 1984. Nitrate reductase and glutamine synthetase activity in leaves and roots of nitrate fed *Helianthus annuus L.* Plant Soil 70, 127-130.

Kaiser, W. M., A. Kandlbinder, M. Stoimenova, and J. Glaab. 2000. Discrepancy between nitrate reduction rates in intact leaves and nitrate reductase activity in leaf extracts: What limits nitrate reduction in situ?. Planta 210, 801-807.

Kaiser, W. M., H. Weiner, and S. C. Huber. 1999. Nitrate reductase in higher plants: A case study for transduction of environmental stimuli into control of catalytic activity. Physiol. Plant. 105, 385-390.

Khamis, S., and T. Lamaze. 1990. Maximal biomass production can occurincorn(*Zea mays*) in the absence of NO_3^- accumulation in either leaves or roots. Physiol. Plant. 78, 388-394.

Kielland, K. 1994. Amino acid absorption by arctic plants: implications for plant nutrition and nitrogen cycling. Ecology 75, 2373-2383.

Kim, Y. S. 2002. The effect of mixed amino acid on nitrate uptake in rice, pea, cucumber and red pepper. Master of Agriculture Degree Thesis, Department of Agricultural Chemistry, College of Agriculture and Biotechnology, Chungnam National University, Daejon, Korea.

King, B. J., M. Y. Siddiqi, T. J. Ruth, R. L. Warner, and A. D. M. Glass. 1993. Feedback regulation of nitrate influx in barley roots by nitrate, nitrite and ammonium. Plant Physiol. 102, 1279-1286.

Kucharski, J., and G. Nowak. 1994. The effect of L-tryptophane on yield bean and activity of soil microorganisms. Acta Microbiol Pol. 43, 381-388.

Lancien, M., P. Gadal, and M. Hodges. 2000. Enzyme redundancy and the importance of 2-oxoglutarate in higher plant ammonium assimilation. Plant Physiol. 123, 817–824.

Leacox, J. D., and J. P. Syvertsen. 1995. Nitrogen Uptake By Citrus Leaves. J. Am. Soc. Hortic. Sci. 120, 505–509.

Lejay, L., P. Tillard, F. D. Olive, M. Lepetit, S. Filleur, and F. Daniel–Vedele. 1999. Molecular and functional regulation of two NO_3^- uptake systems by N–and C–status of Arabidopsis plants. Plant J. 18, 509–519.

Lewis, O. A. M., D. M. James, and E. J. Hewitt. 1982. Nitrogen assimilation in barley (*Hordeum vulgare* L. cv. Mazurka) in response to nitrate and ammonium nutrition. Ann. Bot. 49, 39–49.

Li, X. Z., and A. Oaks. 1993. Induction and turnover of maize nitrate reductase: Influence of NO_3^-. Plant Physiol. 102, 1251–1257.

Li, X. Z., D. E. Larson, M. Glibetic, and A. Oaks. 1995. Effect of glutamine on the induction of nitrate reductase. Physiol Plant 93, 740–744.

Liu, X. Q., Y. S. Kim, and K. S. Lee. 2005. The effect of mixed amino acids on nitrate uptake and nitrate assimilation in leafy radish. Kor. J. Environ. Agr. 24, 245–252.

Lohaus, G., M. Burba, and H. W. Heldt. 1994. Comparison of the contents of sucrose and amino acids in the leaves, phloem sap and taproots of high and low sugar-producing hybrids of sugar beet *(Beta vulgaris* L.). J. Exp. Bot. 45, 1097–1101.

Majerowicz, N., G. B. Kerbauy, C. C. Nievola, and R. M. Suzuki. 2000. Growth and nitrogen metabolism of Catasetum fimbriatum (orchidaceae) grown with different nitrogen source. Environ. Exp. Bot. 44, 195–206.

Matsumoto, S., N. Ae, and M. Yamagata. 2000. Possible direct uptake of organic nitrogen from soil by chingensai (*Brassica campestris L.*) and carrot (*Daucus carota L.*). Soil Biol. Biochem. 32, 1301–1310.

Matt, P., M. Geiger, P. Walch–Liu, C. Engels, A. Krapp, and M. Stitt. 2001. Elevated carbon dioxide increases nitrate uptake and nitrate reductase activity when tobacco is growing on nitrate, but increases ammonium uptake and inhibits nitrate reductase activity when growing on ammonium nitrate. Plant Cell Environ. 24, 1119–1137.

Miflin, B. J., and D. Z. Habash. 2002. The role of glutamine synthetase and glutamate dehydrogenase in nitrogen assimilation and possibilities for improvement in the nitrogen utilization of crops J. Exp. Bot. 53, 979–987.

Muller, B., and B. Touraine. 1992. Inhibition of NO_3^- uptake by various phloem–translocated amino acids in soybean seedling. J. Exp. Bot. 43, 617–623.

Murphy, A. T., and O. A. M. Lewis. 1987. Effect of nitrogen feeding source on the supply of nitrogen from root to shoot and the site of nitrogen assimilation in maize (*Zea mays* L. cv. R201). New Phytol. 107, 327–333.

Näsholm T., K. Huss-Danell, P. Hogberg. 2000. Uptake of organic nitrogen in the field by four agriculturally important plants pecies. Ecology, 81, 1155–1161.

Näsholm, T., A. Ekblad, A. Nordin, R. Giesler, M. Hogberg, and P. Hogberg. 1998. Boreal forest plants take up organic nitrogen. Nature 392, 914–916.

Neelam, A., A. C. Marvier, J. L. Hall, and L. E. Williams. 1999. Functional characterization and expression analysis of the amino acid permease RcAAP3 from castor bean. Plant Physiol. 120, 1049–1056.

NIAST. 1998. Method of soil and plant analysis. National Institute of Agriculural Science and Technology, RDA, Suwon, Korea.

Oaks, A. 1986. Biochemical aspects of nitrogen metabolism in a whole plant context. In: Lambers, H, Neeteson, J J, Stulen, I eds. , Fundamental, ecological and agricultural aspects of nitrogen metabolism in higher plants., Martinus Nijhoff Publishers, Dordrecht, Boston, Lancaster, pp 133-151.

Oaks, A. 1994. Primary nitrogen assimilation in higher plants and its regulation. Can. J. Bot. 72,739-750.

Oaks, A., I. Stulen, and I. Boesel. 1979. Influence of amino acids and ammonium on nitrate reduction in corn seedlings. Can. J. Bot. 57, 1824-1829.

Oaks, A., M. Aslam, and I. Boesel. 1977. Ammonium and amino acids as regulators of nitrate reductase in corn roots. Plant Physiol. 59, 391-394.

Oaks, A., S. Sivasankar, and V. J. Goodfellow. 1998. The specificity of methionine sulfoximine and azaserine inhibition in plant tissues. Phytochemistry 49, 355-357.

Ortiz-Lopez, A., H. C. Chang, and D. R. Bush. 2000. Amino acid transporters in plants. Biochim. Biophys. Acta 1465, 275-280.

Owen, A. G., and D. L. Jones. 2001. Competition for amino acids between wheat roots and rhizosphere microorganisms and the role of amino acids in plant N acquisition. Soil Biol. Biochem. 33, 651-657.

Padgett, P. E., and R. T. Leonard. 1993. Regulation of nitrate uptake by amino acids in maize cell suspension culture and intact roots. Plant Soil 155/156, 159-161.

Padgett, P. E., and R. T. Leonard. 1996. Free amino acid levels and the regulation of nitrate uptake in maize cell suspension cultures. J. Exp. Bot. 47, 871-883.

Pal'ove-Balang, P. 2002. Role of nitrogen metabolites on the regulation of nitrate uptake in maize seedlings. Acta Biol. Szegediensis. 46, 177-178.

Pal'ove-Balang, P., and I. Mistrik. 2002. Control of nitrate uptake by phloem-translocated glutamine in *Zea mays L.* seedlings. Plant Biol. 4, 440-445.

Peeters, K. M. U., and A. J. Van Laere. 1994. Amino acid metabolism associated with N-mobilization from the flag leaf of wheat (*Triticum aestivum L.*) during grain development. Plant Cell Environ. 17, 131-141.

Person, J., and T. Näsholm. 2002. Regulation of amino acid uptake in conifers by exogenous and endogenous nitrogen. Planta 215, 639-644.

Persson, J., and T. Näsholm. 2001. Amino acid uptake: a widespread ability among boreal forest plants. Ecol Lett 4, 434-438.

Persson, J., P. Hogberg, A. Ekblad, M. N. Hogberg, A. Nordgren, and T. Nasholm. 2003. Nitrogen acquisition from inorganic and organic sources by boreal forest plants in the field. Oecologia 137, 252-257.

Pessarakli, M. 2002. Handbook of plant and crop physiology. 2nd Edition. Marcel Dekker, Inc, New York, USA. pp.385-394.

Popova, O. V., K. J. Dietz, and D. Golldack. 2003. Salt-dependent expression of a nitrate transporter and two amino acid transporter genes in *Mesembryanthemum crystallinum*. Plant Mol. Biol. 52, 569-578.

Raad, T. K., D. A. Lipson, and R. K. Monson. 1999. Soil amino acid utilization among species of cyperaceae: plant and soil pro-cesses. Ecology 80, 2408-2419.

Radin J. W. 1975. Differential regulation of nitrate reductase induction in roots and shoots of cotton plants. Plant Physiol. 55, 178-182.

Radin, J. W. 1977. Amino acid interactions in the regulation of nitrate reductase induction in cotton roots tips. Plant Physiol. 60, 467-469.

Rodgers, C. O., and A. J. Barneix. 1993. The effect of amino acids and amides on the regulation of nitrate uptake by wheat seedlings. J. Plant Nutr. 16, 337–348.

Rufty, Jr. T. W., C. T. MacKown, and R. J. Volk. 1990. Alterations in nitrogen assimilation and partitioning in nitrogen stressed plants. Physiol. Plant. 79, 85–95.

Ruiz, J. M., R. M. Rivero, P. C. Garcia, M. Baghour, and R. Romero. 1999. Role of $CaCl_2$ in nitrate assimilation in leaves and roots of tobacco plants (*Nicotiana tabacum L.*). Plant Sci. 141, 107–115.

Sarwar, M., and W. J. Frankenberger. 1994. Influence of L-tryptophan and auxins applied to the rhizophere on the vegetative growth of *Zea mays* L.. Plant Soil 160, 97–104.

Schobert, C., and E. Komor. 1990. Transfer of amino acids and nitrate from the roots into the xylem of Ricinus communis seedlings. Planta 181, 85–90.

Siddiqi, M. Y., A. D. M. Glass, and T. J. Ruth. 1991. Studies of the uptake of nitrate in barley III. Compartmentation of NO- 3. J. Exp. Bot. 42, 1455–1463.

Sivasankar, S., S. Rothstein, and A. Oaks. 1997. Regulation of the accumulation and reduction of nitrate by nitrogen and carbon metabolites in maize seedlings. Plant Physiol. 114, 583–589.

Spalding, R. F., D. G. Watts, J. S. Schepers, M. E. Burbach, M. E. Exner, R. J. Poreda, and G. E. Martin. 2001. Controlling Nitrate Leaching in Irrigated Agriculture. J. Environ. Qual. 30, 1184–1194.

Srivastava, H. S., and R. P. Singh. 1987. Role and regulation of L–glutamate dehydrogenase activity in higher plant. Phytochem. 26, 597–610.

Suzuki, A., A. Oaks, J. Jacquot, J. Vidal, and P. Gadal. 1985. An electron transport system in maize roots for reactions of glutamate synthase and nitrite reductase: Physiological and immunochemical properties of the electron carrier and pyridine nucleotide reductase. Plant Physiol. 78, 374–378.

Ta, T. C., and K. W. Joy. 1984. Transamination, deamination, and the utilisation of asparagine amino nitrogen in pea leaves. Can. J. Bot. 63, 881–884.

Taiz, L., and E. Zeiger. 2002. Plant Physiology. 3rd Edition. Sinauer Associates, Inc., Publisher, Sunderland, Massachusetts. pp.278-281

Tillard, P., L. Passama, and A. Gojon. 1998. Are phloem amino acids involved in the shoot to root control of NO_3^- uptake in *Ricinus communis* plants? J. Exp. Bot. 49, 1371–1379.

Tischner, R., B. Waldeck, S. Goyal, and W. Rains. 1993. Effect of nitrate pulses on the nitrate-uptake rate, synthesis of mRNA coding for nitrate reductase, and nitrate reductase activity in the roots of barley seedlings. Planta 189, 533–537.

Vessey, J. K., and D. B. Layzell. 1987. Regulation of assimilate partitioning in soybean. Initial effects following change in nitrate supply. Plant Physiol. 83, 341–348.

Vicentz, M., T. Moureaux, M. T. Leydecker, H. Vaucheret, and M. Caboche. 1993. Regulation of nitrate and nitrite reductase expression in *Nicotiana plumbaginifolia* leaves by nitrogen and carbon metabolites. Plant J. 3, 315–324.

Virtanen, A. I., and H. Linkola. 1946. Organic nitrogen compounds as nitrogen nutrition for higher plants. Nature 158, 515.

Vose, P. B. 1980. Introduction to nuclear techniques in agronomy and plant biology. Pergamon Press Ltd. pp. 268–270.

Wang, H., L. Wu, and Q. Tao. 2004. Influence of partial replacement of nitrate by amino acids on nitrate accumulation of pakchoi (*Brassica chinensis L.*) (Chinese). China Environ. Sci. 24, 19–23.

Wang, H., L. Wu, and Q. Tao. 2004. Nitrate accumulation and variation of nutrient quality in pakchoi after application of several amino acids in summer and autumn (Chinese). J. Agro-Environ. Sci. 23, 224–227.

Winter, H., G. Lohaus, and H. W. Heldt. 1992. Phloem transport of amino acids in relation to their cytosolic levels in barley leaves. Plant Physiol. 99, 996–1004.

Wray, J. 1989. Molecular biology, genetics and regulation of nitrite reduction in higher plants. Physiol. Plant. 89, 607–612.

Yemm, E. W., and E. C. Cocking. 1955. The determination of amino acids with ninhydrin. Analyst 80, 209–213.

Section 4

Crop Response to Temperature

8

Plant Temperature for Sterile Alteration of Rice

Chuan-Gen Lű
Jiangsu Academy of Agricultural Sciences, Nanjing, P.R. China

1. Introduction

Temperature affects not only the growth rate, but also reproductive development of rice. When temperature was higher than a critical point value during 5-15 days before heading, rice thermo-sensitive genic male sterile (TGMS) line showed pollen sterility. Otherwise, it would be fertile (Lu *et al* 2001). The relationship between temperature and plant growth or reproductive development is usually studied by using a thermometer screen at a weather station placed at height of 150 cm (T_A), although a few studies were also performed to describe more direct and accurate effects of micro-climate temperature at rice canopy on plant growth and reproductive development.

Plant temperature (T_p) is regulated by various factors including solar radiation, cloud cover, wind speed, soil heat flux, and transpiration of plant. Besides, temperature and flow velocity of irrigated water have significant effects on rice T_p. Many methods and simulations on T_p have been reported (Cellier 1993, Cui *et al* 1989, Ferchinger 1998, Hasegawa 1978, Leuning 1988a, 1988b, Lu *et al* 1998, Tetsuya *et al* 1982, Van *et al* 1989, Wei *et al* 1981, Zhao *et al* 1996). T_p was used as the index of water supply regime for wheat or corn (Cheng *et al* 2000, Huang *et al* 1998, Shi *et al* 1997, Yuan *et al* 2000), freezing injury or grain filling rate of wheat (Feng *et al* 2000, Li *et al* 1999, Liu *et al* 1992, Xiang *et al* 1998, Xu *et al* 2000). In recent studies, it was found that the fertility alteration of rice TGMS line was sensitive to T_p (Lu *et al* 2007, Xu *et al* 1996). Lu *et al* (2004, 2007) found that the sterility of TGMS was simulated more accurately by temperature at stem part of 20 cm height or air temperature (T_a) around the part when compared with T_A at a weather station (Lu *et al* 2004, 2007, Zou *et al* 2005). So far, little is known about how T_p is regulated by microclimate or irrigated water (Hu *et al* 2006).

2. Plant temperature and its simulation model of thermo-sensitive male sterile rice

The present chapter was performed to investigate the temporal and spatial distribution of T_p and its relationships with microclimate of canopy and irrigated water by using a TGMS line under irrigated and non-irrigated conditions. Two models were established to understand how T_p is regulated by environments.

A TGMS rice line, Peiai64S, was used as plant material. Flowing irrigated water depth of 10-15 cm was treated, and no irrigated (keeping humid) was treated as control.

T_p and microclimatic factors were determined as below:

T_p: PTWD-2A sensors were inserted in stem sheaths at heights of 10 cm, 20 cm, 30 cm and 40 cm, respectively, to measure T_p.

Air, water and soil temperatures: PTWD-2A sensors were placed at 10 cm and 5 cm under the ground, and 5 cm, 20 cm, 40 cm, 60 cm, 100 cm, 150 cm above the ground, respectively, to measure temperatures of soil, water and air.

Wind speed: EC-9S sensors were used to measure wind speed at height of 150 cm.

All above data were obtained using TRM-ZS1, automatically collecting every 10 seconds and storing every 10 minutes.

When 50% of flag leaves appeared perfectly, five plants were sampled and dissected at height of each 10 cm to determine layered LAI. Layered LAI was determined with traditional method, i.e., calculated according to the dry matter of harvested layer and the specific leaf area of each layer from sampled leaves (SLA).

2.1 Change of rice T_p

2.1.1 Daily change of T_p

Fig. 1 showed the daily changes of air temperature at height of 150 cm (T_A) and T_p at heights of 20 cm and 40 cm (T_{p20} and T_{p40}) under non-irrigated condition. The data were collected from random 30 days, during which 11 d were sunshine, 9 d were cloudy, and 10 d were overcast. T_p showed different value from T_A, although they had a similar trend in daily change. T_A and T_p could be simulated by using same parameters and a similar equation.

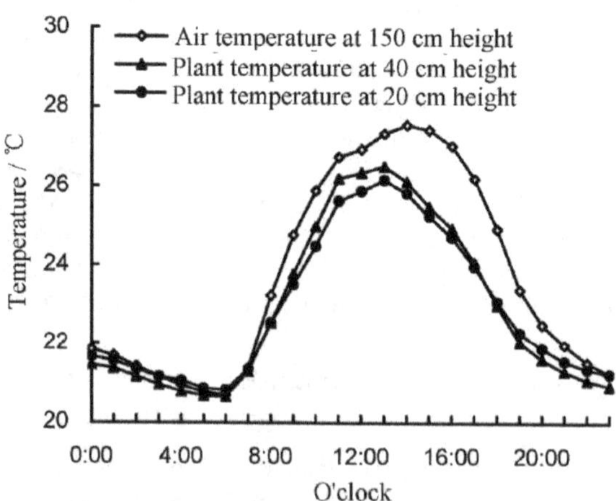

Fig. 1. Daily change of air temperature at 150 cm and plant temperature at 20 cm, 40 cm heights.

Calculated the data of Fig. 1, result showed that, during 08:00-20:00, T_p was significantly lower than T_A. T_{p20} was lower than T_A by an average of 1.44°C and the largest margin of 2.3°C, while T_{p40} was lower than T_A by an average of 1.25°C and the largest margin of 2.1°C.

During 21:00-07:00, T_{p20} and T_{p40} were lower than T_A only by 0.27°C and 0.06°C, respectively.

Fig. 1 and its simulation equations (1), (2), (3) showed that, during daytime (06:00-18:00), the variation of both T_p and T_A fit the sinusoid curve, but the coefficient showed $T_A > T_{p40} > T_{p20}$. During night time (18:00-06:00), the variation of both T_p and T_A fit the exponent curve, but the coefficient showed $T_A > T_{p40} \approx T_{p20}$. The simulation equations and their effects were described as follows：

$$T_{p20} = \begin{cases} 20.8 + 5.3\sin[\pi(t-5.78)/14.44] & 06:00 \le t \le 18:00 \\ \{19.54 + 2.3\exp[-(t+6)/4]\}/0.94 & t < 06:00 \\ \{19.54 + 2.3\exp[-(t-18)/4]\}/0.94 & t > 18:00 \end{cases} \quad R^2 = 0.956^{**} \tag{1}$$

$$T_{p40} = \begin{cases} 20.7 + 5.8\sin[\pi(t-5.78)/14.44] & 06:00 \le t \le 18:00 \\ \{19.38 + 2.3\exp[-(t+6)/4]\}/0.94 & t < 06:00 \\ \{19.38 + 2.3\exp[-(t-18)/4]\}/0.94 & t > 18:00 \end{cases} \quad R^2 = 0.957^{**} \tag{2}$$

$$T_A = \begin{cases} 20.7 + 6.8\sin[\pi(t-5.78)/16.44] & 06:00 \le t \le 18:00 \\ \{19.32 + 4.2\exp[-(t+6)/4]\}/0.94 & t < 06:00 \\ \{19.32 + 4.2\exp[-(t-18)/4]\}/0.94 & t > 18:00 \end{cases} \quad R^2 = 0.972^{**} \tag{3}$$

Differences (visual temperature difference from thermometer, >0.2°C) were detected in value and time between the maximum T_p and the maximum T_A. The daily maximum value was 26.1°C for T_{p20}, and 26.5°C for T_{p40}, whereas the maximum value was 27.5°C for T_A. The maximum value of T_p occurred at 13:00, 1 h earlier than that of T_A. Their minimum values, however, both appeared at 06:00. The fluctuation in daily change also showed significant difference: T_A (6.8°C) > T_{p40} (5.8°C) > T_{p20} (5.3°C).

2.1.2 Difference of T_p at vertical height

Fig. 2 showed T_p and T_a at different heights observed at 06:00 and 13:00 under non-irrigated condition (30 d, same as Fig.1). Although the differences between T_p and T_a were not significant, T_p at 13:00 at heights of 10 cm, 20 cm, 30 cm and 40 cm were all higher than T_a at corresponding heights without exception. T_{p40} and T_{p20} were higher by 0.24°C and 0.60°C, respectively. The decreased rate of T_p (0.1°C /10 cm) was lower than that of T_a (0.41°C /10 cm). At 06:00, T_p at 30-40 cm was higher than T_a, whereas it showed an opposite difference at 10-20 cm.

2.2 Effects of environmental factors on T_p

Solar radiation and irrigated water were two main heat sources affecting rice T_p. Cloud cover, wind speed and LAI of canopy also affect T_p by regulating heat transmission or radiation intensity. Temperature and flowing speed of irrigated water were two main factors regulating T_p.

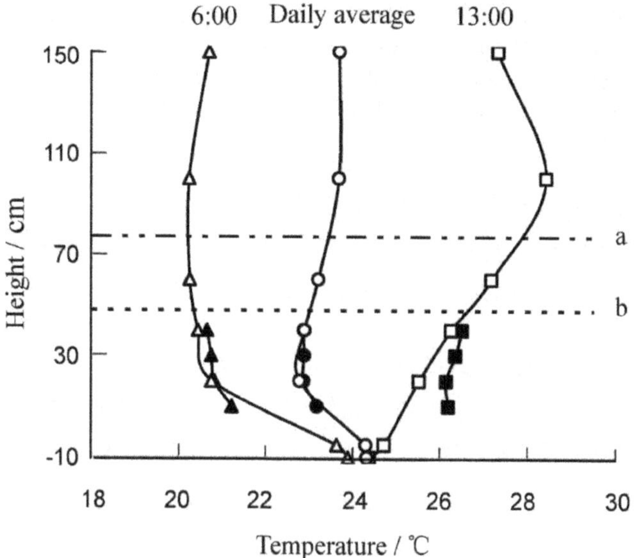

Fig. 2. Plant (▲●■) and air (△○□) temperatures at different heights. a and b lines denotes plant height and layer of maximum leaf density, respectively.

2.3 Effects of T_a on T_p

2.3.1 Change of T_a and its effects on T_p

Fig. 3 showed daily changes of T_p and T_a at heights of 20 cm and 40 cm under sunshine (11 d) and cloudy (10 d) days. During 06:00-13:00 in sunshine days, T_p was increased earlier than T_a by 1 h, and T_p was maximized at 13:00 (28.1°C for T_{p20}, and 28.7°C for T_{p40}). In contrast, T_a was increased later than T_p by 1 h, and T_a was maximized at 14:00 (27.4°C for T_{a20}, and 28.5°C for T_{a40}). Besides, the raising intensity was T_p stronger than T_a by 0.7°C (20 cm) and 0.2°C (40 cm). T_p showed close to or even little lower than T_a during night time (18:00-06:00). The significant difference between T_a and T_p during daytime might be caused by the larger absorption of solar radiation by plant than air. When the heat was absorbed, the plant released its energy in long wave, which resulted in an increase of T_a around the plant. After 13:00, along with the diminishing of solar radiation, T_p began to decrease but T_a was reacted dully (decreased one or two hours later under sunny days). Under cloudy days, T_p was higher than T_a all the day. At heights of 20 cm and 40 cm, T_p was 0.4°C higher on average, and the maximum one by 0.5°C for T_{p20} and 0.4°C for T_{p40}, and the minimum one by 0.3°C for T_{p20} and 0.4°C for T_{p40}, respectively. This suggested that there was a weak exchange between plant and air on cloudy days.

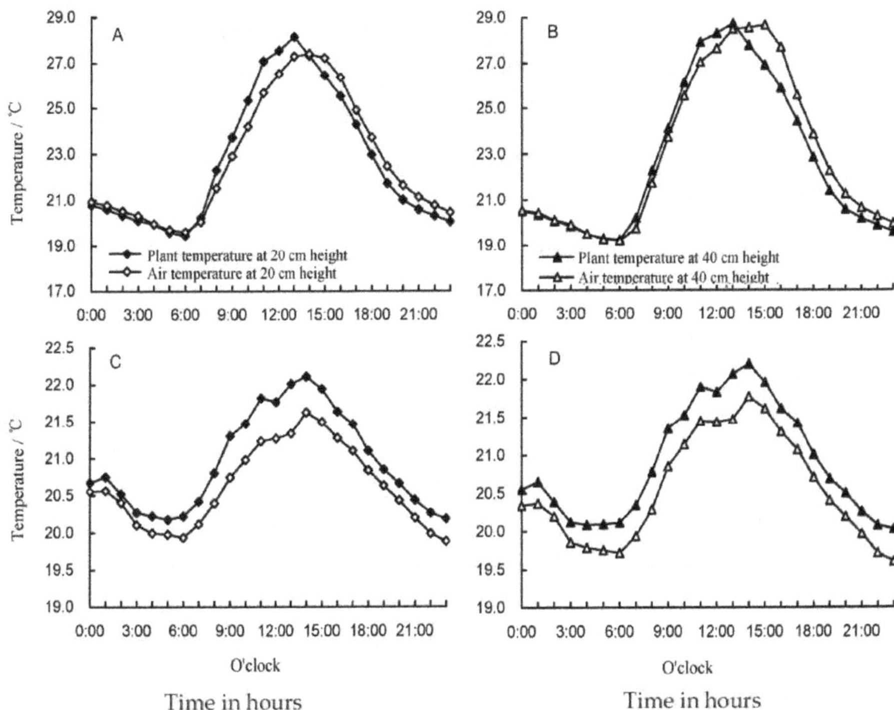

A and B denote sunny day with sunshine time >8 h; C and D denote overcast day with no sunshine time.

Fig. 3. Daily change of air and plant temperature at 20 cm and 40 cm heights.

2.3.2 Statistical relation of T_p to T_A

In the present chapter, T_p at 20 cm and 40 cm, and T_A at 150 cm observed at 06:00 and 13:00 were used to establish linear regression equations. Results showed that the regression coefficient of T_{p40} was lower than that of T_{p20}, confirming that solar radiation was the key resource of heat in plant.

06:00 T_p: 20 cm: $T_{p20} = 0.9949\, T_A + 0.1741 \quad R^2 = 0.992^{**}, \quad n = 30$

40 cm: $T_{p40} = 0.9670\, T_A + 0.8885 \quad R^2 = 0.995^{**}, \quad n = 30$

13:00 T_p: 20 cm: $T_{p20} = 0.9338\, T_A + 2.3121 \quad R^2 = 0.973^{**}, \quad n = 30$

40 cm: $T_{p40} = 0.9199\, T_A + 2.3416 \quad R^2 = 0.982^{**}, \quad n = 30$

Daily average: 20 cm: $T_{p20} = 0.9338\, T_A + 1.6011 \quad R^2 = 0.996^{**}, \quad n = 30$

40 cm: $T_{p40} = 0.9205\, T_A + 1.7991 \quad R^2 = 0.997^{**}, \quad n = 30$

2.4 Effect of wind speed

Fig. 4 showed the relationship between T_{p20} and wind speed at 150 cm (V). The equation was: $T_{p20} = -14.411V + 32.622$ ($R^2=0.334^*$, $n=39$). The result indicated that wind speed on the top of canopy had a significant effect on T_p.

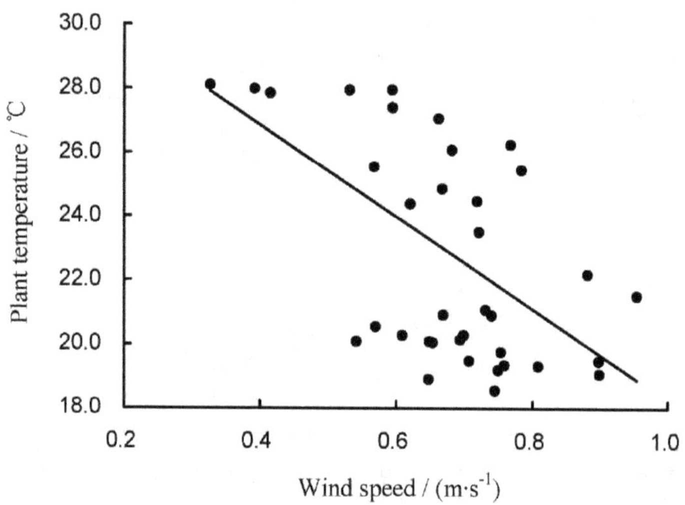

Fig. 4. Relationship between plant temperature at 20 cm and wind speed at 150 cm.

2.5 Effects of irrigated water

Irrigated water was another important effect regulating rice T_p. Water temperature (T_w) and T_a were varying in different manners, which resulted in a remarkable difference in their effects on T_p.

2.5.1 Temporal change of effects of irrigated water on T_p

Fig. 5 showed daily varying curves of T_p (30 d) at 20 cm and 40 cm of plants grown with and without irrigated water. The average temperature of inflow water was 26.9°C (21.5-34.0°C), and T_A was 23.7°C (13.5-36.5°C). Irrigated water increased T_p at 20 cm and 40 cm by 0.7°C and 0.5°C, respectively, or by 0.35°C and 0.25°C, respectively, for per 1°C of water-air temperature margin. The effect of increased T_p by irrigated water was higher at night than at daytime. T_{p20} was increased by 1.01°C during 19:00-05:00, and by 0.37°C during 06:00-18:00, but decreased by 0.19°C during 11:00-14:00. At 40 cm, T_{p40} was increased by 0.77°C during 19:00-05:00, and by 0.20°C during 06:00-18:00, but decreased by 0.25°C during 11:00-13:00. The results showed that irrigated water had more significant effects on T_p during night than daytime. In daytime, solar radiation partly withstood the effects of water irrigated. At the noon when solar radiation was the maximum, irrigated water decreased T_p.

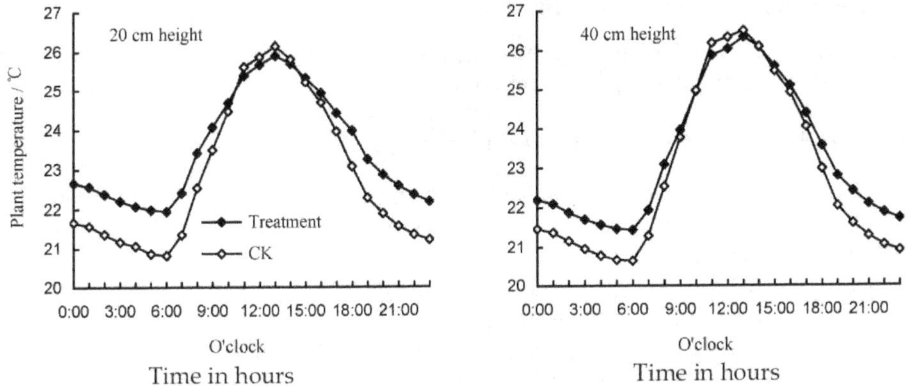

Fig. 5. Daily change of plant temperature at 20 cm and 40 cm under irrigated treatment and non-irrigated conditions.

2.5.2 Effects of irrigated water on T_p at vertical height

T_p (30 d, same as Fig. 1) at four heights at 06:00 and 13:00 was compared between irrigated and non-irrigated conditions (Table 1). The effect of irrigated water on T_p was decreased lower along with the increase of height. It was a difference by 2.5°C at 10 cm, 0.6°C at 20 cm, 0.5°C at 30 cm, and 0.4°C at 40 cm. At 06:00, irrigated water could increase T_p (increasing rate of 3.7°C at 10 cm), but the increasing rate was lowered along with the increasing of plant height. Around noon, T_p showed 0.7°C increase at 10 cm but a decrease at 20-40 cm. The effect of irrigated water on T_p was stronger at the lower site than at the higher site in plants.

Item		06:00	13:00	Daily average
	Irrigated 10 cm	24.7±9.4	26.9±7.6	25.7±8.3
T_{p10}	Non-irrigated	21.2±8.5	26.2±8.0	23.2±7.6
	Difference	3.5±3.9	0.7±3.5	2.6±2.2
	Irrigated 10 cm	21.9±8.7	25.9±7.9	23.5±8.0
T_{p20}	Non-irrigated	20.8±8.4	26.1±8.1	22.9±7.5
	Difference	1.1±1.5	-0.2±1.9	0.7±1.0
	Irrigated 10 cm	21.8±8.4	26.0±7.4	23.4±7.6
T_{p30}	Non-irrigated	20.8±8.3	26.4±8.6	22.9±7.6
	Difference	1.0±2.3	-0.3±2.0	0.5±1.0
	Irrigated 10 cm	21.4±8.3	26.3±8.3	23.3±7.7
T_{p40}	Non-irrigated	20.7±8.4	26.5±8.6	22.9±7.6
	Difference	0.8±1.0	-0.2±1.0	0.5±0.5

Note: Difference: difference of plant temperature between irrigated water of 10 cm depth and non-irrigated.

Table 1. Comparison of plant temperature observed at four heights at 06:00 and 13:00 between irrigated and non-irrigated conditions

2.5.3 Effects of irrigated water temperature on T_p

Irrigated water temperature also regulated T_p, due to the difference between water and air temperature. The simulated equation between water-air temperature difference ($T_{w\text{-}A}$) and average T_A of 117 days was： $T_{w-A} = \pm 25.8\left[1-(\frac{T_A-10.6}{19.0})^{0.048}\right]^{0.5}$ (R^2=0.422**, n=117), based on the weather record (T_A from 17.4-30.5°C with the average of 23.7±8.6°C, T_w from 22.2-31.9°C with the average of 26.9±6.7°C). It implied that, when T_A=29.6°C, then $T_{w\text{-}A}$=0°C；when T_A>29.6°C, $T_{w\text{-}A}$ was lower than T_A, irrigated water decreased T_p. Conversely, when T_A<29.6°C, irrigated water increased T_p. When T_A decreased from 29.6°C, $T_{w\text{-}A}$ would be decreased in power. When T_A decreased from 27.4°C to 22.0°C (decreased by 5.4°C), $T_{w\text{-}A}$ was enlarged from 2°C to 4°C. When T_A decreased from 16.5°C to 12.9°C (decreased only by 3.6°C), $T_{w\text{-}A}$ was enlarged from 6°C to 8°C. Obviously, the regulatory effect would be enlarged under a lower T_A.

Fig. 6 showed the relationship in T_p difference (at 20 cm and 40 cm) between irrigated and non-irrigated (ΔT_p) and the difference between water-air temperature ($T_{w\text{-}A}$, average value of 720 samples) (equation 4 and 5). The results showed that, in a range of (-5.45°C)-(+10.32°C) for $T_{w\text{-}A}$, ΔT_p and $T_{w\text{-}A}$ showed a significant conic relationship. When $T_{w\text{-}A}$= -5°C, ΔT_p at 20 cm and 40 cm was -1.10°C and -0.71°C, respectively. When $T_{w\text{-}A}$=5°C , ΔT_p at 20 cm and 40 cm was 0.84°C and 0.66°C, respectively. When $T_{w\text{-}A}$=10°C, ΔT_p at 20 cm and 40 cm was 1.42°C and 1.26°C, respectively.

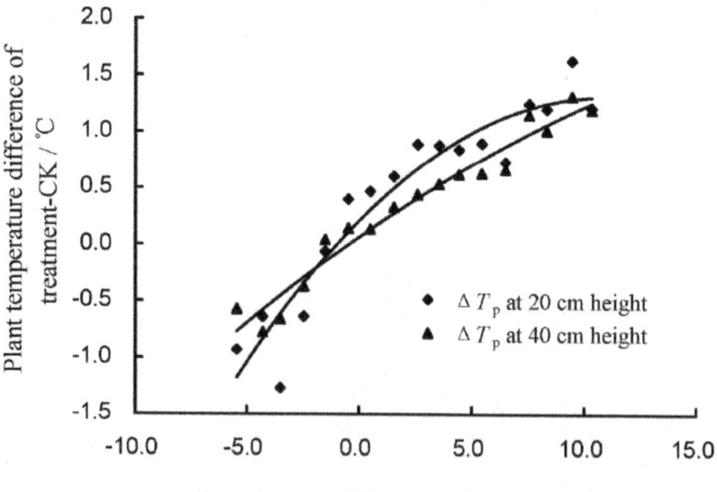

Fig. 6. Relationship between plant temperature difference of irrigated-non-irrigated (ΔT_p) and temperature difference of water-air ($T_{w\text{-}A}$).

$$20\text{ cm}: \Delta T_{p20} = -0.0051\, T_{w\text{-}A}^2 + 0.1934\, T_{w\text{-}A} \quad R^2 = 0.870^{**}, n = 17 \tag{4}$$

$$40\text{ cm}: \Delta T_{p40} = -0.0011\, T_{w\text{-}A}^2 + 0.1368\, T_{w\text{-}A} \quad R^2 = 0.950^{**}, n = 17 \tag{5}$$

2.5.4 Effects of flowing speed of irrigated water on T_p

Flowing speed of irrigated water was also an important factor affecting T_p. The margin of T_w between inflow and outflow (50 m between them) denotes the flowing speed of irrigated water. The equation of T_p established by margin of T_w between inflow and outflow of 13 days (with T_A=17.4-22.5°C) showed that T_p was affected significantly by flowing speed of irrigated water (Fig.7).

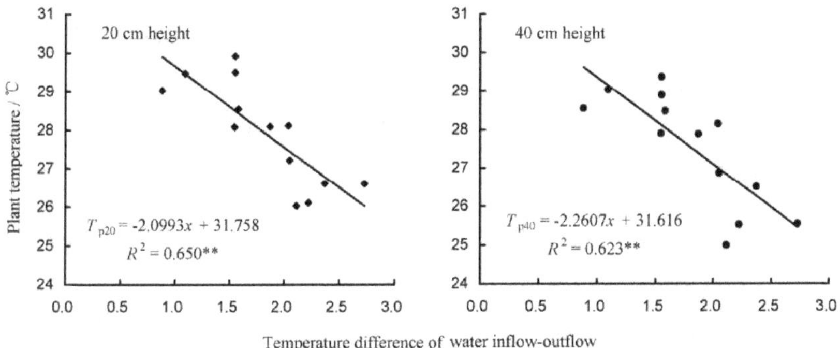

Fig. 7. Relationship between plant temperature and temperature difference of water inflow-outflow.

2.6 Effects of LAI of canopy on T_p

Rice layered LAI was decreased along with the increase of plant height. The relationship between T_p at 06:00 and 13:00 and their corresponding LAI at 5 heights was shown in Fig. 8. At 13:00, T_p was increased (actual value was increased from 26.9°C to 28.4°C) along with the decrease of the accumulated LAI. At 06:00, T_p was decreased (actual value was decreased from 20.6°C to 19.7°C) along with the decrease of the accumulated LAI. It indicated that the canopy absorbed solar radiation at daytime and released the heat energy during night, which regulated T_p.

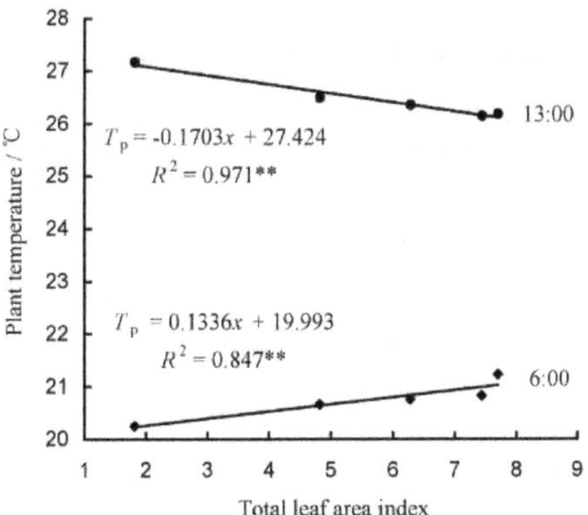

Fig. 8. Relationship between plant temperature (T_p) and LAI (x) at 06:00 and 13:00.

2.7 Simulation model of rice T_p by environmental factors

2.7.1 Analysis on correlations between T_p and environmental factors

Table 2 listed the correlations between T_p and weather or field environmental factors. It implied that T_p was regulated by various factors of canopy.

Item	T_{p20}	T_{p40}	T_{a20}	T_{a40}	150 cm Wind speed (V)	Inflow tempe-rature (T_{in})	Outflow tempe-rature (T_{out})	T_A	Sunshine hours (S)
T_{p20}	1	0.997**	0.985**	0.990**	-0.360**	0.944**	0.973**	0.981**	0.260*
T_{p40}		1	0.982**	0.995**	-0.366**	0.944**	0.971**	0.988**	0.270*
T_{a20}			1	0.973**	-0.392**	0.941**	0.971**	0.971**	0.320*
T_{a40}				1	-0.359**	0.950**	0.975**	0.994**	0.296*
150 cm Wind speed (V)					1	-0.330*	-0.372**	-0.390**	-0.200
Inflow water temperature (T_{in})						1	0.977**	0.943**	0.362**
Outflow water temperature (T_{out})							1	0.967**	0.316*
T_A								1	0.269*
Sunshine hours (S)									1

* Sample no. 60, $\alpha_{0.05}$=0.259, $\alpha_{0.01}$=0.335.

Table 2. Correlations between plant temperature and weather or field environmental factors

2.7.2 Models of T_p at heights of 20 cm and 40 cm

Four indices were chosen to establish models for T_p at heights of 20 cm and 40 cm in the present study: average temperature of inflow and outflow water $[(T_{in}+T_{out})/2]$, temperature margin between inflow water and air at 150 cm (T_{in}-T_A), wind speed at 150 cm (V), and sunshine hours (S). Results showed that both of two equations showed significant high R^2 (equations 6 and 7 and Fig. 9). Among the five parameters, T_{in} and T_{out} can be obtained by an actual measure, and T_A, V and S can be obtained from local weather station.

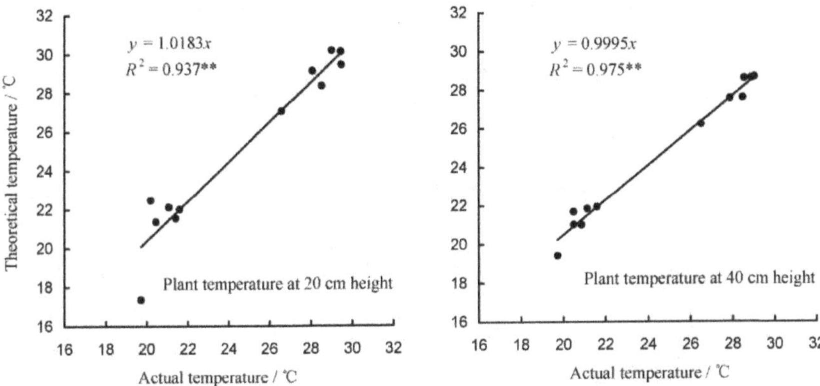

Fig. 9. Simulation effect of daily plant temperature at 20 cm and 40 cm heights.

Validation test showed that, the theoretical and practical values of T_{p20} and T_{p40} had significant correlations, and the slopes of the linear equation were 1.0183 and 0.9995, close to 1 (Fig. 9). The relative error $[(\sum \mid \text{practical}-\text{theoretical} \mid /\text{practical})/n]$ of T_{p20} and T_{p40} were 3.93% and 2.95%, respectively.

T_p of 20 cm:

$$Tp20 = 0.964(T_{in} + T_{out})/2 - 0.803(T_{in} - T_A)V + 0.085(12 - S) \quad R2 = 0.993^{**}, n = 61 \qquad (6)$$

T_p of 40 cm:

$$Tp40 = 0.937(T_{in} + T_{out})/2 - 0.595(T_{in} - T_A)V + 0.063(12 - S) \quad R2 = 0.998^{**}, n = 15 \qquad (7)$$

2.8 Function for rice plant temperature

It could be traced root as early as 1960's when field microclimate regulated by irrigated water was researched. The detailed procedure, however, was not described until two-line hybrid rice was applied (Xiao *et al* 2000, Zhou *et al* 1993). The sterility of rice TGMS line was controlled by temperature. A credible alteration point in temperature was important not only for selection and identification of TGMS lines, but for monitoring sterility alteration, determining effective methods to keep sterility of such TGMS, and increasing seed production of two-line hybrid rice (Lu *et al* 2004, 2007). In the practice of past two decades, techniques commonly used temperatures from thermometer-screen which was located in a 25×25 m^2 green plot and 150 cm height in local weather station (T_A). The sensitive part of rice plant to temperature, however, was in its canopy (Lu *et al* 2004, 2007, Zou *et al* 2005). We have noticed that the sterility of rice TGMS lines was affected directly by T_p rather than T_A. Therefore, it would be more accurate to monitor sterility of TGMS by using T_p in sensitive organs (Lu *et al* 2007). T_p was the final consequence of various environmental factors including air, water, soil, and the heat exchange among them. When attacked by low temperature during sensitive stage, how does the plant response, and how are agronomical methods used for safeguarding the sterility of rice TGMS? Such issues should be addressed by further studies. It will be too late to guarantee seed production of TGMS rice if only traditional methods are used. Simulation models of rice T_P established in this chapter and its regulation by environmental factors such as by inflow and outflow temperature of irrigated water, and T_A, wind speed, sunshine hours from local weather station have provided a more effective method for seed production of two-line hybrid rice.

3. Plant temperature for sterile alteration of a thermo-sensitive genic male sterile rice, Peiai64S

Rice thermo-sensitive genic male sterile (TGMS) line showed their sterile alteration along with the temperature change. An exact parameter to indicate sterile alteration was useful not only for breeding and identification of such TGMS, but was also helpful to determine and estimate the sterility security of TGMS, and select suitable methods for safeguarding the sterility of TGMS in two line hybrid rice seed production. Up to date, the forecast of TGMS sterile alteration was only based on the daily average temperature determined by local weather station (screen temperature at a height of 150 cm in a 25 m × 25 m green plot), and used three days average temperature as the alteration point temperature (from sterility transformed to partial fertility, seed setting rate from zero increased to 0.5%) (Lu *et al* 2001, Yao *et al* 1995). For instance, a TGMS, Peiai64S, set its alteration point temperature to be 23.5-24°C for average temperature with three days duration. When the three days average temperature was higher than the point, Peiai64S will exhibit sterile, otherwise partial fertility (Liao *et al* 2000, Lu *et al* 1999, Zou *et al* 2003). However, rice production practices showed that such parameter revealed its shortages in the following points: I: Temperature forecasted by a weather station was the same value within a county (or a city), and it could not express the difference of microclimate of individual field. Furthermore, the ground of screen in the weather station was different from rice field. It would cause the difference of temperature (Xie *et al* 2001). II: The sensitive part of rice to the environment condition was lower than the height of screen (Xu *et al* 1996), which would cause the difference of temperature as well. III: It was confirmed that when TGMS was attacked by lower temperature weather, it was useful of water irrigated to increase field or plant temperature for maintaining its sterility (Lu *et al* 2004, Xiao *et al* 1997, 2000, Zou *et al* 2005). However, it is difficult to estimate the increased degree of temperature by such forecasted screen temperature. IV: Seed production practices of two-line hybrid rice showed that under the same lower temperature weather, individual field exhibited diversified seed purity for the differences of microclimate. It is inaccurate to estimate fertility by screen temperature. V: The alteration point temperature for fertility used only the upper point temperature (seed setting rate from zero to 0.5%) and no scale for lower and optimum points. Also, there is no research report for such item so far. The author consider that the fertility of TGMS was affected by plant temperature, which was caused by all environmental conditions including air, water, soil, wind and so on. During the fertility sensitive period, for rice TGMS of each seed production field, the damaged degree, and the effect of measures against lower temperature must be estimated immediately. If these are estimated only by the pollen fertility even seed set, it will be late for taking measures to avoid such damage of lower temperature. Thus, it is important to establish an effective estimating and adjusting method for safeguarding seed production of two-line hybrid rice.

The temperature indices of sterile alteration must include two aspects as parameter of type and scale. The parameter type denoted the type of temperature (screen temperature of average or maximum or minimum one, otherwise as plant temperature of stem or leaf). The temperature scale will include the upper point (seed setting rate from zero increased to 0.5%), the optimum (show highest seed setting rate), and the lower point (seed setting rate returned to zero again) of temperature, which were considered as the three basic temperature points. The present chapter was aimed to establish such three basic temperature points.

A lower thermo-sensitive genic male sterile line, Peiai64S, which was widely used in two-line hybrid rice breeding, was chosen as plant material. During the fertility-sensitive period of all treatments, the microclimate and plant temperature were regulated by irrigated water. Three irrigated depths of 5 cm, 10 cm, and 15 cm, each of these with flowing and staying-water were used. Flowing irrigated water of 15 cm depth was also treated. During the fertility sensitive stage, the stem temperature at plant height of 20 cm was measured continually per 10 s with an improved needle thermocouple sensor under various types of weather. At the fertility sensitive stage, the plant and panicle height were determined, and furthermore the distance between the last two leaves. For all the sowing date treatments, their self-fertilized seed setting rate was measured by 30 paper bag insulated panicles.

3.1 Temperature differences between rice field and screen of weather station

Table 3 shows the temperature differences between the screen of weather station and rice field at heights of 150 cm, 100 cm, 60 cm, 40 cm, and 20 cm. The four periods in Table 3 are: I: from 20 August to 29 September in 2004 and from 9 August to 30 September in 2005. II: the earliest lower temperature days, 3 to 5 September in 2004. III: the earliest lower temperature days, 18 to 20 August in 2005. IV: the highest temperature day of 16 August, 2005. Table 3 shows that under various weather conditions, the temperatures of rice field at each height were different from the screen one of the weather station.

Tempe-rature	I		II		III		IV	
Location	Tempe-rature	Difference #	Tempe-rature	Difference	Tempe-rature	Difference	Tempe-rature	Difference
Screen	24.6±5.82		23.0		22.7		32.0	
150cm	23.91±5.82	0.69**	22.89	0.11	22.32	0.38	31.05	0.95
100cm	23.50±6.80	1.10**	22.79	0.21	22.21	0.49	30.79	1.21
60cm	23.77±5.94	0.83**	22.86	0.14	22.33	0.37	30.68	1.32
40cm	23.42±5.88	1.18**	22.69	0.31	22.18	0.52	30.21	1.79
20cm	22.97±5.29	1.63**	22.46	0.54	22.25	0.45	29.44	2.56

Temperature difference between screen and various heights in rice field. ** Different at $p<0.01$.

Table 3. Temperatures at different locations in rice field and their difference when compared to that in weather station

Owing to the difference in the underlay surface and growing plant, the temperature at each canopy height was lower than that of the screen of weather station in the four periods. For the period I, which represented the various weather conditions, the air temperature at height of 150 cm showed 0.69°C lower than the screen temperature of the weather station, which was at the same height. At heights of 100 cm, 60 cm, 40 cm, and 20 cm, owing to the energy absorbing and reflecting, temperature difference was enlarged at the four heights. The average temperature at heights of 100 cm, 60 cm, 40 cm, and 20 cm of period I was significantly lower than screen one by 1.10°C, 0.83°C, 1.18°C, and 1.63°C, respectively. A tendency of larger difference along with the height increase was seen, and the largest one was seen at 20 cm.

3.2 Fertility sensitive position and its height

It was reported that developing panicle was the fertility sensitive part of rice (Xu *et al* 1996). Thus, panicle height was important for determining the fertility sensitive position and the water depth for irrigation to regulate the temperature of the fertility sensitive part. The preceding studies proved that the sensitive stage was around the stage of meiosis of pollen mother cells (panicle developing stage IV - VI), and the visible morphological trait was at ±5 cm distance between the last upper two leaves (DL) (Lu *et al* 2001). Fig. 10 shows the correlation between the panicle height, length, and the DL. Result showed that during the stage, panicle height from the base was 15.3-21.9 cm with an average value of 18.5 cm, and meanwhile, the panicle length was 9.7-16.8 cm with an average value of 13.2 cm when DL was ±5 cm. It indicated that height of 20 cm was the suitable location for fertility sensitivity.

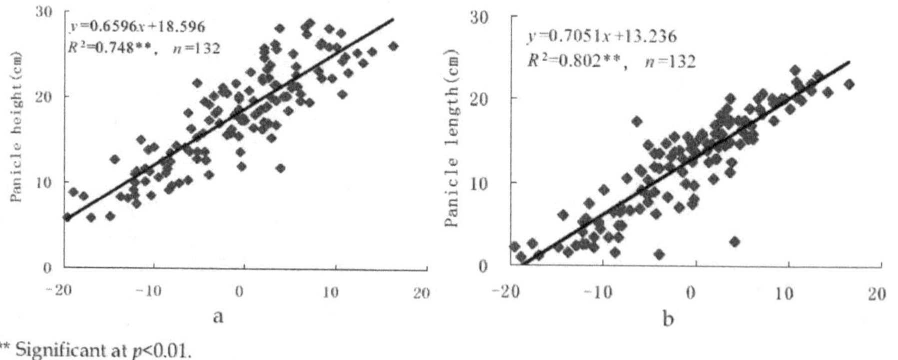

Fig. 10. Correlation of distance between the uppermost two leaves and panicle height (A) or length (B) of Peiai64S.

Table 4 shows the correlation coefficients between the self-fertilized seed setting rate and air temperature at various plant heights or screen one with a liner, conic, and present model (See the part '3.3 Model of fertility-temperature'). The result showed that the coefficients between the self-fertilized seed setting rate and stem or air temperature at 20 cm height were the highest and showed significant correlation.

It indicated that height of 20 cm was the location expressing the highest correlation coefficient. The result happened to meet the height of panicle at that time. Thus, 20 cm was taken as the suitable location for expressing the fertility of TGMS.

Item	20 cm Stem tempe-rature	20 cm Air tempe-rature	40 cm Air tempe-rature	60 cm Air tempe-rature	100 cm Air tempe-rature	150 cm Air tempe-rature	Screen tempe-rature
Line model	-0.316**	-0.410**	-0.275*	-0.225*	-0.224*	-0.200	-0.180
Conic model	0.318**	0.424**	0.275*	0.240*	0.238*	0.227*	0.208*
Present model	0.853**	0.778**	0.503**	0.314**	0.362**	0.369**	0.265*
Sample No.	66	85	85	85	85	85	89

*, ** Significant at $p<0.05$ and $p<0.01$, respectively.

Table 4. Correlation of the seed setting rate of Peiai64S and temperature at different heights in field

3.3 Model of fertility-temperature

3.3.1 Fertility sensitive stage

Determining fertility sensitive stage was the base to establish a model of fertility-temperature. The fertility sensitive stage can be determined by distance between the last upper two leaves or can be checked by the length of panicle. However, a method by growth days was popular for its advantage in quantitative analysis (Lu *et al* 2001). The method was given a range of growing days and it will cause difficulty for calculating temperature accumulation (Lu *et al* 2001), since it does not consider the growing difference owing to temperature. The chapter determined the initiative date (total days as well) by effective accumulated temperature >24. The method was set as：

>24°C°C effective accumulated temperature (Total sensitive days)	Initiative days from heading
<10°C.d	15
10-20°C.d	12
>20°C.d	10

3.3.2 Fertility-temperature model

By analyzing the relationship between the fertility and temperature, it was found that the fertility showed a rule of recovery-increase-decrease along with temperature decreasing. Thus, the chapter set the base of the model as: when the temperature decreased to an upper limit (self-fertilized seed setting rate of 0.5%), the TGMS was regarded as fertility, and at the optimum temperature, it showed the highest fertility. When the temperature decreased furthermore to a lower limit, the self-fertilized seed setting rate returned to zero.

Thus, equation (8) was established as:

$$P = P_0\left(\frac{T - T_L}{T_0 - T_L}\right)^A\left(\frac{T_H - T}{T_H - T_0}\right)^B \tag{8}$$

Here, P indicated self-fertilized seed setting rate, while P_0 indicated its maximum value. T denoted the averaged temperature of the sensitive stage, and T_H, T_L indicated the upper and lower limit temperature, respectively, when fertility was zero. T_0 denoted the optimum temperature when the TGMS showed maximum fertility. A and B were undetermined parameters and were both$>$0. Equation (8) shows the following parameters:

The derivative equation:

$$P' = P_0\frac{A(T - T_L)^{A-1}(T_H - T)^B}{(T_0 - T_L)^A(T_H - T_0)^B} - P_0\frac{B(T - T_L)^A(T_H - T)^{B-1}}{(T_0 - T_L)^A(T_H - T_0)^B}$$

i. When P' =0, the optimum temperature T_0 was calculated, and the highest self-fertilized seed setting rate, $P_{max}=P_0$
ii. When $T \geq T_H$ and $T \leq T_L$, then $P_{min}=0$

3.3.3 Analysis of temperature indices for sterile alteration by the model

Table 5 shows the simulation of stem or air temperatures with self-fertilized seed setting rate by three parameters as average, maximum, and minimum temperatures, with three durations of 1, 3, 5 days, and by 0.1°C of temperature step length for the simulation. Result showed that there were significant correlations between self-fertilized seed setting rate and the temperatures of 20 cm stem, air temperature or 150 cm air temperature with the three temperatures and three durations. It also implied that the effect of average temperature was better than that of the maximum or minimum temperature, and three-day duration was better than one-day or five-day durations. The result also showed that the stem and air temperatures at 20 cm were better than 150 cm air temperature, and the stem temperature at 20 cm showed the best effect of simulation. Thus, stem temperature at 20 cm was selected as the best simulating parameter for the model.

Equations (9), (10), and (11) are the statistic models of the self-fertilized seed setting rate simulated with the actual self-fertilized seed setting rate and temperatures of 20 cm stem, air, and 150 cm air. With the self-fertilized seed setting rate of 0.5% as initial (upper limit) and that of zero as the lower limit, and with stem temperature at 20 cm for three days, using equation (9), upper limit temperature and lower limit temperature was determined to be 22.8°C and 21.7°C, respectively. When compared with air temperature at 150 cm, there was 1.2°C and 1.1°C lower, respectively. With air temperature at 20 cm for three days, using equation (10), upper limit temperature and lower limit temperature was determined to be 23.2°C and 21.5°C, respectively. When compared with air temperature at 150 cm, there was 0.8°C and 1.3°C lower, respectively.

Location			A	B	T_L	T_0	T_H	P_0	R	n
20 cm Stem tempe-rature (T_{p20})	One day	Average	1	0.24	21.6	22.3	22.5	10.82	0.755**	68
		Maximum	1	0.55	27.9	28.1	28.2	10.82	0.371**	68
		Minimum	1	1.12	17.0	17.7	18.5	10.82	0.964**	34
	Three days	Average	1	0.37	21.7	22.5	22.8	10.82	0.853**	66
		Maximum	1	0.55	28.5	28.7	28.8	10.82	0.470**	66
		Minimum	1	1.33	17.0	17.5	18.1	10.82	0.550**	66
	Five days	Average	1	19.35	22.2	22.4	26.0	10.82	0.586**	64
		Maximum	1	0.62	28.0	28.2	28.3	10.82	0.344**	64
		Minimum	1	0.63	18.7	18.9	19.0	10.82	0.610**	64
20 cm Air Tempe-rature (T_{a20})	One day	Average	1	1.17	21.9	22.1	22.3	10.82	0.715**	83
		Maximum	1	1.50	26.6	26.8	27.1	10.82	0.360**	83
		Minimum	1	27.27	17.8	18.0	23.0	10.82	0.659**	76
	Three days	Average	1	12.14	21.5	21.8	26.0	10.82	0.778**	85
		Maximum	1	0.60	26.8	27.0	27.1	10.82	0.356**	85
		Minimum	1	5.02	17.7	17.9	19.1	10.82	0.684**	83
	Five days	Average	1	0.88	22.0	22.3	22.6	10.82	0.491**	86
		Maximum	1	0.97	26.7	26.8	26.9	10.82	0.408**	86
		Minimum	1	0.56	19.1	19.3	19.4	10.82	0.361**	85
150 cm (ck) Air Tempe-rature (T_A)	One day	Average	1	0.22	22.9	23.4	23.5	10.82	0.559**	41
		Maximum	1	0.67	29.5	29.6	29.7	10.82	0.280	41
		Minimum	1	0.55	17.7	17.9	18.0	10.82	0.426**	37
	Three days	Average	1	12.61	22.8	23.0	26.0	10.82	0.594**	42
		Maximum	1	0.34	28.5	28.8	28.9	10.82	0.548**	42
		Minimum	1	15.56	17.5	17.8	22.9	10.82	0.686**	41
	Five days	Average	1	9.15	22.9	23.2	26.0	10.82	0.504**	43
		Maximum	1	0.08	27.1	28.3	28.4	10.82	0.223	43
		Minimum	1	0.31	18.7	19.0	19.1	10.82	0.623**	42

Table 5. Simulation of stem or air temperature with three parameters and three durations

$$T_{p20} = 10.82 \times (\frac{T - 21.7}{22.5 - 21.7})(\frac{22.8 - T}{22.8 - 22.5})^{0.37} \quad (R=0.853^{**} ; n=66) \tag{9}$$

$$T_{a20} = 10.82 \times (\frac{T - 21.5}{21.8 - 21.5})(\frac{26.0 - T}{26.0 - 21.8})^{12.14} \quad (R=0.778^{**} ; n=85) \tag{10}$$

$$T_{A} = 10.82 \times (\frac{T - 22.8}{23.0 - 22.8})(\frac{26.0 - T}{26.0 - 23.0})^{12.61} \quad (R= 0.594^{**} ; n=42) \tag{11}$$

3.4 Stem and air temperatures at 20 cm

Considering energy exchange and balance, stem and air temperatures in rice canopy were connected with two energy sources: solar radiant energy and water heat energy. To avoid trouble in actual operation, the chapter established effective statistic models as equations (12) and (13), which included the screen temperature (T_A), cloud cover (N, from 1 to 10), and the water temperature at inflow (T_{in}) and outflow (T_{out}). Only by determining T_{in} and T_{out}, stem or air temperature at 20 cm could be calculated by screen temperature and cloud cover, which were both offered by the local weather station. It was helpful to estimate the damage of lower temperature weather and the effects of the adjusting measures in actual seed production.

$$T_{p20} = \frac{N}{15.2}\frac{(T_{in} + T_{out})}{2} + (1 - \frac{N}{15.2})T_A \quad (R=0.812^{**}, n=46) \tag{12}$$

$$T_{a20} = (1 - \frac{N}{34.9})\frac{(T_{in} + T_{out})}{2} + \frac{N}{34.9}T_A \quad (R=0.975^{**}, n=46) \tag{13}$$

3.5 Techniques by used plant temperature to safeguard sterile of Peiai64S

Peiai64S was a widely used TGMS in two-line hybrid rice breeding. Its fertility was controlled by temperature during its sensitive stage. Its sterility usually fluctuated owing to the frequent fluctuation caused by monsoon, which resulted in damage to the seed purity of seed production in China southern rice area (Lu *et al* 2001, Yao *et al* 1995). The first so-called super hybrid rice in China, Liangyoupeijiu, which was released by the authors' research group, was popularized over 7 million ha, and has been the major planted rice for its largest planting area of China from 2002. However, in lower reaches of Yangtze River, in the past five years, there was twice lower temperature weather (daily averaged temperature for three days lower than 24°C) in August, during which was the sterility sensitive period of TGMS, that caused damage to the seed production of Liangyoupeijiu and other two-line hybrid rice. It was a hidden problem for seed production of two-line hybrid rice. In the seed production practices of Liangyoupeijiu, it was found that when attacked by lower temperature weather, the seed purity exhibited a large difference even if the TGMS was in a same weather condition, owing to the individual differences at landform and water treatment and so on. Researches showed that the fertility of rice TGMS was affected directly by plant temperature, which was infected by the microclimate of the field (Hu *et al* 2006). When attacked by lower temperature weather, by irrigating warmer water from river or deep pool, plant temperature would be increased by 2°C, which was effective for safeguarding the sterility of TGMS (Lu *et al* 2004, Zou *et al* 2005). So far, the forecast of sterile alteration is only based on the temperature information from weather station; there is lack of any study on field microclimate or plant temperature, and especially no research has focused on the temperature scale of plant or air around it. In researches of wheat, some researchers used plant temperature as the parameter for freeze injury and grain growing speed (Feng *et al* 2000, Liu *et al* 1992).

The chapter put forward a method to conclude the fertility of TGMS by stem or air temperature at 20 cm height, which takes various factors including microclimate and location of field into account. It is more direct and exact than the traditional method. In rice production, we can use it to estimate any field or representative plot of a large field, and to monitor directly the result of regulation for safeguarding seed production in two-line hybrid rice. The technique is: when attacked by lower temperature weather with average temperature lower than 24°C during 5-15 days before TGMS heading, using infrared or thermosensor temperature indicator to determine plant stem temperature at 20 cm height or air temperature around it at 02:00, 08:00, 14:00, 20:00 (or only 08:00 and 20:00) every day. If the averaged value is lower than the line of 22.8°C for stem temperature or air temperature is lower than 23.2°C, it implies that the TGMS will transform its sterility to fertility. For safeguarding its sterility, it is necessary to irrigate by warmer water higher than 25°C, and by depth of 15 cm, until the temperature is higher than the above index.

The present chapter also established a statistic model for stem or air temperature at 20 cm height. By the inflow and outflow water temperatures of any field, and screen temperature and cloud cover from the local weather station, stem or air temperature at 20 cm height can be concluded. For an application example, if one day, the average temperature and cloud cover was 22°C and 9, respectively, and the actual water temperature of the field was 23°C, by (12) and (13), stem or air temperature at 20 cm height was calculated to be 22.55°C and 22.74°C, respectively, both of which were lower than the above temperature index. By irrigating warmer water from river, the inflow and outflow water temperature were measured as 26°C and 24°C, by (12) and (13), the stem and air temperature at 20 cm height was calculated to be 23.74°C and 24.23°C, respectively, since both are higher than the above temperature index, it will be effective for safeguarding the sterility of TGMS.

4. Abbreviation

Tp: plant temperature. T_{p10}, T_{p20}, T_{p30}, T_{p40} denotes rice stem temperature at plant heights of 10cm, 20cm, 30cm and 40cm, respectively.
ΔT_p: plant temperature difference between water irrigated and non-irrigated.
T_{a20}, T_{a40} and T_a: air temperature at heights of 20cm, 40cm and 150cm, respectively.
T_w: water temperature.
T_{in}: temperature of inflow water.
T_{out}: temperature of outflow water.
T_{w-A}: temperature difference between water temperature and air temperature at height of 150cm.
LAI: leaf area index.
DL: distance between the last upper two leaves of rice plant.

5. Notes

The most data were published in the journals of <Agricultural Sciences in China>, 2007, 6(11):1283-1290, and <Rice Science>, 2008, 15(3):223-231.

6. References

Cellier P. Estimating the temperature of a maize apex during early growth stages. *Agric Forest Meteor*, 1993, 63: 35-54.

Cheng W D, Yao H G, Zhao G P, Zhang G P. Application of canopy temperature in detecting crop moisture. *Chin Agric Sci Bull*, 2000, 16(5): 42-44. (in Chinese)

Cui L S, Lou X D. Statistical character of wheat leaf temperature in Qinghai tableland. *Meteor*, 1989, 15(10): 53-56. (in Chinese with English abstract)

Feng Y X, He W X, Rao M J, Zhong X L. Relationship between frost damage and leaf temperature with winter wheat after jointing stage. *Acta Agron Sin*, 2000, 26(6): 707-712. (in Chinese with English abstract)

Ferchinger G N. Simulating surface energy fluxes and radiometric surface temperature for two arid vegetation communities using the SHAW model. *J Appl Meteor*, 1998, 37: 449-460.

Hasegawa T. Agroclimatological studies of C sub (3) plants and C sub (4) plants, Pt. 4, Diurnal variations of leaf temperature and transpiration rates of rice and Japanese barnyard millet. *J Agric Meteor*, 1978, 34(3): 119-124.

Huang L, Leng Q, Bai G C, Hua B G. Appling the method of the distinguishing analysis to characterizing the relationship between the temperature dispersion on the leaves lower surface and the water status in sweet potato seeding. *J Biom*, 1998, 13(3): 388-393. (in Chinese with English abstract)

Hu N, Lu C G, Zou J S, Yao K M. Research on plant temperature of TGMS and its application. *Chin J Ecol*, 2006, 25(5): 512-516. (in Chinese with English abstract)

Leuning R. Leaf temperature during radiation frost, Part I. Observations. *Agric Forest Meteor*, 1988a, 42(2): 121-133.

Leuning R. Leaf temperature during radiation frost, Part II. A steady state theory. *Agric Forest Meteor*, 1988b, 42(2): 135-155.

Liao F M, Yuan L P. Study on the fertility expression of photo-thermo-sensitive genic male sterile rice Peiai64S at low temperature. *Sci Agric Sin*, 2000, 33(1): 1-9. (in Chinese with English abstract)

Li S P, Cai S Z, Fu X H. Effect of leaf temperature on differentiation and formation of Lichi floral buds and its control. *Chin J Trop Crops*, 1999, 20(4): 38-43. (in Chinese with English abstract)

Liu R W, Dong Z G. Effect of leaf temperature on grain milking in wheat. *Chin J Agrom*, 1992, 13(3) : 1-5. (in Chinese with English abstract)

Lu C G, Zou J S, Yao K M. Techniques for safeguarding seed production of two-line hybrid rice, *Rev Chin Agric Sci Techn*, 2004, 6 (supple) : 41-44. (in Chinese with English abstract)

Lu C G, Zou J S, Hu N, Yao K M. Plant temperature for sterile alteration of a temperature sensitive genic male sterile rice, Peiai64S. *Agric Sci China*, 2007, 6(11):1283-1290.

Lu P L, Ren B H, Yu Q. The observation and simulation of dew formation over maize canopy. *Acta Ecol Sin*, 1998, 18(6): 615-620. (in Chinese with English abstract)

Lu X G, Yao K M, Yuan Q H, Cao B, Mou T M, Huang Z H, Zong X M. An analysis of fertility of PTGMS rice in China. *Sci Agric Sin*, 1999, 32(4): 6-13. (in Chinese with English abstract)

Lu X G, Yuan Q H, Yao K M, Zong X M. *Ecological adaptability of photo-sensitive male sterile rice*. Beijing: China Meteorology Press. 2001. (in Chinese)

Shi P H, Mei X R, Leng S L, Du B H. Relationship between canopy temperature and water condition of winter wheat farmland ecosystem. *Chin J Appl Ecol*, 1997, 8(3): 332-334. (in Chinese with English abstract)

Tetsuya H, Daijiro I. Leaf temperature in relation to meteorological factors, Pt. 2, Leaf temperature variation with air temperature and humidity. *J Agric Meteor*, 1982, 38(3): 269-277.

Van A A, Griend D. Water and surface energy balance model with a multiplayer canopy representation for remote sensing purpose. *Water Resour Res*, 1989, 25(5): 949-971.

Wei L, Jiang A L, Jiang S K. Research on method of leaf temperature in field. *Chin J Agrom*, 1981, 6: 37-41. (in Chinese with English abstract)

Xiang X Q, Chen J, Chen G X, Liu Y D. Study on the leaf temperature, transpiration and photosynthesis in the summer period of Kiwifruit. *J Fruit Sci*, 1998, 15(4): 368-369. (in Chinese with English abstract)

Xiao G Y, Deng X X, Tang L, Tang C D. Approaches and methods for overcoming male sterility fluctuation of PTGMS lines in rice. *Hyb Rice*, 2000, 15(4)：4-5. (in Chinese with English abstract)

Xiao G Y, Yuan L P. Effects of water temperature on male sterility of the thermo-sensitive genic male sterile rice lines under the simulated low air temperature conditions appeared occasionally in high summer. *Chin J Rice Sci*, 1997, 11(4): 241-244. (in Chinese with English abstract)

Xie J H, Huang P B, Huang P J, Long Z Y, Jiang D X. Comparison of climatic factors between farming fields and meteorological station in early autumn seed production of two-line hybrid rice. *Hyb Rice*, 2001, 16(3): 32-35. (in Chinese with English abstract)

Xu M L, Zhou G Q. Studies on thermo-sensitive part of Peiai64S relating to its fertility expression. *Hyb Rice*, 1996, 11(2)：28-30. (in Chinese with English abstract)

Xu W J, Ta Y E, Ba H R. The explanatory simulation of the dewdrop state in vineyard. *Fores Sci Techn*, 2000, 25(4): 49-52. (in Chinese with English abstract)

Yao K M, Chu C S, Yang Y X, Sun R L. A preliminary study of the fertility change mechanism of the photoperiod (temperature period) sensitive genic male sterile rice (PSGMR). *Acta Agron Sin*, 1995, 21(2) :187-197. (in Chinese with English abstract)

Yuan G F, Tang D Y, Luo Y, Yu Q. Advances in canopy-temperature-based crop water stress research. *Adv Earth Sci*, 2000, 16(1): 49-54. (in Chinese with English abstract)

Zhao L X, Jing J H, Wang S T. Studies on water transports in the soil-plant-atmosphere continuum on Weibei rainfed highland, Shanxi Province-Effects of ecological environment on leaf temperature of winter wheat. *Acta Bot Boreali-Occidentalia Sin*, 1996, 16(4): 345-350. (in Chinese with English abstract)

Zhou C S, Liu J B. Studies of the cold water irrigation technique for multiplication of lower critical temperature TGMS rice, *Hyb Rice*, 1993, 8 (2): 15-17. (in Chinese with English abstract)

Zou J S, Lu C G, Yao K M, Hu N, Xia S J. Research on theories and techniques of irrigation for safeguarding seed production of two-line hybrid rice. *Agric Sci Chin*, 2005, 38 (9): 1780-1786.

Zou J S, Yao K M, Deng F P. Analysis on fertility characteristics of Peiai64S and technology for its safely applying. *Acta Agron Sin*, 2003, 29 (1) : 87-92. (in Chinese with English abstract)

Section 5

Crop Protection

9

Insect Pests of Green Gram *Vigna radiata* (L.) Wilczek and Their Management

R. Swaminathan[1], Kan Singh[1] and V. Nepalia[2]

[1]*Department of Entomology, Rajasthan College of Agriculture, Maharana Pratap University of Agriculture and Technology, Udaipur, Rajasthan,*

[2]*Department of Agronomy, Rajasthan College of Agriculture Maharana Pratap University of Agriculture and Technology, Udaipur, Rajasthan, India*

1. Introduction

1.1 Importance of the crop and its cultivation

Pulses, the food legumes, have been grown by farmers since millennia providing nutritionally balanced food to the people of India (Nene, 2006) and many other countries in the world. The major pulse crops that have been domesticated and are under cultivation include black gram, chickpea, cowpea, faba bean, grass pea, green gram, horse gram, lablab bean, lentil, moth bean, pea and pigeon pea. The probable geographical origin of the more common pulses has been reported as:

Crop	Geographical origin and domestication
Black gram	Indian subcontinent
Chickpea	Turkey Syria
Cowpea	West Africa
Faba bean	West Africa
Grass pea	Southern Europe
Green gram	Indian subcontinent
Horse gram	Indian subcontinent
Lablab bean	Indian subcontinent
Lentil	Southwest Asia (Turkey-Cyprus)
Moth bean	Indian subcontinent
Pea	Southern Europe
Pigeonpea	India

Source: Nene, 2006

A low input, short duration, high value crop, mung bean fits very well into rice-wheat cropping systems and other crop rotations. It fixes nitrogen in the soil, requires less irrigation than many crops to produce a good yield, and helps maintain soil fertility and texture. Including green gram to the cereal cropping system has the potential to increase farm income, improve human health and soil productivity, save irrigation water and promote long term sustainability of agriculture (Chadda, 2010).

1.2 The plant

Vigna radiata (L.) Wilczek [Synonyms: *Phaseolus radiatus* L. (1753), *Phaseolus aureus* Roxb. (1832)], often known as green gram/mung bean, is native to India and Central Asia. It has been grown in these regions since prehistoric times (Vavilov, 1926) and as an important legume crop in India throughout the year. The Sanskrit name for green gram is *mudgaparni* or *mashaparni* as per ancient Indian literature the *Yajurveda* (c. 7000 BC). While green gram cultivation spread over to many countries, especially in tropical and subtropical Asia, black gram (*Vigna mungo*) cultivation has remained more or less confined to South Asia. The progenitor of these pulses is believed to be *Vigna trilobata*, which grows wild in India (Nene, 2006). The more common vernacular names include: mung bean, green gram, golden gram (En). Haricot mungo, mungo, ambérique, haricot doré (Fr). Feijão mungo verde (Po). Mchooko, mchoroko (Sw) (Mogotsi, 2006).

The genus *Vigna* comprises about 80 species and occurs throughout the tropics. *Vigna radiata* belongs to the subgenus *Ceratotropis*, a relatively homogenous and morphologically and taxonomically distinct group, primarily of Asian distribution. Other cultivated Asiatic *Vigna* species in this subgenus include *Vigna aconitifolia* (Jacq.) Maréchal (moth bean), *Vigna angularis* (Willd.) Ohwi & Ohashi (adzuki bean), *Vigna mungo* (L.) Hepper (black gram or urd bean), *Vigna trilobata* (L.) Verdc. (pillipesara) and *Vigna umbellata* (Thunb.) Ohwi & Ohashi (rice bean). Hybrids have been obtained between many of these species. The species have often been confounded, especially *Vigna radiata* and *Vigna mungo*. The wild types of mung bean, which are usually smaller in all parts than cultivated types, are usually classified into 2 botanical varieties: - var. *sublobata* (Roxb.) Verdc., occurring in India, Sri Lanka, South-East Asia, northern Australia (Queensland), in tropical Africa from Ghana to East Africa, southern Africa and Madagascar; - var. *setulosa* (Dalzell) Ohwi & Ohashi, with large, almost orbicular stipules and dense long hairs on the stem, and occurring in India, China, Japan and Indonesia. The cultivated types of mung bean are grouped as *Vigna radiata* var. *radiata*, although a classification into cultivar groups would be more appropriate. Two types of mung bean cultivars are usually distinguished, based mainly on seed colour: - golden gram, with yellow seeds, low seed yield and pods shattering at maturity; often grown for forage or green manure; - green gram, with bright green seeds, more prolific, ripening more uniformly, less tendency for pods to shatter. Two additional types are recognized in India, one with black seeds and one with brown seeds (Mogotsi, 2006).

1.3 Uses

Mature mung bean seeds or flour enter a variety of dishes such as soups, porridge, snacks, bread, noodles and even ice-cream. In Kenya mung bean is most commonly consumed as whole seeds boiled with cereals such as maize or sorghum. Boiled whole seeds are also fried with meat or vegetables and eaten as a relish with thick maize porridge ('ugali') and pancakes ('chapatti'), whereas consumption of split seeds (dhal) is common among people of Asian descent. In Ethiopia the seeds are used in sauces. In Malawi the seeds are cooked as a side dish, mostly after removing the seed coat by grinding. In India and Pakistan the dried seeds are consumed whole or after splitting into dhal. Split seeds are eaten fried and salted as a snack. The seeds may also be parched and ground into flour after removing the seed coat; the flour is used in various Indian and Chinese dishes. The flour may be further processed into highly valued starch noodles, bread, biscuits, vegetable cheese and extract for the soap industry. Sprouted mung bean seeds are eaten raw or cooked as a vegetable; in French they are erroneously called '*germes de soja*', in English 'bean sprouts'. Immature pods and young leaves are eaten as a vegetable. Plant residues and cracked or weathered seeds are fed to livestock. Mung bean is sometimes grown for fodder, green manure or as a cover crop. The seeds are said to be a traditional source of cures for paralysis, rheumatism, coughs, fevers and liver ailments (Mogotsi, 2006).

1.4 Nutritional facts

Green gram is an important source of easily digestible high quality protein for vegetarians and sick persons. It contains 24 per cent protein, 0.326 per cent phosphorus, 0.0073 per cent iron, 0.00039 per cent carotene, 0.0021 per cent of niacin and energy 334 cal/100g of green gram. The gap between realizable and actual yields needs to be bridged up with appropriate technologies. The latter basically revolves around two aspects: production and protection. Although the former has been a subject of greater emphasis at all levels of strategy, yet the latter continues to suffer from neglect and a sort of apathy under the cover of ignorance, economic status of the farmers and many more factors. In India, Acharya (1985) has pointed out that plant protection remains a most neglected aspect in pulse cultivation; further stating that only 5 to 6 per cent of the growers use plant protection measures in only 1.5 per cent of the total area under this crop. In view of the above facts, for the control of various pests in green gram, only certain insecticides have been recommended by several workers.

The composition of mature mung bean seeds per 100 g edible portion is: water 9.1 g, energy 1453 kJ (347 kcal), protein 23.9 g, fat 1.2 g, carbohydrate 62.6 g, dietary fibre 16.3 g, Ca 132 mg, Mg 189 mg, P 367 mg, Fe 6.7 mg, Zn 2.7 mg, vitamin A 114 IU, thiamin 0.62 mg, riboflavin 0.23 mg, niacin 2.3 mg, vitamin B_6 0.38 mg, folate 625 µg and ascorbic acid 4.8 mg. The essential amino-acid composition per 100 g edible portion is: tryptophan 260 mg, lysine 1664 mg, methionine 286 mg, phenylalanine 1443 mg, threonine 782 mg, valine 1237 mg, leucine 1847 mg and isoleucine 1008 mg. The starch consists of 28.8per cent amylose and 71.2per cent amylopectin. Mung bean seed is highly digestible and low in antinutritional factors. It causes less flatulence than the seed of most other pulses, making it suitable for children and older people. Mung bean starch is considered to have a low glycaemic index, i.e. to raise the blood sugar level slowly and steadily. The composition of sprouted mung bean seeds per 100 g edible portion is: water 90.4 g, energy 126 kJ (30 kcal), protein 3.0 g, fat 0.2 g, carbohydrate 5.9 g, dietary fibre 1.8 g, Ca 13 mg, Mg 21 mg, P 54 mg, Fe 0.9 mg, Zn 0.4 mg, vitamin A 21 IU, thiamin 0.08 mg, riboflavin 0.12 mg, niacin 0.75 mg, vitamin B_6 0.09 mg, folate 61 µg and ascorbic acid 13.2 mg. The essential amino-acid composition per 100 g edible portion is: tryptophan 37 mg, lysine 166 mg, methionine 34 mg, phenylalanine 117 mg, threonine 78 mg, valine 130 mg, leucine 175 mg and isoleucine 132 mg. Sprouting especially leads to an increased ascorbic acid concentration. Mung bean hay contains: moisture 9.7 per cent, crude protein 9.8 per cent, fat 2.2 per cent, crude fibre 24.0 per cent, ash 7.7 per cent, N-free extract 46.6 per cent, digestible crude protein 7.4 per cent, total digestible nutrients 49.3per cent. Aqueous extracts of mung bean seeds have shown in-vivo hypotensive and hepatoprotective effects in rats. Extracts from mung bean seeds and husks have shown antioxidative effects.

The chemical composition of green gram (dal) has been worked out as:

Calorific value (cal./ 100g)	Crude protein (%)	Fat (%)	Carb-ohyd-rate (%)	Ca (mg/ 100g)	Fe (mg/ 100g)	P (mg/ 100g)	Vitamine(mg/100g)		
							B_1	B_2	Niacine
334	24.0	1.3	56.6	140	8.4	280	0.47	0.39	2.0

Source: Pulse Crops, by B.Baldev, S.Ramanujam and H.K.Jain, PP. 563.

1.5 Cultivation in Kenya

Green gram is a warm-season crop and grows mainly within a mean temperature range of 20–40°C, the optimum being 28–30°C; hence, can be grown in summer and monsoon season/autumn in warm temperate and subtropical regions and at altitudes below 2000 m in the tropics. It is sensitive to frost. It is mostly grown in regions with an average annual rainfall of 600–1000 mm, but it can do with less. It withstands drought well, by curtailing the period of flowering and maturation, but it is susceptible to water-logging. High humidity at maturity causes damage to seeds leading to seed discoloration or sprouting while still in the field. Green gram cultivars differ markedly in photoperiod sensitivity, but most genotypes show quantitative short-day responses, flower initiation being delayed by photoperiods longer than 12–13 hours. The crop grows well in a wide range of soil types, but prefers well-drained loams or sandy loams with pH (5–) 5.5–7 (–8). Some cultivars are tolerant to moderate alkaline and saline soils (Mogotsi, 2006).

Green gram is propagated by seed. There is no seed dormancy, but germination can be affected by a hard seed coat. Green gram seeds are broadcast or dibbled in hills or in rows. Recommended sowing rates are 5–30 kg/ha for sole crops, and 3–4 kg/ha under intercropping. Recommended spacing ranges within 25–100 cm × 5–30 cm; whereas, for the more modern cultivars ripening in 60–75 days, maximum yields are obtained at plant densities of 300,000–400,000 plants/ha. The later-maturing traditional cultivars generally need wider spacing. Recommended spacing for sole crop of green gram in Kenya are 45 cm between rows and 15 cm within the row, with a seed rate of 6–10 kg/ha and a sowing depth of 4–5 cm. Mung bean can be grown mixed with other crops such as sugar cane, maize, sorghum or tree crops in agroforestry systems. Short-duration mung bean is often relay-cropped to make use of a short cropping period. In Kenya mung bean is usually intercropped with maize, sorghum or millet; it is occasionally grown in pure stands or intercropped with other pulses. The usual practice here is to place 1–2 rows of mung bean between rows of a cereal, or to plant mung bean in the cereal row. In pure stands, 1–2 weedings are necessary during the early stages of growth. In Kenya weeding is done using hoes and machetes. Farmers do not normally apply any inorganic fertilizer to a mung bean crop. Mung bean uses residues from fertilizer applications to the main crops in the system, though it responds well to phosphorus. Nutrient removal per ton of seed harvested (dry weight) is 40–42 kg N, 3–5 kg P, 12–14 kg K, 1–1.5 kg Ca, 1.5–2 kg S and 1.5–2 kg Mg. The nutrient removal is much higher when crop residues are removed to be used for fodder. In its major area of cultivation, the monsoon tropics, mung bean is mainly grown as a rainy season crop on dryland or as a dry-season crop after the monsoon in rice-based systems on wetland, making use of residual moisture or supplementary irrigation. In some areas where adequate early rains occur, an early-season crop can be grown before the monsoon. In semi-arid regions of Kenya with 600–800 mm rainfall evenly distributed over 2 rainy seasons, 2 mung bean crops are grown per year. In the Wei Wei Integrated Development Project in Sigor, Kenya, mung bean is grown under irrigation. In India mung bean is often sown as a fallow crop on rice land as a green manure (Mogotsi, 2006).

1.6 Cultivation in India

1.6.1 Historical background

India has been universally accepted as the original home of these two pulse crops. While green gram was spread to many countries, especially in tropical and subtropical Asia, black gram has remained more or less confined to South Asia. Currently, green gram is being grown in USA.

1.6.2 Area and production

India is reportedly the largest pulse growing country in the world both in terms of area as well as production covering 43.30 per cent of land area under pulses with 33.15 per cent production. In another report, it has been described that India is the largest producer and consumer of pulses in the world accounting for 33 per cent of world's area and 22 per cent of world's production of pulses. Green gram is one of the most widely cultivated pulse crops after chickpea and pigeonpea. The major producing states in India are Andhra Pradesh, Orissa, Maharashtra, Madhya Pradesh and Rajasthan accounting for about 70 per cent of total production.

Reliable production statistics for mung bean are difficult to obtain, as its production is often lumped together with that of other *Vigna* and *Phaseolus* spp. India is the main producer, with an estimated production in the late 1990s of about 1.1 million t. China produced 891,000 t (19per cent of total pulse production in China) from 772,000 ha in 2000. No mung bean production statistics are available for Africa. China exported 110,000 t in 1998, 290,000 t in 1999 and 88,000 t in 2000. All mung bean produced in India is for domestic consumption. In most parts of Africa where there are Asian communities, mung bean food products are sold in the cities (Mogotsi, 2006).

In India it is cultivated in two seasons: *kharif* and summer. However, peak market arrivals are from September to October (*kharif*) and June to July (summer). Green gram is mostly grown as a *kharif* crop (monsoon season) in the states of Rajasthan, Maharashtra, Gujarat, Karnataka, Andhra Pradesh, Madhya Pradesh and Uttar Pradesh; but, in Tamil Nadu, Punjab, Haryana, Uttar Pradesh and Bihar it is grown as a summer crop. Green gram is a short duration crop; the canopy closes in earlier than in cereal crops. Therefore, the critical period for crop-weed interference is limited to the first 30 days after sowing. However, the presence of weeds during this period may lead to 20 to 40 percent reduction in yield. The most phenomenal way of reducing weed infestation include preventive measures like use of clean crop seed and pre-planting destruction of existing weeds by tillage. In the standing crop, one inter-row hoeing and weeding at 25 to 30 days after sowing is good enough to check potential losses. Sometimes in the monsoon crop, due to incessant rains, tillage operations are not feasible; hence, the use of pre-plant incorporation of fluchloralin or pendimethalin at the rate of 0.75 to 1.0 kg per hectare is recommended.

1.6.3 Cropping systems

The common crop rotations followed in India include: green gram - mustard; green gram - safflower; green gram - linseed; and green gram - wheat. In *kharif*, intercropping with maize, pearl millet, sesame, pigeon pea, and cotton is common. Spring or summer green gram is grown as a catch crop. The crop sequences that have been successful are green gram - maize - wheat, green gram - rice - wheat, green gram - maize - toria - wheat, green gram - maize - potato - wheat. In spring planted sugarcane, it is also grown as an intercrop. During *rabi*, it is grown in rice fallows of southern and south eastern region (ICAR, 2006).

The area (m-ha), production (m-t) and Productivity (kg/ha) in India

Years	Area	Production	Productivity
2004-05	3.34	1.06	415
2005-06	3.10	0.95	428
2006-07	3.19	1.12	440
2007-08	3.43	1.52	510
2008-09	3.30	1.24	425

Source: Directorate of pulses development Bhopal

2. Insect pests and their associated natural enemies

2.1 Qualitative and quantitative abundance

Methodology used to quantify major insect pests of green gram and the associated natural enemies:

Study on population dynamics of insect pests and natural enemies:

i. Population of jassids (nymphs & adults) and white flies (adults) can be estimated by the visual count technique during early hours of the day from requisite plants per replication (usually counting from 1/5th of total plant population per plot), selected at random and tagged. The top, middle and bottom parts of the tagged plants should be given due consideration. Alternatively, the sudden trap method using a cubical iron-frame trap of 45cm x 45cm base and 60cm height clothed in high density polythene can be used to trap the adult jassids and whiteflies for easy counting. The nymphs can be counted adopting the visual/sight-count technique from the plants by gently turning the leaves. Similarly, for aphids, the nymphs and adults can be counted on the plants directly taking observations from at least a 10cm top shoot/twig for sampling. Yellow pan-traps or sticky traps can also be used for counting the jassids and whiteflies.

The population of jassds and whiteflies can be estimated by the sudden trap method using a cubical iron-frame trap of 45cm x 45cm base and 60cm height clothed in high density polythene or covered by muslin cloth on three sides, while one side and the top can be clothed with high density polythene for viewing the adults captured in the trap. This is possible when the crop is 35 to 40 day-old; thereafter, we have to rely on visual count technique as it is not feasible to use the sudden trap because as per agronomical practices the spacing is recommended as 30cm between the rows and 10cm between two pants in a row; though, we may go for a spacing of 45cm (row to row) by 15cm (plant to plant in a row). Yes, an aspirator should be used to suck the adults for a reliable count. The few adults that might escape while counting from each tagged plant or replicate is a common error and will be taken care of if the sample size is more. The jassid nymphs move sideways but do not leave the foliage while observing, hence their count is reliable.

Data from visual counts can be homogenized by following the square root transformation method, especially when zero counts are recorded. We add 0.5 or 1.0 to the visual count data and find the square root before analysis that has to be retransformed after analysis for interpretation. In case of percentage data, if the percentage values range between 30 and 80, usually arc sine (angular) transformation would not be required; however, if the percentage values happen to be less than 30 and/ or more than 80, then angular transformation is required before analysis of data.

Green gram being an indeterminate plant, vegetative growth and flowering go together when the plant is 35 to 40 days of age; hence, using a suction trap (battery operated/electric) might not be successful, as it will cause more harm to the plant and at the same time disturb the insects. We can use the suction trap if the field/plot size is large so that sampling can be taken from distant areas within the field being observed without disturbing the insect species on adjacent plants.

ii. The associated natural enemies like syrphid flies, coccinellids and others, can be recorded by the visual/sight count technique from the same number of plants per replication (1/5th of total population) randomly tagged, during early hours of the day. Observations can be taken to study the predator-prey function under caged conditions.

iii. Observations can be taken on a weekly basis or a 10-day basis and the prevailing abiotic conditions of the atmosphere can also be recorded accordingly to work out the correlation coefficients between the populations and the abiotic factors of the environment.

It is true that correlations do not establish the cause and effect relationships and must be interpreted with caution. Therefore, further working out the simple regression lines or linear regressions through regression equations enables us to know the effect of the abiotic factors (independent variable) on the population (dependent variable). Data so collected and collated over many years (at least for 5 years in succession) will give a good understanding of the population trend.

The equation of a line of regression (Y on X) is given as:

$$Y = a + bX$$

The data can be entered and processed in "MS-Excel" using the correlation function and make a chart in excel using the custom type Classic Combination Chart (either line-column on two axes or lines on two axes) that also enables to have the regression equation and coefficient of determination (R^2).

iv. Estimation of the population density of insect pests and their natural enemies in the different treatments can be made and expressed as a percentage after comparing the data from the control treatment or the standard check. Wherever applicable, diversity indices can be computed using suitable techniques (Shanon-Weiner or Simpson Diversity Index).

The following mathematical/ statistical analysis can be made towards estimating the species richness and diversity:

Mean density:

$$\text{Mean density} = \Sigma \frac{Xi}{N} \times 100$$

Where,

Xi = No. of insects or natural enemies in i^{th} sample
N = Total number of plants sampled.

Shannon-Weiner diversity index (H'):

Shannon-Weiner diversity index

$$(H') = -\Sigma\ pi\ ln\ pi$$

Where, pi = the decimal fraction of individuals belonging to i^{th} species.

However, along with the Shanon Diversity Index the Simpson's Index can also be computed.

Simpson's index is calculated using the equation:

$$Ds = \frac{N(N-1)}{\sum n(n-1)}$$

N= Total number of individuals of all species
n= Number of individuals of a species

v. To record the incidence of blister beetles at flowering stage the numbers of beetles per plant for a fixed time interval during the morning hours (8 to 10am) or evening hours (3 to 5pm) of the day can be observed on the randomly tagged plants. However, some species visit during early hours while others late; hence, a preliminary observation on their behaviour shall become essential before standardizing the methodology for blister beetle counts.

vi. To note the damage of pod borers, the numbers of healthy and damaged pods can be counted from a known (pre-decided) sample of pods (say 100) taken from the tagged plants and the data expressed as a percentage of the total. Usually the pods are split open to record the species of the borer under study.

vii. For the estimation of the population of soil dwelling predators, especially carabids, pitfall traps (500ml capacity glass jars) should be laid out in each replication and at least 3 traps should be randomly placed in each plot of 18 sq. m. (6m X 3m as length X breadth). For instant killing of the predatory insects and to avoid cannibalism (as in carabid grubs), ethylene glycol or formalin (1-2%) can be used in the traps. Comparisons among the treatments can be accounted for.

2.2 Loss estimation and establishment of economic threshold for the pod borer

In order to asses the losses caused by insect pests of green gram the paired plot experiment, as suggested by Leclerg (1971), can be adopted. The method involves growing the crop in 26 plots, each measuring preferably 6m X 3m. Each plot should be separated by a buffer strip of one meter all around. One set of plots has to be kept protected from insect infestation by regular need-based application of recommended insecticides. The other set of plots has to be exposed to natural infestation and thus called unprotected. Observations on the plant height, number of primary branches, pod length, pod and grain damage (%), and any other yield attributing parameter recorded from five randomly selected plants from each plot at maturity should be taken. Loss in yield can be calculated by comparing the yield obtained from protected and unprotected plots using the following formula:

$$\text{Loss in yield}(\%)=\frac{X_1-X_2}{X_1}\times 100$$

Where

X_1 = Yield in treated plot
X_2 = Yield in untreated plot

The yield data can be analyzed statistically and significance tested using the "t" test.

$$\text{Standard deviations (s)}=\frac{\text{Sum of square of the deviation from the mean difference}}{\text{Number of paired plots}-1}$$

$$\text{Standard error of mean difference (Sd)}=\frac{\text{Standard deviation (s)}}{\text{Number of paired plots (n)}}$$

$$t=\frac{Y_1-Y_2}{sd}$$

Where

Y_1 = Average yield in treated plot; Y_2 = Average yield in untreated plot

sd = Standard error of mean difference

2.3 Determination of economic threshold level for the lycaenid pod borer

In order to calculate the economic injury level for the pod borer, losses in grain weight due to various levels of larval density of the pod borer has to be estimated. Green gram can be sown in pots of suitable size and the neonate larvae can be released on the developing tender pods or flowers at different population densities (1, 2, 3 and 4 larvae per plant or in a geometrical progression as 2, 4, 8 and 16). A no-larval release control should also be taken side by side on the pot plants. The plants should be caged properly and the treatments replicated. Observations on the number of healthy and damaged pods, and grain weight per plant should be recorded. Taking the reduction in yield due to different levels of larval density release, the regression analysis can be worked out to quantify the damage. The economic injury level for the pod borer on green gram can be determined by using the method suggested by Hammond and Pedigo (1982).

$$\text{Gain threshold (G.T.)} = \frac{\text{Management Cost (Rs/ha)}}{\text{Marketed value of Mungbean (Rs/kg)}} = \text{kg/ha}$$

$$\text{Economic injury level (EIL)} \frac{\text{Gain threshold (kg/ha)}}{\text{Loss per insect (kg/insect)}} = \text{insect/ha}$$

The economic threshold level can be calculated by the method suggested by Johnston and Bishop (1987). They established economic threshold level as the population of economic injury level minus the increase in population of the pest concerned per day. The increasing rate of larval population under natural field conditions can be determined by recording the weekly population of the pod borer during larval activity. The rate of increase in population can be calculated arithmetically.

2.4 Farmscaping in green gram with annual marigold and niger

Early flowering marigold variety must be sown in well prepared, raised nursery beds at least 45 days before transplanting. Niger crop has to be directly sown 28 to 30 days before sowing the main crop of green gram. In short, sowing should be adjusted in such a manner that flowering of niger/marigold and green gram should coincide so that nectar/pollen feeding natural enemies would be attracted to the farmscape plants. The sowing operations for green gram and the different farmscape plants should be as:

1. In the green gram and niger farmscaping (3: 1 ratio), niger sowing is done first and followed by sowing of green gram a month later.
2. In the green gram and marigold farmscaping (3: 1 ratio), transplanting of marigold on ridges should be carried out 45 days after sowing green gram.

The row to row distance and plant to plant spacing for green gram can be 60cm and 10cm, respectively or 45cm and 10cm, respectively; whereas, mature seedlings of marigold are to be transplanted in between two rows of green gram at a distance of 30cm.

3. Pest management strategies

3.1 The pest insects

An estimated 200 insect pests that belong to 48 families in Coleoptera, Diptera, Hemiptera, Hymenoptera, Isoptera, Lepidoptera, Orthoptera, Thysanoptera, and 7 mites of the order Acarina are known to infest green gram and black gram. Under severe cases stem fly may alone cause more than 90 per cent damage with a yield loss of 20 per cent (Talekar, 1990). The galerucid beetle, *Madurasia obscurella* causes damage up to 20 – 60 per cent. Whitefly, a potential vector of mungbean yellow mosaic virus (MYMV), can cause losses ranging from 30-70 per cent. The major insect pests, particularly those often cited, have been enlisted in Table - 1.

The insect pests that infest green gram are better classified according to their appearance based on crop phenology. Accordingly, they can be: (1) stem feeders, (2) foliage feeders, (3) pod feeders, and (4) pests of stored grains; which are also convenient to access their economic importance so as to devise suitable management measures.

At the seedling stage are the agromyzid flies, also known as bean flies (possibly few species), *Melanagromyza* (*Ophiomyia*) *phaseoli* (Tryon) being of more common occurrence. *Ophiomyia phaseoli* larva is a cortex feeder and pupates in the cortex mostly at the root-shoot junction. Sometimes pupae can be seen sticking under the membranous epidermis. In India the girdle beetle, *Oberiopsis brevis* (Swedenbord), a major pest of soybean, sometimes infests mungbean locally (Talekar, 1990).

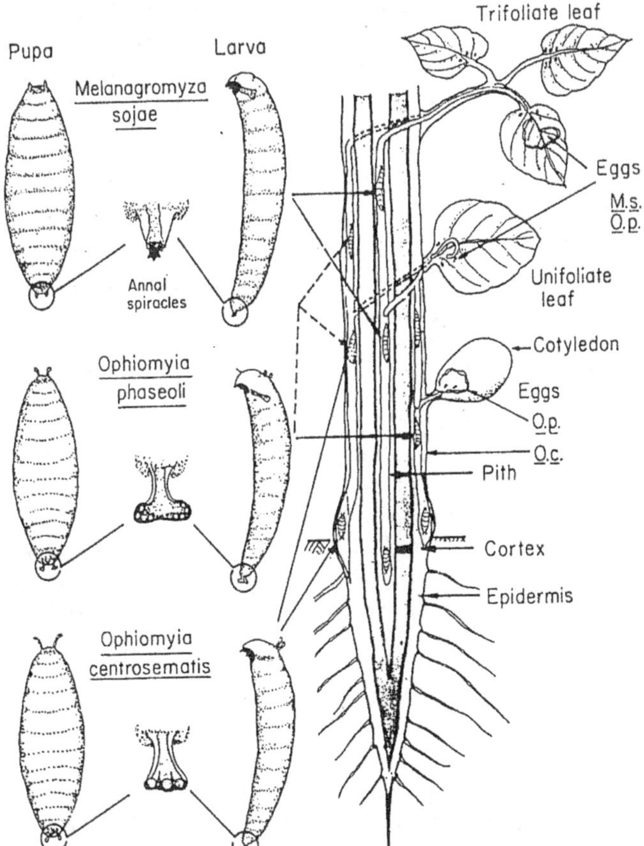

(*M. s.* = *M. sojae*, *O. p.* = *O. phaseoli*, *0. c.* = *O. centrosematis*; please note, *O. phaseoli* does not lay eggs in the cotyledons of green gram) [Ref.: Talekar, 1990]

Fig. 1. Location of ovipositional and larval feeding sites in soybean plant

The foliage feeders, especially defoliators that belong to Lepidoptera and Coleoptera include: the leaf folder, *Lamprosema indica* (F.); caterpillars of *Spodoptera exigua* (Hubner), *Anticarsia irrorata* (F.) the tobacco caterpillar, *Spodoptera litura* (F.), the hornworms, *Agris convolvuli* (L.) and *Acherontia styx* (Westwood); the Bihar hairy caterpillar, *Spilosoma obliqua* (Walker), the tussock caterpillars, *Euproctis fraterna* (Moore), *Dasychira mendosa*; the weevils, *Cyrtozemia dispar* Pascoe, *Myllocerus undecimpustulatus maculosus* Desbr., *Myllocerus discolor* Boheman, *Myllocerus viridanus* Boheman, *Episomus lacerata* Fabr.; the hada beetle, *Henosepilachna* spp., chrysomelid (leaf) beetles, *Monolepta signata*, and the grasshopper, *Attractomorpha crenulata crenulata*.

Among the sap feeding insects the more common are aphids, especially black bean aphids, *Aphis craccivora* Koch; jassids, *Empoasca kerri* Pruthi; white flies, *Bemisia tabaci* Gennadius, thrips belonging to genus *Megalurothrips* and *Caliothrips indicus* Bagnall; the plant bugs, *Riptortus* spp., *Nezara viridula* L., *Plautia fimbriata* (Fabricius) and the pod bug, *Clavigralla* spp. They cause significant damage to green gram foliage and pods; besides causing damage to other related legumes. It was observed that green gram cultivated in the vicinity of pigeon pea was heavily infested and rather preferred by the pigeon pea pod bug, *Clavigralla* spp. (Swaminathan, *et al.*, 2007). A linear relationship was observed between pod feeder infestation and seed loss, with the rate of seed loss being greater for *Riptortus linearis* and *Nezara viridula* than for *Maruca testulalis* (Hussain and Saharia, 1994).

The blister beetles (species of *Mylabris*) cause serious damage to the flowers, especially to the second and third flush during August - September months in most green gram cultivation areas in India.

The key pod borers include the lepidopteran caterpillars - the spotted pod borer, *Maruca testulalis* (Geyer) [*Maruca vitrata*] and the spiny pod borer, *Etiella zinckenella* Tretsche; however, the blue butterflies, *Lampides boeticus* Linnaeus and *Catechrysops cnejus* Fabricius; the gram caterpillar, *Helicoverpa armigera* (Hubner) have also been reported among the major pests.

The primary insect pests of stored green gram include species of bruchids belonging to the genus *Callosobruchus*. The annual yield loss is estimated to be 20 per cent in pigeonpea, 15 per cent in chickpea and 30 per cent in black gram and green gram. On an average 2.5 to 3.0 million tonnes of pulses are lost annually due to pests (Ali, 1998). Damage due to bruchids, *Callosobruchus chinensis* (L.) begins right from the field; adults emerging from the stored seeds lay eggs on healthy grains. The field infestation ranges from 7.8–9.9 per cent (Banto and Sanchez, 1972) and 100 per cent destruction of seeds occurred at 9.9 per cent field infestation.

3.2 Pest management strategies

3.2.1 Organic approach

Of late, use of various cultural practices and framscaping for the management of insect pests of green gram seems to gain importance. Adjusting the sowing dates, use of resistant varieties and growing inter or trap crops can be followed depending on the availability and effectiveness in a particular location. Since use of bio-control agents has not been successful in these crops although it is a viable alternative despite the record of several natural enemies in the field, their augmentation through farmscaping is a viable option.

Farmscaping is an ecological approach to pest management; comprising the use of hedgerows, insectary plants, cover crops, and water reservoirs to attract and support populations of beneficial organisms such as insects, bats, and birds of prey. Such mini-livestock requires adequate supplies of nectar, pollen, and herbivorous insects and mites as food to sustain and increase their populations. The best source of these foods is flowering plants. Flowering plants are particularly important to adults of the wasp and fly families, which require nectar and pollen sources in order to reproduce the immature larval stages that parasitize or prey on insect pests. However, using a *random* selection of flowering plants to increase the biodiversity of a farm may favor pest populations over beneficial organisms. It is important to identify those plants, planting situations, and management practices that best support populations of beneficial organisms.

There are many approaches to farmscaping: some farmers, after observing a cover crop harboring beneficial insects, plant strips of it in or around their crop fields. The advantages of this kind of approach are that it is simple to implement, is often very effective and the farmer can modify the system after observing the results. Problems arise when the beneficial insect habitat, without the knowledge of the cultivator, also harbors pest species. In other instances the beneficials may not exist in numbers sufficient to control pest populations, *especially during the time when pest populations generally increase*. Predator/prey population balances are influenced by the *timing* of availability of nectar, pollen and alternate prey/hosts for the beneficials; therefore, essentially efforts must be made to for have year-round beneficial organism habitat and food sources. The beneficial habitat season may be extended by adding plants that bloom sequentially throughout the growing season or the whole year (Rex Dufor, 2000).

The mechanisms by which insectary plantings can help natural enemies of crop pests and other beneficial arthropods are complex, and their effectiveness can vary greatly from site to site depending on the specific situation. For this reason, it is especially important that insectary plantings are planned and assessed on a case-by-case basis, and integrated into whole-farm plans for pest management and other farm operations. Insectary plantings that are well thought out can maximize the benefits to natural enemies and minimize the benefits to pest species (Pfiffner and Wyss 2004, Quarles and Grossman 2002).

The goal of farmscaping is to prevent pest populations from becoming economically damaging. This is accomplished primarily by providing habitat to beneficial organisms that increase ecological pressures against pest populations. Farmscaping requires a greater investment in knowledge, observation, and management skill than conventional pest management tactics, while returning multiple benefits to a farm's ecology and economy. However, farmscaping alone may not provide adequate pest control. It is important to monitor pest and beneficial populations so that quick action can be taken if beneficials are not able to keep pest populations in check. Measures such as maintaining healthy soils and rotating crops are complementary to farmscaping and should be integrated with farmscaping efforts. Bio-intensive Integrated Pest Management (IPM) measures, such as the release of commercially-reared beneficials (applied biological control) and the application of soft pesticides (soaps, oils, botanicals) can be used to augment farmscaping efforts.

In a case study on the impact of farmscaping in greengram on the major insect pests and their natural enemy complex at the College farm, Udaipur, India, a comparison of the seasonal mean abundance of the major foliage feeding and pod damaging insect pests showed a significant difference among the treatments. The Shanon Weiner diversity index was the maximum under green gram + marigold weeded and unweeded farmscape conditions being 0.7936 and 0.7790. The sole crop of green gram had the lowest diversity index of 0.6622 for weeded and 0.6863 for unweeded conditions. Comparisons made for the associated natural enemy complex in the different treatments showed that the farmscape treatment green gram + niger under unweeded conditions had the highest Shanon Weiner diversity index of 1.5932 followed by that for green gram + marigold under unweeded conditions with an index of 1.5716. Green gram sole crop had the lowest diversity indices being 1.2882 and 1.3854 under weeded and unweeded conditions, respectively. Niger, by virtue of being taller than green gram, acts as a physical barrier to blister beetle infestation on green gram floral parts. Some blister beetles may happen to alight on niger flowers and cause some damage, thereby safeguarding damage to green gram. Marigold is preferred by *Helicoverpa armigera* (Hubner) for laying eggs; thereby, the main crop of mung bean/green gram significantly escapes the pest infestation (Unpublished data - Swaminathan, 2011).

3.3 Cultural practices

Different cultural practices have been advocated from time to time; however, these traditional practices and those improved happen to vary from place to place and have responded in a varied manner. Intercropping green gram with cereals/millets (maize, sorghum and pearl millet) is often in vogue. Green gram is sown by keeping the row to row spacing at 30 cm and plant to plant distance at 10 cm. In the inter-cropped system, green gram and maize (in 1: 1 ratio) are sown in alternate rows at a distance of 30cm apart. In spring planted sugarcane, 1 or 2 rows of green gram can be planted in between the sugarcane rows. Intercropping of green gram can also be done in *Mentha*. Similarly, in the newly planted poplar crop and in horticultural plants or orchards (papaya, pomegranate) intercropping green gram is a viable option.

Weed-free crop of green gram harboured lower populations of major insect pests, while weedy crop was conducive to their population build-up. With respect to insect infestation, keeping the field weed-free throughout the crop period was equivalent to removal of weeds up to the vegetative-3 stage of the crop. The effectiveness of weeding however varied according to the pest species (Rekha Das Dutta, 1997).

Showler and Greenberg (2003) observed that the presence of weeds in cotton was associated with greater populations of 9 of the 11 prey arthropod groups, and 9 of the 13 natural enemy arthropod groups counted. These trends were mostly evident late in the season when weed biomass was greatest. Weed-free cotton harboured more cotton aphids (*Aphis gossypii*), early in the season and silver leaf whiteflies (*Bemisia argentifoli*) later in the season than weedy cotton on some of the sampling dates. Diversity (Shannon's index) within the selected arthropod groups counted was significantly greater in DVAC samples from the weed foliage than from weed-free cotton plants during both years, and diversity on weedy cotton plants was greater than on weed-free cotton plants during 2000. Boll weevil oviposition injury to squares was unaffected by weeds, but the higher weed-associated predator populations mainly occurred after most squares had become less vulnerable bolls. Weed competition resulted in lower lint yields of 89 and 32 per cent in 2 years.

Some of the more recent literature on the impact of intercropping on insect pest situation has been reviewed herein. Populations of *O. phaseoli* on *V. mungo* and *B. tabaci* on cowpea increased when these crops were intercropped with maize. The incidence of yellow mosaic was lower in intercrops of *V. radiata* with maize and sorghum than in monocultures. Conversely, pod borer damage to *V. radiata* was lower in monocultures than in intercrops. There was no significant difference in populations of *A. soccata* and *C. partellus* on pure and intercrops (Natarajan *et al.*, 1991). Rekha Das Dutta (1996) observed that intercropping *Vigna radiata* with maize resulted in reduced populations of the pests *viz.*, *Monolepta signata, Aphis craccivora, Nacolea vulgaris, Nezara viridula* and *Riptortus linearis* on *V. radiata* than when intercropped with other legumes like *Vigna umbellata* (rice bean), *Glycine max* (soybean), *Vigna mungo* (black gram) and *Arachis hypogea* (groundnut). Intercropping maize and sorghum along the periphery significantly reduced the whitefly (*Bemisia tabaci*) population and the damage caused by the pod borers (*Maruca testulalis* [*M. vitrata*] and *Lampides boeticus*). All intercrops resulted in increased yields over the sole crops; however, maize and sorghum intercropped along the periphery was more promising (Dar *et al.*, 2003).

Various forms of farmscaping in the form of permanent hedgerows or temporary insectary strips in vegetable fields to increase the activity of beneficial insects have shown that data on the effectiveness of these practices is sparse at best, as is information on the best plant species to use. The primary pest target is often aphids. The use of sweet alyssum (*Lobularia maritima*) provides long periods of flowering and fits into most grower operations, yet was chosen originally for its ability to attract and provide resources to hymenopteran aphid parasitoids. Now that the aphid species of concern has shifted from green peach aphid (*Myzus persicae*) to lettuce aphid, the natural enemy of greatest importance has also shifted to hoverfly (Diptera: Syrphidae) larvae (Chaney, 2004).

Higher numbers of arthropod pests were observed in onion plants 30 m from the marigold strip, while higher numbers of predators and parasitoids were found at 5 m distance. Species richness and Shannon's diversity index were higher at 5 m from marigold. Therefore, marigold rows next to onion fields resulted in higher number of entomophagous species, potentially enhancing the natural control of onion pests (Silveira et al., 2009).

Evaluating the suitability of some farmscaping plants as nectar sources for the parasitoid wasp, *Microplitis croceipes* (Hymenoptera: Braconidae), Nafziger and Fadamiro (2011) observed that the greatest longevity (~16 days) was recorded for honey-fed wasps (positive control). Buckwheat significantly increased the lifespan of female and male wasps by at least two-fold as compared to wasps provided water only (longevity=3-4 days). Licorice mint significantly increased female longevity and numerically increased male longevity. Sweet alyssum slightly increased longevity of both sexes though was not significantly different from the water only control. Females had a significantly longer longevity than males on all the diet treatments. The greatest carbohydrate nutrient levels (sugar content and glycogen) were recorded in honey-fed wasps followed by wasps fed buckwheat, whereas very little nutrients were detected in wasps provided sweet alyssum, licorice mint or water only. However, female wasps were observed to attempt to feed on all three flowering plant species.

Insect pest tolerant/resistant varieties have been evaluated often, but there is no single variety of green gram that might offer resistance to the major insect pests; however, some varieties are less preferred than others. Of 20 cultivars of green gram (*Vigna radiata*) screened in the field in Madhya Pradesh, India, for resistance to 8 species of insect pests, PDM-84-139 and ML-382 were promising against *Caliothrips indicus*, an unidentified chrysomelid and a galerucid beetle, BM-112 against *Raphidopalpa* sp. [*Aulacophora* sp.] and TAM-20, PDM-84-143 and Pusa-105 against *Aphis craccivora*, *Amrasca kerri* [*Empoasca kerri*] and *Myllocerus undecimpustulatus* (Devasthali and Joshi, 1994). Green gram cultivar, MV 1-6 was relatively less susceptible to both paddy grasshopper and cotton grey weevil. MI 7-21 was found to be promising against pea thrips, semilooper and cotton grey weevil but was most susceptible to paddy grasshopper. MI-131-(Ch) was less attacked by blue beetle. The variety MI-67-9 was less infested by bean aphid but was most susceptible to blue beetle. Infestation of jassid was comparatively less in varieties MI-67-3 and MI-29-22 (Devasthali and Saran, 1998).

3.4 Bio-pesticide use

Pesticides of biological origin offer good response in the management of some of the major insect pests of green gram. The fungus, *Nomuraea rileyi* (2 x 106 spores/ml) has been reported highly virulent under laboratory trials resulting in approximately 97.5, 93.33, 80.0 and 100.0 per cent mortality of *Thysanoplusia orichalcea*, *Spodoptera litura*, *Spilosoma obliqua* and *Helicoverpa armigera*, respectively (Ingle *et al.*, 2004). The fungus, *Paecilomyces lilacinus* (0.02%) caused higher reduction in the larval population of *Lampides boeticus*, followed by *Verticillium lecani* (0.02%) and *Beauveria bassiana* (0.02%). While comparing the *neem* products, *neem* (*Azadirachta indica*) oil (0.05%) was better than *neem* seed kernel extract (5%) in reducing the pod borer larval population (Arivudainambi and Chandar, 2009).

3.5 Integrated approach

Integrated management strategies involve the use of resistant varieties, use of disease free seeds, manipulation of cultural practices, management of vectors, and biological and chemical control methods (Raguchandar *et al.*, 1995; Vidhyasekaran and Muthamilan, 1995). In a 2-year study (2001 and 2002), the maximum yield of maize and green gram in the intercropped pattern (1: 1 ratio) and that as sole crop of green gram, as well as the maximum rupee equivalent yield value was recorded for the management schedule comprising release of the green lace wing, *Chrysoperla carnea* at 25 DAS, spray of *Azadirachta indica* oil at 40 DAS and a contact insecticide, endosulfan at 55 DAS (Kan Singh *et al.*, 2009). Earlier, Kan Singh (2002) observed that for every unit increase in the larval density there was a significant and subsequent decrease in the number of pods per plant. The linear relationship between larval density and the reduction in number of pods per plant caused by borer damage was positive and significant for both the years. The increased reduction in number of pods as a result of increased larval density was significant. Likewise, for every unit increase in larval density there was a significant reduction in number of grains, which was reflected in the losses caused. The estimated losses to grains were the maximum at a larval density of 4 per plant, 78.87 and 68.01 per cent for *kharif* 2001 and 2002, respectively. Obviously, the linear relationship between larval density and reduction in yield was significantly positive.

Insect Pest	Taxonomic Position Order/Family	Plant Part Damaged
Alcidodes collaris Pasc.	Coleoptera Cuculionidae	Stem
Amsacta albistriga W.	Lepidoptera Arctiidae	Foliage
Anarsia spp.	Lepidoptera Gelechiidae	Shoot webber feeding within
Anoplocnemis phasiana Fabr.	Hemiptera Coreidae	Pods
Anticarsia irrorata (F.)	Lepidoptera Noctuidae	Foliage
Aphis craccivora Koch.	Hemiptera Aphididae	Leaves
Apion ampulum Fst.	Coleoptera Apionidae	Buds, flowers and pods
Aspongopus janus Fabr.	Hemiptera Dinidoridae	Plant parts
Attractomorpha crenulata Fabr.	Orthoptera Pyrgomorphidae	Leaves
Callosobruchus spp.	Coleoptera Bruchidae	Pods
Catechrysops cnejus Fabricius	Lepidoptera Lycaenidae	Flowers and pods
Chrotogonus spp.	Orthoptera Pyrgomorphidae	Pods
Colemania sphenarioides B.	Orthoptera Pyrgomorphidae	Leaves
Coptosoma cribria Fabr.	Hemiptera Plataspidae	Plant parts, leaf axils, shoots
Cyrtozemia dispar Pascoe	Coleoptera Curculionidae	Foliage
Dolycoris indicus Stal.	Hemiptera Pentatomidae	Plant parts, pods
Empoasca kerri Pruthi *Empoasca* spp.	Hemiptera Jassidae	Leaves
Eucosma melanaula Meyr	Lepidoptera Gelechiidae	Flowers and pods
Helicoverpa armigera (Hubner)	Lepidoptera Noctuidae	Pods
Herse convolvuli L.	Lepidoptera Sphingidae	Foliage
Hyalospila leuconeurella Rag.	Lepidoptera Pyralidae	Seeds in pods

Insect Pest	Taxonomic Position Order/Family	Plant Part Damaged
Lampides boeticus (L.)	Lepidoptera Lycaenidae	Flowers and pods
Lamprosema indicata Fabr.	Lepidoptera Pyralidae	Tender foliage
Liogryllus bimaculatus DeGeer (= *Gryllus bimaculatus*)	Orthoptera Gryllidae	Pods
Maruca testulalis G. (= *Maruca vitrata*)	Lepidoptera Pyralidae	Flowers and pods
Melanagromyza phaseoli Coq.	Diptera Agromyzidae	Stem
Mylabris pustulata Thunberg	Coleoptera Meloidae	Buds and flowers
Myllocerus spp.	Coleoptera Cuculionidae	Foliage
Nacolea vulgalis Gn.	Lepidoptera Pyralidae	Tender foliage
Nezara viridula Linneaus	Hemiptera Pentatomidae	Plant parts, pods
Pachytychius mungosis Marsh.	Coleoptera Cuculionidae	Seeds
Piezodorus rubrofasciatus Fabr. *P. hybneri* Gmel	Hemiptera Pentatomidae	Plant parts, pods
Plautia fimbriata (Fabricius)	Hemiptera Pentatomidae	Plant parts, pods
Riptortus pedestris Fabr., *R. linearis*, *R. fuscus*	Hemiptera Coreidae	Pods
Spilosoma obliqua (W.)	Lepidoptera Arctiidae	Foliage
Spilostethus pandurus (Scopoli)	Hemiptera Lygaeidae	Shoots/leaves
Spodoptera litura (Fabr.)	Lepidoptera Noctuidae	Foliage
Thrips (*Megalurothrips*) *distans* Ky.,	Thysanoptera Thripidae	Tender leaves and flowers

Note: The spider mites also happen to be serious arthropod pests of green gram, foliage especially during the warmer months of the year.

Table 1. Record of insect pests that infest green gram

Plate 1 (a). Major insect pests of green gram at Udaipur (India)

Plate 1 (b). Major insect pests of green gram at Udaipur (India)

Plate 1 (c). Major insect pests of green gram at Udaipur (India)

Plate 2 (a). Farmscaping in green gram with marigold

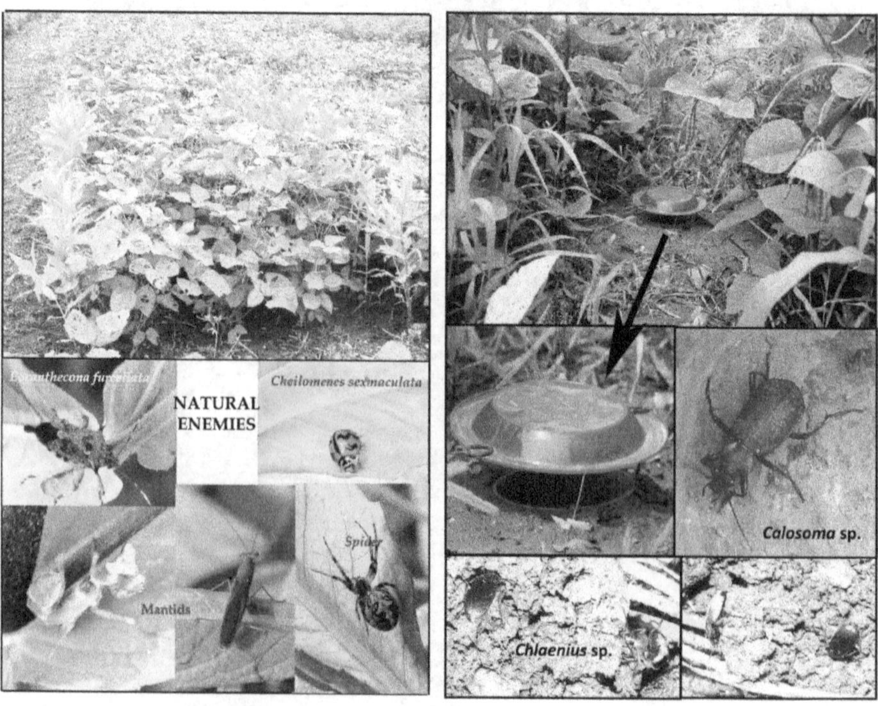

Plate 2 (b). Farmscaping in green gram with niger

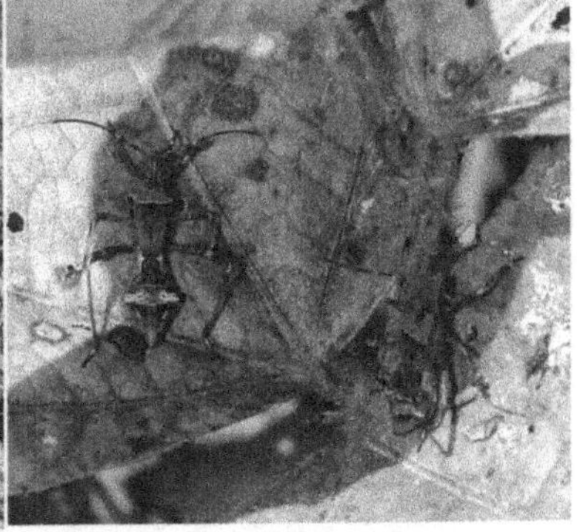

Infestation by *Riptortus pedestris* in sole green gram

Plate 2 (c). The Plate has two different aspects: (1) Benefits of farmscaping in green gram: Niger intercrop helps to escape blister beetle (*Mylabris* spp.) infestation in green gram. (2) Severe pest infestation in sole green gram: blister beetle (*Mylabris* spp.) and the coreid bug (*Riptortus pedestris*).

4. References

Acharya, S.S. 1985. *Prices and Price Policy for pulses and cereals*. Sukhadia University, Udaipur XVIII+439+A 135 pp.

Ali, M. 1998. Research, Development and Management for production of pulses. In: IPM system in Agriculture. Vol.4. Pulses (eds. R.K.Upadhyay, K.G. Mukerji and R.L.Rajak) Aditya Books Private Limited, New Delhi. pp 1-40.

Arivudainambi, S. and Chandar, A. V. 2009. Management of pulses blue butterfly, *Lampides boeticus* L. in green gram. *Karnataka Journal of Agricultural Sciences*, 22: 624-625.

Baldev, B., Ramanujam, S. and Jain, H. K. 2000. Pulse Crops. Oxford IBH Publishing Company Pvt. Ltd., New Delhi, pp. 229-258.

Banto,S.M. and F.F..Sanchez. 1972. The biology and chemical control of *Callasobruchus chinensis* (Linn) (Coleoptera:Bruchidae) *Philippines Entomologist*, 2: 167–182.

Chaney, W. E. 2004. Insectary plants for vegetable crops. *California Conference on Biological Control IV*, Berkeley, California, USA, 13-15 July, 2004. 53 - 54.

Dar, M. H. Rizvi, P. Q. and Naqvi, N. A. 2003. Effect of intercropping on the major insect pests of greengram and blackgram. *Shashpa*, 10: 85-87.

Devesthali , S. and Joshi, M. 1994. Infestation and varietal preference of insect pests in green gram. *Indian Agriculturist*, 38: 263-272.

Devesthali, S. Saran, R. K. 1998. Relative susceptibility of new cultivars of green gram (*Vigna radiata* L. Wilczek) to insect pests at Indore (M.P.). *Indian Agriculturist*, 42: 261-266.

Dixit, G. P. 2005. Project Coordinator's Report. Annual Group Meet (*kharif*, 2005), Indian Institute of Pulses Research, Kanpur, India.

Dufour, R. 2000. Farmscaping to enhance biological control. ATTRA, Fayetteville, AK Online: http://www.attra.ncat.org/attra-pub/PDF/farmscaping.pdf

Fowler, J. Cohen, L. and Jarvis, P. 1998. *Practical Statistics for Field Biology* (2nd Edition), John Wiley & Sons, Chichester, West Sussex, England.

Hammond, R.B. and Pedigo, L.P. 1982. Determination of yield loss relationship for two soybean defoliators by using simulated insect defoliation technique. *Journal of Economic Entomology*, 75: 102-107.

Handbook of Agriculture 2006. Indian Council of Agricultural Research, New Delhi.

Hussain, S. and Saharia, D. 1994. Linear model for predicting seed loss in green gram due to pod feeders. *Journal of the Agricultural Science Society of North East India*, 7: 98-99.

Hussain, S. and Saharia, D. 1994. Linear model for predicting seed loss in green gram due to pod feeders. *Journal of the Agricultural Science Society of North-East India*, 7: 98-99.

Ingle, Y. V. Aherkar, S. K. Lande, G. K. Burgoni, E. B. and Autkar, S. S. 2004. Natural epizootic of *Nomuraea rileyi* on lepidopterous pests of soybean and green gram. *Journal of Applied Zoological Researches*, 15: 160-162.

Johnston, R.L. and Bishop, G.W. 1987. Economic injury levels and economic threshold for cereal aphids on spring planted wheat. *Journal of Economic Entomology*, 80: 478-482.

Kan Singh 2003. Estimation of losses, management of insect pests of mung bean [*Vigna radiata* (L.) Wilczek] and determination of economic threshold Level against the lycaenid borer (*Lampides boeticus* Linnaeus), Ph. D. Thesis submitted to Maharana Pratap University of Agriculture and Technology, Udaipur, India.

Leclerg, E.L. 1971. Field experiments for assessment of crop losses. *In: Crop loss assessment methods FAO manual on the evaluation and prevention of losses by pests, diseases and weeds.* Edited by Chirappa, L. 2.1/1.2. 1/11.

Mogotsi, K.K., 2006. *Vigna radiata* (L.) R.Wilczek. [Internet] Record from Protabase. Brink, M. & Belay, G. (Editors). PROTA (Plant Resources of Tropical Africa/ Ressources végétales de l'Afrique tropicale), Wageningen, Netherlands. http://database.prota.org/search.htm.

Nafziger, T. D. and Jr. Fadamiro, H. Y. 2011. Suitability of some farmscaping plants as nectar sources for the parasitoid wasp, *Microplitis croceipes* (Hymenoptera: Braconidae): effects on longevity and body nutrients. *Biological Control*, 56: 225 - 229.

Natarajan, N. Rao, P. V. S. and Gopal, S. 1991. Effect of intercropping of pulses in cereals on the incidence of major pests. *Madras Agricultural Journal*, 78: 59-67.

Nene, Y. L. 2006. Indian pulses through the millennia. *Asian Agri-history* 10: 179-202.

Pfiffner, L., and Wyss. E. 2004. Use of sown wildflower strips to enhance natural enemies of agricultural pests, pp. 165-186. *In* G. M. Gurr, S. D. Wratten and M. A. Altieri [eds.], Ecological Engineering for Pest Management: Advances in Habitat Manipulation for Arthropods. CSIRO, Melbourne.

Quarles, W., and Grossman, J. 2002. Insectary plants, intercropping and biological control. *IPM Practitioner* 24: 1-11.

Raguchander, T. Rajappan, K. and Prabakar, K. 1995. Evaluation of tale based product of *Trichoderma viride* of the control of blackgram root rot. *Journal of Biological Control*, 9: 63-64.

Rekha Das Dutta, S. K. 1996. Effect of intercropping on infestation of insect pests of green gram. *Journal of the Agricultural Science Society of North East India*, 9: 220-223.

Rekha Das Dutta, S. K. 1997. Effect of weeding on infestation of insect pests of green gram. *Journal of the Agricultural Science Society of North East India*, 10: 136-139.

Showler, A. T. and Greenberg, S. M. 2003. Effects of weeds on selected arthropod herbivore and natural enemy populations and on cotton growth and yield. *Environmental Entomology*, 32: 39 - 50.

Silveira, L. C. P., Berti Filho, E., Pierre, L. S. R., Peres, F. S. C. and Louzada, J. N. C . 2009. Marigold (Tagetes erecta L.) as an attractive crop to natural enemies in onion fields. *Scientia Agricola*, 66: 780 - 787.

Singh, K. Sharma, U. S. Swaminathan, R. and Dashora, P. K. 2009. Management of insect pests of *Vigna radiata* (L.) Wilczek. *Applied Ecology and Environmental Research*, 7: 99-109.

Swaminathan, R., Hussain, T. and Bhati, K. K. 2007. Influence of crop diversity on host preference by major insect pests of *kharif* pulses. *Indian Journal of Applied Entomology* 21: 59-62.

Talekar, N. S. 1990. Agromyzid flies of food legumes in the tropics. Wiley Eastern Limited, New Delhi.

Vavilov, N.I. 1926. Studies on the origin of cultivated plants. *Bulletin of Applied Botany and Plant Breeding,* 16, No.2.

Vidhyasekaran, P. and Muthamilan, M. 1995. Development of formulation of *Pseudomonas fluorescens* for the control of chickpea wilt. *Plant Disease,* 79: 782-786.

10

Infrared Spectroscopy Applied to Identification and Detection of Microorganisms and Their Metabolites on Cereals (Corn, Wheat, and Barley)

Cécile Levasseur-Garcia

Université de Toulouse, Institut National Polytechnique de Toulouse, Ecole d'Ingénieurs de Purpan, Département Sciences Agronomiques et Agroalimentaires, Toulouse, France

1. Introduction

Over the last several years, mycotoxins, which are metabolites secreted by fungi, have been the subject of numerous studies. These eukaryotes play a major ecological role in the life cycle of plants. Indeed, for some fungi, the role of saprophyte places them at the heart of ecosystem dynamics [Alexopoulous et al., 1996].

Some 350 mold species produce a large range of secondary metabolites (over 300, of which ~30 are toxic) [Fremy et al., 2009] and represent a potential danger for animal and human health and cause significant losses for the cereals industry [Le Bars et al., 1996]. The effects of mold are well illustrated by decreases in crop and livestock yields, public health problems, or write-offs on the international cereal market [Le Bars, et al., 1996]. The United Nations Food and Agriculture Organization estimates annual global losses from mycotoxins at 1 billion tons of foodstuffs [Fao, 2001]. The primary organisms impacted by mycotoxins are plants. Currently, about 25% of agricultural crops worldwide are contaminated by these metabolites [Charmley et al., 2006].

In response to these significant economic and health risks, global non-tariff barriers (i.e., specific food-safety standards imposed on imported products) were erected to control commercial trade based on the mycotoxic quality of foodstuffs. These measures generate significant economic and material losses for countries that export contaminated foodstuffs, either because their cargo is refused or because of a reduction in prices. To limit these consequences, farmers and the food industry strive to reduce the presence of mycotoxins in their products. Therefore, producers and processors are searching for alternative analytical methods to determine, in a quick, simple, and inexpensive manner, the risk of their products containing fungi or mycotoxins. The use of infrared spectroscopy—a mature technology—to monitor foodstuffs could respond to this need.

In this chapter, we focus on mycotoxins found mainly in wheat, barley, and corn and that have been studied in the international literature; namely, deoxynivalenol, fumonisins, and aflatoxin B1.

2. Advantages of using infrared spectroscopy to manage fungal and mycotoxic risk for wheat, barley, and corn

Fungus can be detected by microbiological methods involving visual, microscopy, and microbial-cultural methods.

Conventional methods of mold detection are based on direct observation by eye or by microscope of thalli, contaminated foodstuffs, or microbial cultures. These methods are time consuming and require viable samples and a good deal of expertise. Counting methods are difficult to apply to fungi because, during their reproduction, a spore generates a mycelium that can in turn divide itself into tens of individuals. Furthermore, a fungal contamination may not be visible at the surface of grains [Hirano, et al., 1998, Pearson, et al., 2001].

Other methods are based on molecular biology or on the detection of antigens specific to given molds. Organisms, either dead or alive, can be detected by the polymerase chain reaction (PCR) by copying a large number of DNA sequences that are originally present in small quantities (with a multiplicative factor on the order of 10^9). By amplifying certain genes of toxigenic strains, PCR serves as a tool to determine the risk. Various researchers have tested PCR to detect *Fusarium* contamination in corn [Jurado et al., 2006; Jurado et al., 2005; Nicolaisen et al., 2009]. These methods are rapid, sensitive, and can be automated. They are good qualitative methods (e.g., good selectivity) but offer only average precision in quantitative terms (they are called "semiquantitative"). These techniques are thus very reliable, provided the fungal strain to be detected is known beforehand, and so are used as referential methods. With such methods, a grain is deemed of suitable microbiological quality if less than 10 000 germs of the storage flora per gram of grain are detected.

New approaches are based on detecting constituents and fungal metabolites. Such approaches exploit the fact that molds have specific characteristics that distinguish them from other eukaryotes. These characteristics include the regulation of certain enzymes, the synthesis of lysine amino acid by a particular metabolic route, extremely structural characteristics (e.g., the Golgi apparatus), and genetic characteristics (e.g., haploid). From among these attributes, two types of compounds can be used as indicators of a fungal contamination.

The secreted compounds are synthesized compounds such as soluble carbohydrates (e.g., disaccharide trehalose and polyhydric alcohols such as mannitol or arabitol) or products of the metabolization of complex carbons such as volatile aldehydes, alcohols, ketones, spores, primary metabolites, secondary metabolites (i.e., volatile compounds). The last item gives rise to the characteristic fungal odor and is often detected by an electronic nose. For nonvolatile compounds, other tools such as infrared spectroscopy seem better suited.

The structural compounds of mold can also be used for their detection. The main polysaccharides of the cell wall of mold are the α et β (1-3) glucans, as well as chitin. Ergosterol is a component of fungal cell membranes.

Chitin may absorb infrared light, making it useful for infrared spectroscopy [Nilsson, et al., 1994; Roberts et al., 1991]. The main inconvenience in using this component as an indicator of fungal contamination is that chitin is not limited to fungi; it is found in insects, diatoms, arachnids, nematodes, crustaceans, and several other living organisms [Muzzarelli, 1977]. In addition, it may take different forms, each of which requires a specific detection technique. Roberts et al. [Roberts, et al., 1991] estimates the quantity of mold on barley by detecting this molecule but also detects glucans by near-infrared spectroscopy.

Ergosterol, however, is more specific to mold. This molecule, which may still be called provitamin D2, is a C24-methylated sterol (and is part of the subgroup of organic compounds that are soluble in lipids) and is found in the cell membranes of yeasts and filamentous fungi. This molecule is not found in animal cells [Verscheure et al., 2002] and is in the minority among the sterols found in higher plants [Pitt et al., 1997] and insects [Weete, 1980]. Griffiths et al. [Griffiths et al., 2003] demonstrated that ergosterol is the primary sterol found in molds: it represents 95% of the total sterols, with the remaining 5% being ergosterol precursors from *Leptosphaeria maculans*. This specificity makes this molecule a potential tracer of fungal activity. It is generally agreed that the ergosterol content of grains must be less than a given threshold; the limit for corn is 8 µg/g.

3. Infrared spectroscopy to detect fungal and mycotoxic contamination of wheat, barley, and corn

3.1 Background and methods

The first application of infrared spectroscopy to detect microorganisms dates from the 1950s [Miguel Gomez et al., 2003]. In these applications, the spectrometers were calibrated depending on the method of dosing the fungi or mycotoxins. In the 1980s, Fraenkel *et al.* [Fraenkel et al., 1980] and Davies *et al.* [Davies et al., 1987] published their first works on the detection by infrared spectroscopy of fungal contamination (*Botrytis cinerea* and *Alternaria tenuissima*), but the application of this tool to detecting mold really grew in the 1990s. This growth was due to the fact the existing agronomic models required collecting a significant amount of data in the field, making this approach unsuitable for routine use. In addition, industry required nondestructive techniques to assess the health safety of crops. Therefore, several research teams used infrared spectroscopy to detect mold and mycotoxins on cereals, which could be done concomitantly with the quantification of other parameters such as protein content, humidity, etc.

One method proposed to determine the fungal or mycotoxin content is to quantify the total fungal biomass. Toward this end, ergosterol is used as a fungus marker [Castro et al., 2002; Saxena et al., 2001; Seitz et al., 1977; Seitz et al., 1979]. Very often, this type of study is coupled with a study of the mycotoxin content and fungal units (colony-forming units or CFU). Indeed, the quantity of fungi is not proportional to the quantity of mycotoxins; it is possible to have small quantities of fungi but large quantities of mycotoxins, and vice versa. Indeed, fungi may disappear after secreting its toxins, either because of the evolution of the mycoflora or because of the application of chemical treatments. In addition, certain strains are more toxic than others. Two conclusions exist from the work on this subject: some researchers find a correlation between the mycotoxin content, the ergosterol content, and/or the fungal units [Lamper et al., 2000; Le Bouquin et al., 2007; Miedaner et al., 2000; Seitz et al., 1977; Wanyoike et al., 2002; Zill et al., 1988] whereas the others find no correlation or cannot make categoric conclusions [Beyer et al., 2007; Diener et al., 1982; Gilbert et al., 2002; Nowicki 2007; Penteado Moretzsohn De Castro et al., 2002; Perkowski,et al., 1995].

Covering the last 20 years, we count over 20 articles dealing with the use of infrared spectroscopy (primarily near-infrared) to detect molds and mycotoxins in wheat, barley, and corn. Because some of the work in infrared spectroscopy deals both with the detection and the identification of mycotoxins, we separate the articles into three groups. Table 1 is for molds, Table 2 compiles the trials dealing with deoxynivalenol (DON), fumonisins (FUMs), and B1 aflatoxins (AF1). Finally, Table 3 contains articles in which the authors worked

Molds	Matrix	Infrared Specificities	Performances	Characteristic wavenumbers and wavelengths	Reference
F. verticillioides **and** ***A. flavus***	corn	FTIR-photoacoustic and FTIR-diffuse reflectance 400-4000 cm^{-1}		* Amide I [1650 cm^{-1}] and amide II [1550 cm^{-1}] = increase in protein or acetylated amino sugar content * Peak [3400 cm^{-1}]: shift = increase of NH absorptions with respect to OH absorptions * [2855 and 2925 cm^{-1}] = increase of lipid contents * [400 to 1500 cm^{-1}] = change in carbohydrate composition * Conclusion: the best for establishing a regression is to use the amide II peak	Greene, 1992
A. flavus	112 corn	FTIR-photoacoustic 4000-600 cm^{-1}	Classification between uninfected and infected corn: - training: 100% of correct classification - test: 94% of correct classification	On infected grains: CO_2 evolution [2366 cm^{-1}] Ester peak shift [1256 to 1240 cm^{-1}] COOH band elevation [3200 to 2200 cm^{-1}] CH_2 peak ratio [2853 cm^{-1}/2923 cm^{-1}] OH, NH_2 peak shift [3400 to 3360 cm^{-1}] Protein (amide II) increase [1550 cm^{-1}] NH_2, COOH peak shift [3400 to 3300 cm^{-1}] Carbonyl shoulder height [1725 cm^{-1}] FTIR-PAS measures the level of fungal infection, rather than the level of mycotoxins, because the mycotoxin concentrations of interest and concern are too low to be detected using infrared spectral data.	Gordon, 1998
A. flavus	20 corn	Transient infrared 2000-600 cm^{-1}	Correct classification rate between healthy and infected corn: 85% to 95%	* Decrease in absorption of carbonyl group = consumption of corn lipids by mold * Increase in absorption of amide II group and changes in absorption of C–O group = increase in fungal proteins and carbohydrates. This may also be the manifestation of the damages caused by molds: they expose the grain interior to the infrared, where the protein, lipid, etc. composition differs from that of the grain surface. * Conclusion: the damages suffered by corn seem more significant than the detection of the fungus itself.	Gordon, 1999
Identification of species ***Penicillium, Memnoniella, Fusarium***		FTIR microscopy 4000-600 cm^{-1}		* [1655 and 1546 cm^{-1}]: proteins: amide I and II bands * Shoulder [1750 cm^{-1}]: C=O of lipids * [1465 cm^{-1}]: CH2 of cell lipids * [1450 cm^{-1}]: CH3 of end ethyl groups of proteins * [1396 cm^{-1}]: C=0 of lipids * [1377 cm^{-1}]: CH of CH2 * [peaks at 1237 and 1082 cm^{-1}]: PO_2^- and phospholipids * [1064 cm^{-1}]: carbohydrates	Erukhimovitch, 2005
Aspergillus **and** ***Penicillium*****: identification of 10 different species**		FTIR 4000-500 cm^{-1}		Best results obtained using : [1765-1590 cm^{-1}] [1470-1275 cm^{-1}] [1170-1000 cm^{-1}] [930-715 cm^{-1}] corresponding to amide I and amide II, polysaccharides and microorganism fingerprints.	Fischer, 2006
Different species of fungi	corn	Reflectance 500-1700 nm	Correct classification rate between fungal infected or undamaged kernels: above 96.6%	Use of two wavelengths: 715 and 965nm => fungal damaged kernels are generally discolored, with modifications in oil content.	Pearson, 2006
F. graminearum, F. proliferatum, F. subglutinans, F. verticillioides	PDA	400–2500 nm	Correct classification rate awaiting external validation: 98.8%	Significant spectral differences : [500–950 nm], [1500–1850 nm], and [2100–2300 nm] (between *F. subglutinans*, and *F. proliferatum* and *F. verticillioides*), and [400–800 nm] and [900–1400 nm] (between *F. graminearum* and the 3 others)	Levasseur, 2010

Table 1. Infrared spectroscopy applied to identification of mold species and gena (FTIR spectroscopy =Fourier transform infrared spectroscopy).

Mycotoxins	Matrix /number of samples	Spectral Range	Content Range	Performance	Characteristic wavelength	Reference
Deoxyni-valenol	188 barley	NIR 400–2500 nm	0.3 to 50.8 ppm	R^2 between 0.933 and 0.805 SEP between 3.097 and 5.461 ppm		Ruan, 2002
	14 corn	FTIR with attenuated total reflection (ATR) 4500-650 cm^{-1}		Correct classification rate between blank and contaminated samples: 79 to 100%	* Choice of spectral area: 1800–800 cm^{-1} * Sieving the samples increases the percentage of good classification	Kos, 2007
	340 wheat	NIR	70–11000 ppb	$0.33 < R^2 < 0.801$ 421 ppb<SECV<1900 ppb		Roumet, 2007
	197 wheat (durum + common)	FTIR-NIR 10000-4000 cm^{-1}	0-3000 µg/kg	* Correct classification rate between blank and contaminated wheat (limit at 300 µg/kg DON): 69% * Quantification Training: R^2=0.71 and SEC: 386 µg/kg		De Girolamo, 2009
	wheat	diffuse reflectance 350–2500 nm	0-100.75 mg/kg	SECV=3.61 mg/kg R^2=0.84	* [1400–1900nm]: highest differences between sound and damaged kernels: changes of kernel composition and pigmentation * The SEC is too high in comparison with the EU limit (1.25 mg/kg)	Beyer, 2010
Fumonisin	corn (330 kernels)	NIR transmittance 500–1050nm and NIR reflectance 400-1700 nm	0.5 – 610 ppm	Correct classification rate (distinction between corn containing high *>100 ppm* and low *<10 ppm* levels of fumonisin, classified as positive or negative): from 80 to 91% (use of entire spectra or ratios of two or four wavelengths)	* Models are relying on the damages done by the fungus to the kernel (changes in color, vitreousness, protein structure, or oil content) as fumonisin present at the ppm level does not absorb detectable amounts of NIR energy. * Transmittance: 650 nm, 710 nm, 935 nm, 990 nm * Reflectance: 590 nm, 995 nm, 1200 nm, 1410 nm	Dowell, 2002
Aflatoxin	corn (500 kernels)	NIR transmittance 500–950nm and NIR reflectance 550-1700 nm	1 to >1000 ppb	Correct classification rate (distinction between corn containing high *>100 ppb* and low *<10 ppb* levels of aflatoxins, classified as positive or negative): >95%	* The differences in spectra of contaminated and uncontaminated kernels can be explained by the damages of fungi in the kernel. * The use of ratio of 2 absorbencies gives better results than the use of full spectra - Transmittance: 720/780 nm or 710/760 and 615/645 nm - Reflectance: 735/1005 nm or 1075/1135 nm and 880/1075 nm	Pearson, 2001
	76 corn 76 barley	NIR grating 400–2500 nm FT-NIR: 9000-4000 cm^{-1}	0–50 ppb	* Correct classification rate for samples classified as positive (aflatoxin B1>20 ppb) and negative (aflatoxin B1 <20ppb) : 100% for corn and barley * Quantification - Corn # NIR grating: R^2=0.80 and SECV=0.211 ppb # FTNIR: R^2=0.82 and SECV=0.2 ppb - Barley # NIR grating: R^2=0.85 and SECV=0.176 ppb # FTNIR R^2=0.84 and SECV=0.183 ppb	* Indirect detection of aflatoxins because of excessively small concentrations. Specific bands can be related to fungal infection and fungal cellular compounds : - [480–600nm]: changes in color of fungal infected grains - [870–1200nm], [1750–1800nm] and [2020-2190nm]: deteriorative alterations of kernels	Fernandez-Ibanez, 2009

Table 2. Infrared spectroscopy applied to quantification of levels of deoxynivalenol (DON), fumonisins (FUM), and aflatoxins in cereals (NIR=Near Infrared Spectroscopy / FTIR=Fourier-transform infrared spectroscopy)

Mycotoxin or mold	Matrix /number of samples	Spectral Range	Content Range	Performance	Characteristic wavelength	Reference
Deoxynivalenol and ergosterol	wheat (114 for DON and 46 for ergosterol)	NIR 400-1700 nm	deoxynivalenol: 0-789 ppm ergosterol: 0-1497 ppm	* deoxynivalenol (>5 ppm) R^2=0.64 and SEC=44 ppm R^2=0.66 and SEP=52 ppm * ergosterol (>50 ppm) R^2=0.64 and SEC=108 ppm	* [750, 950, and 1400 nm]: OH * [1200, 1400, and 1650 nm]: CH * [1050 and 1500 nm]: NH	Dowell, 1999
***Fusarium*, ergosterol and deoxynivalenol**	52 corn	FTIR-ATR 4000-650 cm^{-1}	ergosterol : 0.79-947 mg/kg deoxynivalenol: 0.13-2.59 mg/kg	Discrimination of blank and contaminated samples : at least 75% correct classification when concentrations of ergosterol and deoxynivalenol are respectively greater than 8.23 mg/kg and 0.13 mg/kg		Kos, 2002
***F. graminearum*, ergosterol and deoxynivalenol**	14 corn	FTIR-ATR 4500-650 cm^{-1}	ergosterol : 880-3600 µg/kg deoxynivalenol: 310-2596 µg/kg	* Classification : up to 100% * Quantification : - deoxynivalenol : R^2 = 0.66 and SECV: 494.5 µg/kg - ergosterol :R^2 = 0.60 and SECV = 833.7 µg/kg	* Choice of wavelength: 1800–800 cm^{-1} * Changes in protein, carbohydrate and lipid contents	Kos, 2003
Deoxynivalenol and *Fusarium*	50 wheat	NIR 570–1100 nm		R^2=0.98 and SECV = 404 µg/kg	* Some of these factors may come from deoxynivalenol and some from other compounds or effects related to deoxynivalenol and produced by *F. culmorum*	Pettersson, 2003
***Fusarium*, deoxynivalenol and ergosterol**	21 corn	FTIR-ATR and mid-infrared diffuse reflection	ergosterol : 880-3600 µg/kg deoxynivalenol : 310-2600 µg/kg	* Correct classification rate between blank and contaminated samples : - ATR : 100% - Diffuse reflection: 79% * Quantification - deoxynivalenol-ATR: R^2 = 0.94 and SEC = 178.8 µg/kg R^2 = 0.65 and SECV = 438.7 µg/kg - deoxynivalenol-diffuse reflection: R^2 = 0.87 and SEC = 297.3 µg/kg R^2 = 0.51 and SECV = 597.5 µg/kg - ergosterol-ATR : R^2=0.80 and SEC = 593 µg/kg R^2=0.60 and SECV = 864.7 µg/kg - ergosterol- diffuse reflection : R^2=0.72 and SEC = 708.4 µg/kg R^2= 0.36 and SECV = 1102 µg/kg	* Choice of wavelength: 1800–800 cm^{-1}	Kos, 2004
Ergosterol, fumonisin B1, *F. verticillioides*	corn fungi infection (220); *F. verticillioides* (217); ergosterol (160); fumonisin B1 (180)	NIR 400–2500 nm	fungal infection: 28-100% *F. verticillioides*: 2-100% ergosterol: 0.78-41.52 mg/kg fumonisin B1: 0.01-19.6 mg/kg	* fungal infection: R^2=0.80, SEP=6.34 % * *F. verticillioides* R^2=0.78, SEP=9.64% * ergosterol: R^2=0.81, SEP=1.74 mg/kg * fumonisin B1: R^2=0.78, SEP=1.33 mg/kg	* Berardo reports the spectroscopic/chemical relationship for detection of fungal contamination, ergosterol and *F. verticillioides*.	Berardo, 2005
***Fusarium* and deoxynivalenol**	wheat 4800 grains	410–865 nm and 1032–1674 nm		* Correct classification rates of damaged kernels: - only VIS: 94% - only NIR: 97% - NIR+VIS: 86%	* Selected pairs of wavelengths: - only VIS: 500 and 550 nm - only NIR: 1152 and 1248 nm - NIR+VIS: 750 and 1476 nm	Delwiche, 2005
Deoxynivalenol and fumonisin	539 corn for deoxynivalenol 198 corn for fumonisins	1100–2500 nm	deoxynivalenol: 1.04–8.68 ln mg/kg fumonisin: 0.14–6.43 ln mg/kg	* Deoxynivalenol: R^2=0.90 and SEC=0.44 ln mg/kg R^2=0.88 and SECV=0.50 ln mg/kg * Fumonisin: R^2=0.68 and SEC=0.80 ln mg/kg R^2=0.46 and SECV=1.04 ln mg/kg	* Better performances for deoxynivalenol than for fumonisin * Due to the fact that mycotoxins are present in a very low concentration, the potential of NIRS to predict mycotoxins might be based on indirect associations of disease symptoms and spectra.	Bolduan, 2009

Table 3. Infrared spectroscopy applied to fungal and mycotoxic quantification in cereals (NIR=Near Infrared Spectroscopy / FTIR=Fourier-transform infrared spectroscopy / ATR=Attenuated Total Reflection)

simultaneously on the fungal and mycotoxic aspects. Each table lists the matrix studied (wheat, barley, corn), the apparatus, the content ranges, the performance of the models, the principle conclusions, and the characteristic wavelengths.

3.2 Principal conclusions

3.2.1 Identification and quantification of fungi

For identification of fungi, the performances, given in terms of percentage of correct classification, are very satisfactory because they exceed 77%. Each study identifies the peaks or spectral zones related to the growth of fungi or to the damage inflicted on the grains by fungi.

Moreover, the performance of the quantification of ergosterol always gives enticing results.

3.2.2 Quantification of mycotoxins deoxynivalenol, fumonisins, and aflatoxins

Mycotoxins are present in quantities too small (in the order of parts per million) for direct detection. Their detection is thus associated with a complex ensemble of information related to the growth of the fungus on the cereal; notably with modifications of the protein or carbohydrate level (starch, cellulose, etc.).

Regarding the capabilities of infrared spectroscopy to quantify mycotoxins, the conclusions differ from one author to another. In general, when dealing with deoxynivalenol, the performance is higher than when dealing with fumonisins. Yet despite this, even if the quantification of mycotoxins appears possible, it is not sufficiently precise to be used in the field. Indeed, the standard error of prediction (SEP) is too large with respect to the regulatory limits—notably European limits. This could be explained by a magnification of the non-negligible standard errors of the chemical benchmarks from which they are developed. Moreover, to work under conditions of realistic of toxin levels, the main avenues for improvement of these studies may be the following:

- Increasing the number of samples;
- Increasing the annual variability (samples often come from a single harvest);
- Allowing a natural contamination of the grains (artificial contamination does not account for all the natural parameters of contamination, notably so for multicontaminations);
- Accepting a range of mycotoxin levels more adapted to the reality in the field (the ranges are often very narrow);
- Acquiring spectra not grain-by-grain, but from entire lots of grain. Indeed, it is more difficult to assign a global mycotoxin level to a lot, because the distribution of mycotoxins is very heterogeneous, as is the case for the molds that synthesize the mycotoxins;
- Using a set of test samples (the proposed performances are often cross validated and are thus better than they would be for an external test);
- Displaying the ratio of standard error of prediction to sample standard deviation (RPD).

Thus, instead of quantification, several studies propose a classification of cereals samples as a function of the mycotoxin level. This qualitative approach works better and, at least until the quantitative models are improved, seems the most conclusive for applications in realistic conditions. Note also that, even if the SECs (Standard Error of Calibration), SECVs (Standard Error of Cross Validation), and SEPs (Standard Error of Prediction), are improved for the quantifications, these models are developed based on chemical benchmarks that themselves have non-negligible standard errors.

4. Conclusion

Infrared spectroscopy offers multiple advantages. This tool requires no preparation and does not use toxic products. It enables nondestructive analysis of the samples. Moreover, it is a rapid, low-cost technique that can be used online. The sample cell is resistant and inexpensive (glass or quartz). Finally, infrared spectroscopy enables multiparametric analyses and a large range of robust devices are available.

Models to classify cereals samples as a function of the type of fungus present or its level, or the presence of mycotoxins, represent an attractive tool to determine the fungal and mycotoxic risk in the food industry; for example, in the field or in the silo. Regarding the quantification of mycotoxins, chemometrics, which is a field in constant progression, may one day deliver performances that meet agricultural and industrial needs.

5. References

Alexopoulous CJ, Mims CW, Blackwell M. 1996. Introductory Mycology. Chichester: John Wiley & Sons, Inc.

Berardo N, Pisacane V, Battilani P, Scandolara A, Pietri A, Marocco A. 2005. Rapid detection of kernel rots and mycotoxins in maize by near-infrared reflectance spectroscopy. Journal of Agricultural and Food Chemistry 53 (21): 8128-8134.

Beyer M, Klix MB, Verreet JA. 2007. Estimating mycotoxin contents of Fusarium-damaged winter wheat kernels. International Journal of Food Microbiology 119 (3): 153-158.

Beyer M, Pogoda F, Ronellenfitsch FK, Hoffmann L, Udelhoven T. 2010. Estimating deoxynivalenol contents of wheat samples containing different levels of Fusarium-damaged kernels by diffuse reflectance spectrometry and partial least square regression. International Journal of Food Microbiology 142 (3): 370-374.

Bolduan C, Montes JM, Dhillon BS, Mirdita V, Melchinger AE. 2009. Determination of Mycotoxin Concentration by ELISA and Near-Infrared Spectroscopy in Fusarium-Inoculated Maize. Cereal Research Communications 37 (4): 521-529.

Castro MFPMd, Bragagnolo N, Valentini SRdT. 2002. The relationship between fungi growth and aflatoxin production with ergosterol content of corn grains. Brazilian Journal of Microbiology (33): 22-26.

Charmley LL, Trenholm HL, Prelusky DB, Rosenberg A. 2006. Economic losses and decontamination. Natural Toxins 3 (4): 199-203.

Davies AMC, Dennis C, Grant A. 1987. Screening of Tomato Purée for Excessive Mould Content by Near-Infrared Spectroscopy : a Preliminary Evaluation. Journal of Science and Food Agriculture (39): 349-355.

De Girolamo A, Lippolis V, Nordkvist E, Visconti A. 2009. Rapid and non-invasive analysis of deoxynivalenol in durum and common wheat by Fourier-Transform Near Infrared (FT-NIR) spectroscopy. Food Additives and Contaminants Part a-Chemistry Analysis Control Exposure & Risk Assessment 26 (6): 907-917.

Delwiche SR, Gaines CS. 2005. Wavelength selection for monochromatic and bichromatic sorting of fusarium-damaged wheat. Applied Engineering in Agriculture 21 (4): 681-688.

Diener UL, Davis ND. 1982. Aflatoxins in corn. Abstracts of Papers of the American Chemical Society 184 (SEP): 86.

Dowell FE, Pearson TC, Maghirang EB, Xie F, Wicklow DT. 2002. Reflectance and Transmittance Spectroscopy Applied to Detecting Fumonisin in Single Corn Kernels Infected with Fusarium verticillioides. Cereal Chemistry 79 (2): 222-226.

Dowell FE, Ram MS, Seitz LM. 1999. Predicting scab, vomitoxin, and ergosterol in single wheat kernels using near-infrared spectroscopy. Cereal Chemistry 76 (4): 573-576.

Erukhimovitch V, Pavlov V, Talyshinsky M, Souprun Y, Huleihel M. 2005. FTIR microscopy as a method for identification of bacterial and fungal infections. Journal of Pharmaceutical and Biomedical Analysis 37 (5): 1105-1108.

FAO. 2001. Safety Evaluation of Certain Mycotoxins in Food. Document préparé par le Comité mixte FAO/OMS d'experts des additifs alimentaires (JECFA) à sa cinquante-sixième réunion. Étude FAO: alimentation et nutrition no 74, Organisation des Nations Unies pour l'alimentation et l'agriculture, Rome, Italie.

Fernandez-Ibanez V, Soldado A, Martinez-Fernandez A, de la Roza-Delgado B. 2009. Application of near infrared spectroscopy for rapid detection of aflatoxin B1 in maize and barley as analytical quality assessment. Food Chemistry 113 (2): 629-634.

Fischer G, Braun S, Thissen R, Dott W. 2006. FT-IR spectroscopy as a tool for rapid identification and intra-species characterization of airborne filamentous fungi. Journal of Microbiological Methods 64 (1): 63-77.

Fremy JM, AFSSA- Agence Française de Sécurité Sanitaire des Aliments. 2009. Evaluation des risques liés à la présence de mycotoxines dans les chaînes alimentaires humaine et animale. Rapport final de l'Agence française de sécurité sanitaire des aliments (Afssa), mars 2009, 308 pp.

Gilbert J, Abramson D, McCallum B, Clear R. 2002. Comparison of Canadian Fusarium graminearum isolates for aggressiveness, vegetative compatibility, and production of ergosterol and mycotoxins. Mycopathologia 153 (4): 209-215.

Gordon SH, Jones RW, McClelland JF, Wicklow DT, Greene RV. 1999. Transient infrared spectroscopy for detection of toxigenic fungi in corn: potential for on-line evaluation. Journal of Agricultural and Food Chemistry 47 (12): 5267-72.

Gordon SH, Wheeler BC, Schudy RB, Wicklow DT, Greene RV. 1998. Neural network pattern recognition of photoacoustic FTIR spectra and knowledge-based techniques for detection of mycotoxigenic fungi in food grains. J. Food Prot. 61 (2): 221-230.

Greene RV, Gordon SH, Jackson MA, Bennett GA, McClelland JF, Jones RW. 1992. Detection of Fungal Contamination in Corn: Potential of FTIR-PAS and -DRS. Journal of Agricultural and Food Chemistry 40 (7): 1144-1149.

Griffiths KM, Bacic A, Howlett BJ. 2003. Sterol composition of mycelia of the plant pathogenic ascomycete Leptosphaeria maculans. Phytochemistry 62 (2): 147-153.

Hirano S, Okawara N, Narazaki S. 1998. Near infra red detection of internally moldy nuts. Bioscience, Biotechnology, and Biochemistry 62 (1): 102-107.

Jurado M, Vazquez C, Marin S, Sanchis V, Gonzalez-Jaen MT. 2006. PCR-based strategy to detect contamination with mycotoxigenic Fusarium species in maize. Systematic and Applied Microbiology 29 (8): 681-689.

Jurado M, Vazquez C, Patino B, Gonzalez-Jaen MT. 2005. PCR detection assays for the trichothecene-producing species Fusarium graminearum, Fusarium culmorum, Fusarium poae, Fusarium equiseti and Fusarium sporotrichioides. Systematic and Applied Microbiology 28 (6): 562-568.

Kos G, Krska R, Lohninger H, Griffiths PR. 2004. A comparative study of mid-infrared diffuse reflection (DR) and attenuated total reflection (ATR) spectroscopy for the detection of fungal infection on RWA2-corn. Anal. Bioanal. Chem. 378 (1): 159-66.

Kos G, Lohninger H, Krska R. 2002. Fourier transform mid-infrared spectroscopy with attenuated total reflection (FT-IR/ATR) as a tool for the detection of Fusarium fungi on maize. Vib. Spectrosc. 29 (1-2): 115-119.

Kos G, Lohninger H, Krska R. 2003. Development of a method for the determination of Fusarium fungi on corn using mid-infrared spectroscopy with attenuated total reflection and chemometrics. Anal. Chem. 75 (5): 1211-1217.

Kos G, Lohninger H, Mizaikoff B, Krska R. 2007. Optimisation of a sample preparation procedure for the screening of fungal infection and assessment of deoxynivalenol content in maize using mid-infrared attenuated total reflection spectroscopy. Food Addititives & Contaminants 24 (7): 721-729.

Lamper C, Teren J, Bartok T, Komorocsy R, Mesterhazy A, Sagi F. 2000. Predicting DON contamination in Fusarium-infected wheat grains via determination of the ergosterol content. Cereal Research Communications 28 (3): 337-344.

Le Bars J, Le Bars P. 1996. Recent acute and subacute mycotoxicoses recognized in France. Veterinary Research (27): 383-394.

Levasseur C, Pinson-Gadais L, Kleiber D, Surel O. 2010. Near Infrared Spectroscopy used as a support to the diagnostic of Fusarium species. Revue De Médecine Véterinaire 161 (10): 438-444.

Miedaner T, Reinbrecht C, Schilling AG. 2000. Association among aggressiveness, fungal colonization, and mycotoxin production of 26 isolates of Fusarium graminearum in winter rye head blight. Journal of plant diseases and protection 107 (2): 124-134.

Miguel Gomez MA, Bratos Perez MA, Martin Gil FJ, Duenas Diez A, Martin Rodriguez JF, Gutierrez Rodriguez P, Orduna Domingo A, Rodriguez Torres A. 2003. Identification of species of Brucella using fourier transform infrared spectroscopy. Journal of Microbiological Methods 55 (1): 121-131.

Muzzarelli RAA. 1977. Chitin. N.Y. London: Pergamon Press

Nicolaisen M, Supronien S, Nielsen LK, Lazzaro I, Spliid NH, Justesen AF. 2009. Real-time PCR for quantification of eleven individual Fusarium species in cereals. Journal of Microbiological Methods 76 (3): 234-240.

Nilsson M, Elmqvist T, Carlsson U. 1994. Use of near-infrared reflectance spectrometry and multivariate data analysis to detect anther smut disease (Microbotryum violaceum) in Silene dioica. Phytopathology 84 (7): 764-770.

Nowicki T. CCF/CWFHB : Session 2 - Toxicologie, qualité du grain et impact sur l'industrie. http://sci.agr.ca.ecorc/fusarium01/session2b_f.htm[Internet]. 19/07/2009.

Pearson TC, Wicklow DT. 2006. Detection of corn kernels infected by fungi. Transactions of the Asabe 49 (4): 1235-1245.

Pearson TC, Wicklow DT, Maghirang EB, Xie F, Dowell FE. 2001. Detecting aflatoxin in single corn kernels by transmittance and reflectance spectroscopy. Transactions of the Asae 44 (5): 1247-1254.

Penteado Moretzsohn de Castro MF, Bragagnolo N, de Toledo Valentini SR. 2002. The relationship between fungi growth and aflatoxin production with ergosterol content of corn grains. Brazilian Journal of Microbiology 33 (1): 22-26.

Perkowski J, Miedaner T, Geiger HH, Muller HM, Chelkowski J. 1995. Occurence of deoxynivalenol (DON), 3-acetyl-DON, zearalenone, and ergosterol in winter rye inoculated with Fusarium culmorum. Cereal Chemistry 72 (2): 205-209.

Pettersson H, Aberg L. 2003. Near infrared spectroscopy for determination of mycotoxins in cereals. Food Control 14 (4): 229-232.

Pitt JI, Hocking AD. 1997. Fungi and food spoilage. London: Blackie Academic & Professional

Roberts CA, Marquardt RR, Frohlich AA, McGraw RL, Rotter RG, Henning JC. 1991. Chemical and spectral quantification of mold in contaminated barley. Cereal chemistry 68 (3): 272-275.

Ruan R, Li Y, Lin X, Chen P. 2002. Non-destructive determination of deoxynivalenol levels in barley using near-infrared spectroscopy. Applied Engineering in Agriculture 18 (5): 549-553.

Saxena J, Munimbazi C, Bullerman LB. 2001. Relationship of mould count, ergosterol and ochratoxin A production. International Journal of Food Microbiology 71 (1): 29-34.

Seitz LM, Mohr HE, Burroughs R, Sauer DB. 1977. Ergosterol as an indicator of fungal invasion in grains. Cereal Chemistry 54 (6): 1207-1217.

Seitz LM, Sauer DB, Burroughs HE, Mohr HE, Hubbard JD. 1979. Ergosterol as a Measure of Fungal Growth. Phytopathology 69 (11): 1202-1203.

Verscheure M, Lognay G, Marlier M. 2002. Revue bibliographique : les méthodes chimiques d'identification et de classification des champignons. Biotechnology, Agronomy, Society and Environment 6 (3): 131-142.

Wanyoike MW, Walker F, Buchenauer H. 2002. Relationship between virulence, fungal biomass and mycotoxin production by Fusarium graminearum in winter wheat head blight. Journal of plant diseases and protection 109 (6): 589-600.

Weete JD. 1980. Lipid Biochemistry of Fungi and Other Organisms. New York and London: Plenum Press

Zill G, Engelhardt G, Wallnofer PR. 1988. Determination of ergosterol as a measure of fungal growth using SI-60 HPLC. Zeitschrift Fur Lebensmittel-Untersuchung Und-Forschung 187 (3): 246-249.

Section 6

Agriculture and Human Health

11

The Agricultural Landscape for Recreation

Erik Skärbäck[1,*], John Wadbro[1], Jonas Björk[2],
Kim de Jong[2], Maria Albin[2], Jonas Ardö[2] and Patrik Grahn[1]
[1]*SLU Alnarp*
[2]*LU Lund*
Sweden

1. Introduction

Food production is not the only use of our agricultural landscape. The landscape also fulfils basic human needs for recreation outside the urban fringe. This chapter describes a study of how certain qualities in the rural and semi-urban landscape correspond to well-being. The study covers Skåne, the southernmost region in Sweden as well as the most productive agricultural region. Large datasets on the environment - land use, land cover, environmental qualities, impacts, etc. - were associated with results from a major public health survey. The results were published in the Journal of Epidemiology and Community Health (Björk et al., 2008). The chapter explains in more detail the method used to assess environmental qualities using Geographical Information System (GIS), and the relevance of these qualities for well-being and health promotion.

2. Background

A number of interview studies conducted between 1995 and 2005 in landscape architecture/environmental psychology (Grahn, Stigsdotter and Berggren-Bärring, 2005) have revealed eight characteristics of the outdoor environment (Serene, Wild, Lush, Spacious, the Common, the Pleasure garden, Festive/centre and Culture) that correspond to basic human needs. The line of research in which the eight characteristics were discovered has existed for decades. Already in 1989, Kaplan and Kaplan pointed out that sounds of nature can reduce stress and improve well-being. Several investigations have shown that people are often afflicted by illnesses related to stress (Grahn and Stigsdotter, 2003; Ottosson and Grahn, 2005a; Ottosson and Grahn, 2005b; Ottosson and Grahn P. 2008).

Many studies have shown a relationship between urban green areas and health (Hartig et al., 1996; Ottosson and Grahn, 1998). When walking in a natural environment, people's blood pressure drops already after a few minutes (Hartig, 1993). Certain biotopes and habitats seem to have been of great importance during human evolution (Coss, 1991; Ulrich, 1993). When people are stressed or ailing and in pressed situations, the availability of such environments seems to be even more important. If people can visit environments with

---

*Corresponding author

certain characteristics, their blood pressure, pulse, etc., can return to normal more quickly (Ottosson and Grahn, 1998).

That the landscape promotes well-being is a common feeling among most people, but more specific knowledge about different landscape qualities has been lacking, as have evidence from epidemiological research and hard facts that can be used in social economic calculations.

An epidemiological study was enabled by merging data from a large regional public health survey with regional GIS data on landscape, land-use, nature and cultural preservation, etc.

3. Materials and methods

3.1 Public health survey

The public health survey was distributed in 2004 as a postal questionnaire in Skåne, the southernmost region of Sweden as well as the most productive agricultural region. The study population consisted of a total of 855,599 individuals. The population was stratified by gender and geographical area. Samples were randomly selected from the population registry. In total 50,000 questionnaires were sent out to individuals 18-80 years of age who had geocoded residential addresses. The participation rate was 59%. Survey questions posed to respondents included topics such as neighbourhood satisfaction, time spent on moderate physical activity per week, body mass index (BMI), self-rated physical and psychological health at present, and a 36-item short-form health survey item called "vitality". Neighbourhood satisfaction was measured in the survey by the question 'How much do you like the environment you live in?' Participants scored their level of satisfaction on a four-point ordinal scale with an additional 'don't know' option. For individuals using that answer, neighbourhood satisfaction was unknown and was therefore excluded in the analyses that use neighbourhood satisfaction as outcome vatiable. Blank answers to this question, or to any of the other questions that represent outcome variables (i.e. physical activity, length and weight used to calculate BMI and self-rated health), were also excluded in corresponding analyses and for the same reason (Björk et al., 2008).

The questionnaire was not pretested for clarity, but parts of the survey had been used in the year 2000 in the same region. A steering committee, with representatives from the county, municipalities and the research community, drafted the survey questions carefully before launching the questionnaire. Validated questions were used when available.

3.2 Environmental qualities for well-being

A number of studies in the fields of environmental psychology and landscape architecture at SLU, the Swedish University of Agricultural Sciences, have resulted in new assessment criteria for recreational values. This research revealed eight characteristics of outdoor recreation values that correspond to basic human needs: Wild, Lush, Spacious, the Common, the Pleasure garden, Festive/centre and Culture/History (see Table 1). Thus far, these values have mainly been studied on a local scale, in parks, gardens, and small forests, but also at the neighbourhood and urban fringe level. In the present study, the characteristics are applied at the regional level.

1. Serene	A place of peace, silence and care. Sounds of wind, water, birds and insects. No rubbish, no weeds, no disturbing people.
2. Wild	A place of fascination with wild nature. Plants seem self-sown. Lichen and moss-grown rocks, old paths.
3. Lush	A place rich in species. A room offering a variety of wild species of animals and plants.
4. Spacious	A room offering a restful feeling of "entering another world", a coherent whole, like a beech forest.
5. The Common	A green open place allowing vistas and stays.
6. The Pleasure garden	A place of imagination. An enclosed, safe and secluded place where you can relax and be yourself, let your children play freely.
7. Festive/centre	A meeting place for festivity and pleasure.
8. Culture	The essence of human culture: A historical place offering fascination with the course of time.

Table 1. Eight characteristics that meet recreational needs (from Grahn, Stigsdotter and Berggren-Bärring, 2005).

3.3 Assessing characteristics for well-being on a regional scale using GIS

GIS data from the county administration were used to elaborate characteristics. Corine land cover data play a vital role in this classification, as do other data of relevance, such as topography, environmental protection, noise disturbance, etc. CORINE (Coordination of Information on the Environment) is a programme initiated by the European Commission in 1985 to compile information on the environment with regard to certain topics and to ensure that information is consistent and that data are compatible across member states. One main part of the programme is the Corine land cover project covering 12 countries with a working scale of 1:100,000, and the smallest mapping unit 25 hectares. Sweden, however, has used a more detailed working scale with 5 hectares as the smallest mapping unit (Lantmäteriverket, 2003). Since the inventory of Corine land cover data is a European project, data corresponding to the present data should be available for most European countries.

Other data sources for this project are administrated at the County Administration level and deal with, e.g., natural and cultural protection areas, key biotopes, a pasture land inventory, Nature 2000, a beach zone protection plan and a regional inventory of "silent areas" (a large-scale noise impact calculation). In addition to this topographic evaluation, data from the land survey administration were used.

Of significance in assessing characteristics for recreation on a large regional scale is that only existing data sources can be used for evaluation. In the present study, no resources, neither time nor money, have been available for a detailed systematic process of ground truth validation of different classification methods. However, the studied region is well known to members of the research group, which facilitated preliminary and overall checks of the produced classification maps.

After testing different alternatives for identification of the characteristics, including revisions, we ended up with the classification in Table 2. The final maps are presented at the end of the chapter

Serene	Wild	Lush	Space	Culture
Broad-leaved forest (3.1.1) Mixed forest (3.1.3) Pastures (2.3.1) Marshes, mires (4.1.1, 4.2.1) Water courses, Lakes and ponds (5.1)	Forest (3.1) Thickets (3.2.4.1) Bare rock (3.3.2) Inland marshes, mires (4.1.1) Water courses, Lakes and ponds (5.1) each >15 ha, or if <1 km from city	Mixed forest (3.1.3) Open space with little or no vegetation (3.3) (Beaches, dunes, and sand plains 3.3.1) Bare rock) Wetlands (4)	Forest >25 ha (3.1) Natural grassland (3.2.1) Heath land (3.2.2) Open space with little or no vegetation (3.3) Open wetland (4)	Non-urban parks (1.4.2.5)
	Slopes > 10 °	All registered "key biotopes"	Slopes > 10 °	
		Pasture land of regional interest	Farmland pointed out in a preservation plan	Farmland pointed out in a preservation plan
		Biodiversity areas, Bird biotopes ref. Nature 2000		National interests of cultural preservation Ancient remains
		National park	Coastal zone preservation	Nature reservation areas
Excluded areas				
Noise > 30 dB(A) No artillery range	Noise > 40 dB(A) < 800 m distance to wind power aggregates		Noise > 40 dB(A)	

Table 2. Criteria for assessment of the five characteristics, final version. Figures in brackets refer to the land cover nomenclature (http://www.eea.europa.eu/publications/COR0-landcover/page001.html).

In the present regional study, we have limited our investigation to the five characteristics Serene, Wild, Lush, Spacious and Culture, as the three characteristics the Common, the Pleasure garden and Festive/centre require local data from the municipalities not available at the time of the study.

Also we have excluded the four largest cities in Skåne (Malmö, Lund, Helsingborg and Kristianstad) but kept smaller towns and villages. In cities, large parks and other inner city green areas are a main recreational resource for residents. GIS data to elaborate characteristics of these areas are not available on a regional level. Therefore, the present study is limited to the rural and semi-urban areas of Skåne.

3.4 Statistical analysis

This study included 24,819 respondents located using their residential coordinates. Using GIS, the presence/absence of each of the five characteristics within 300 m from each respondent was defined based on the criteria in Table 2. The working process for the assessment of the characteristics is described in *Results* below. Table 3 shows the percentage of the population living close (300 m or 100 m) to the different characteristics. Spearman´s rank correlation coefficient, appropriate for investigating associations between ordinal scales, was used to test associations statistically between the number of characteristics (0-5) present within 300 m or 100 m of the respondent's residence and ordered answers to the survey questions. P-values below 0.05, and equivalently 95% confidence intervals for odds ratios excluding unity in ordinal regression analyses with adjustments for a broad list of individual determinants of health, were regarded as statistically significant (Björk et al. 2008).

Characteristics	% of population	
within	< 300 m	< 100 m
Serene	6	4
Wild	3	1
Lush	24	7
Spacious	10	5
The common	(not in the study)	
The pleasure garden	(not in the study)	
Festive, centre	(not in the study)	
Culture	24	15

Table 3. Percentage of the population that has the different characteristics within 300 and 100 meters, respectively, distance from their residence (Björk et al. 2008). Note that all individuals that have a certain characteristics within 100 meters also have it within 300 meters distance from their residence.

Each respondent's coordinates are defined as the centre point of the complex in which he/she lives. That centre point can be quite far from the position of the person's home, making the 100 m distance incorrect. 300 meters is a fairly normal distance to walk to get to a nature area or a park.

One outcome of the study is that the objectively GIS-assessed availability of the five characteristics near one's residence (< 300 m) is positively associated with neighbourhood satisfaction (Figure 1), moderate physical activity (Figure 2) and, among tenants, low BMI. Thus, figure 2 suggests that individuals spend more time on average on moderately demanding physical activities the more characteristics they have within 300 m from home. This association remained after adjustment for individual (socioeconomic) factors.

The impact of the number of characteristics on BMI was less clear. After adjustment for individual factors associated with BMI, the beneficial effect of the characteristics was present among tenants but not among house-owners. The proportion of obese (BMI > 30 kg/m2) individuals among tenants was 17% in residences with zero characteristics within 300 metres compared with 13% in residences with at least one characteristics present (Björk et al 2008). No clear association between the number of characteristics and self-rated general health was detected after adjustment for individual factors.

The result for neighbourhood satisfaction among people living in flats is remarkable. If all five characteristics exist within 300 m, 70% of the tenants are satisfied with their neighbourhood, whereas a maximum of 50% are satisfied if only one or no characteristics are present. The corresponding figures for house owners are 83 % and 74 %. Consequently, house owners seem to be rather satisfied with having their own garden, while tenants' well-being is highly dependent on having good natural environments or parks within 300 m from home. This provides important input to the current debate on global warming and densification in urban planning. We are supposed to live more densely in cities to minimize commuting to our workplaces, but on the other hand, if we do not have sufficient nature and park qualities close to our homes in the cities, we will need to commute into the rural landscape for recreation, and that also has an impact on the climate effect.

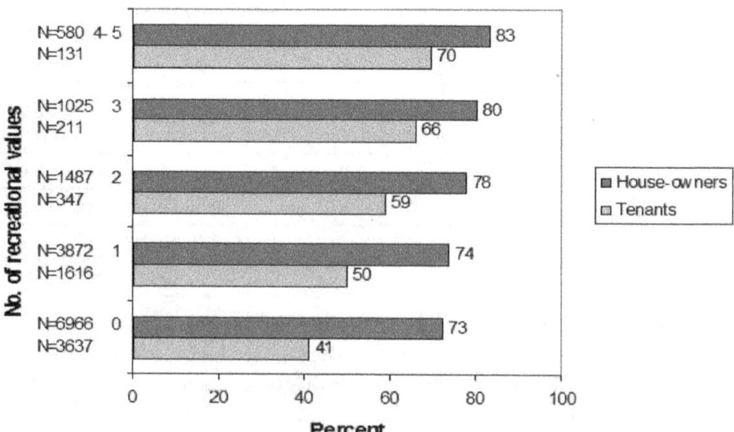

Fig. 1. The relation between the number of recreational values (0–5) of the natural environment within 300 metres distance from the residence and the percentage reporting high neighbourhood satisfaction among house-owners (N = 13,930 answers) and tenants (N = 5,942 answers) (from Björk et al.2008).

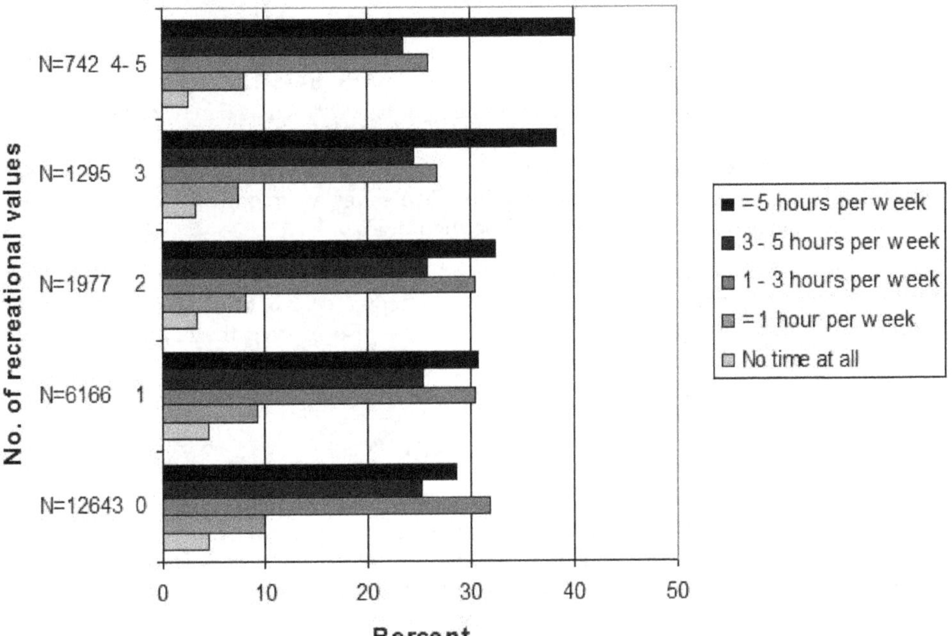

Fig. 2. Time spent on moderate physical activities in relation to the number of recreational values (0–5) of the natural environment within 300 metres distance from the residence (from Björk et al. 2008).

3.5 Results – Working process for the assessment of characteristics

The choice of criteria for elaborating each characteristic using GIS was in some cases a long process in which different combinations of classes of parameters for accessing each characteristic were elaborated successively. These trials generally started with a discussion among the research group members to develop a first set of criteria for identification. The first attempt at classification aimed at achieving an appropriate number of areas for the concerned characteristic, not too many and not too few.

As seen in Table 3, the scope of characteristics close to respondents on average differs substantially. Culture and Lush are most common. Wild and serene are most rare.

The resulting maps were examined by the project staff and compared to their own experiences of certain areas in the studied region. After critical studies of the resulting maps, other classification criteria were tested in a second attempt, and so on in further attempts until the final classification versions were established – those found in Table 2. At the end of the paper, you can see the five maps of the final classification.

3.5.1 Serene

The effort to develop the classification model can be illustrated to some extent by looking at the work with the characteristic Serene. For Serene, we first used a minimum size criterion: Areas should have a minimum size of 20 hectares to be classified as Serene. The reason for this was that an area must be relatively large to be perceived as Serene. From the beginning, we also claimed that the average noise level should not exceed 40 $dB(A)_{24\,h}$, and that a 250 m buffer along roads and dwellings was required. Moreover, we required a buffer of 800 m around large wind power generators and shooting ranges.

Later on, the area size criterion for Serene proved to be useless, as a single large forest often consists of several serene land use areas grouped together, each of which is smaller than 20 hectares. Therefore we rejected that size criterion. We also rejected the minimum distance criterion to roads and wind power generators. The noise criterion, however, was sharpened to 30 $dB(A)_{24h}$. This is perhaps the reason for the relatively rare presence (Table 3) of Serene near the residents.

3.5.2 Wild

Forests, thickets, bare rock, mires and wetlands, lakes and rivers – all larger than 15 ha, if not closer than 1 kilometre from villages and towns. If closer, there is no minimum size for the area, because children seem to sense the Wild characteristic even in very small areas. Also areas steeper than 10° are classified as Wild.

Excluded are areas with a noise level over 40 dB(A), and areas with wind generators within 800 m.

In our first attempts, we set the minimum size of areas to 30 ha and to 5 ha within 1 km from villages and towns. But that reduced the amount of Wild too drastically.

3.5.3 Lush

Mixed forest, marshes and mires, beaches, dunes, sand plains and bear rock. Plus all registered "key biotopes", certain inventories of pasture land, biodiversity areas, bird biotopes of regional interest, also Nature 2000 objects and national parks.

Many of the mixed forests are very small areas, down to 25 x 25 m integrated in the open agricultural landscape, and these are of great importance to biodiversity from an ecological as well as recreational point of view. At our first attempt we included a buffer zone of 100 m. The reason was that fringe zones between different biotopes are very species rich, and we wanted to capture that by including a buffer zone in the biotope area concerned. However, this resulted in areas that were far too large to make the classification useful. One hundred meters is an overestimate of specious rich fringe zones, they are narrower than this.

The best areas cannot be drowned in an overly generous classification, which is a general problem in most landscape analysis. Therefore, this buffer zone was rejected from our classification. Despite this rejection, Lush was classified close to a large percentage of the population.

3.5.4 Space

Beaches, dunes, and sand plains, bare rock, sparsely vegetated areas, burnt areas, natural grassland, moors and heath land, all forests larger than 25 ha. Plus slopes more than 10 degrees (creating viewpoints), farmland pointed out in a preservation plan and coastal zone preservation in a national plan.

Excluded are areas with a noise level over 40 dB(A).

In our first attempt, we used minimum size criteria: Forest 100 ha; natural grassland 20 ha; heath 50 ha. This resulted in very few areas. One observation was that many open space areas are small, but together form large open spaces. Therefore the size criteria for grassland and heath were rejected in our classification. Moreover, forests of different categories are in reality rather small, but together form large forest areas, giving the impression of "entering another world". To address this, we reduced the size criterion for forest to 25 ha.

In our first attempt, we also used noise level criteria for roads 250 m from the residence, but rejected these as being too high in many cases. At first we also used 800 m minimum distance to wind power aggregates as a criterion for obstructing a feeling restfulness, but later rejected this as well.

3.5.5 Culture

Non-urban parks, farmland pointed out in a preservation plan, areas of national interest in cultural preservation, nature reservation areas.

In our first attempt, we also included national interests in recreation. Such areas often overlap national interests in cultural preservation, and are in these cases included anyway. However, they also include large areas for recreation without meeting criteria for the Culture characteristic, but instead for some of the other characteristics, so that dataset was rejected from Culture. Culture was classified as being close to a relatively large percentage of the population (see Table 3).

4. Discussion

When conducting a regional analysis of environmental perception, there are a great many pitfalls to consider. Regional studies require existing datasets mainly produced to show objective data, but perception is a subjective interpretation of the real environment. People´s perceptions also differ from individual to individual due to their previous experiences. This, however, is a general issue independent of the scale of the study. For these kinds of studies, the elaboration of classification criteria always has to be done by representatives of the local community.

The present epidemiological study aimed at finding general associations between well-being and proximity to nature for the population in Skåne. Epidemiological studies provide overall patterns of associations and variations. These kinds of studies do not necessarily give a "correct" picture of individual cases.

One pitfall of using objective data such as forest types, topography, etc., is the issue of interpreting and combining the data so they correspond to common perceptions of nature characteristics. That is what the present paper tries to describe. It is the first study of its kind, and there is a great deal more research to be done to improve the methods. One issue, for example, is that Corine land cover data are derived from an analysis of satellite data, and this is associated with a relatively large percentage of misinterpretation, for some land use types around 30% (Rost & Ahlcrona, 2005).

Because the book focuses on agriculture, we would like to comment additionally on the agricultural landscape. Skåne is in the southwestern half of the region dominated by agriculture. The soil is fertile. Avenues and other landscape elements are impressive. A rather flat zone 30 kilometres wide and following the southern and western coast is the most productive part (the Baltic moraine clay), and is at the same time the most urbanized area, with many roads and railroads creating noise. To some extent this zone is spacious and includes long views, but it is not particularly accessible for walking, and walking promotes a feeling of restfulness. More accessible is the diagonal zone of Skåne from southeast to northwest. This zone has more hills and small scale farmland and is more accessible and silent, offering a restful feeling of "entering another world", a coherent whole. Both zones are rich in impressive visible cultural history, as can be seen in the last map.

There is a need for validation of this kind of analysis. High correlations between the prevalence of GIS-evaluated characteristics and people's responses concerning health and well-being indicate that development of parameters for GIS evaluation is on the right path. Overall, we found a high correlation between people´s perception of a cosy atmosphere and the prevalence of the characteristics. The Skåne study shows that the rural landscape, both arable land and forests, plays an important role in health promotion.

The Swedish National Institute of Public Health now recommends these characteristics as a checklist for green planning (FHI, 2009). The eight characteristics have been implemented in a number of planning projects for housing and infrastructure in Sweden (Skärbäck, 2007).

5. Conclusion

In the study by Björk et al. (2008) green characteristics of the nearby natural environment was shown to be positively associated with neighbourhood satisfaction and physical activity. The association with neighbourhood satisfaction was especially marked among tenants, and a beneficial effect on BMI was also noted in this group. No evident effect of the green characteristics on self-rated health was detectable. The cross-sectional design limits definite conclusions regarding cause-effects. A further limitation is that the study was restricted to rural and semi-urban areas and the generalizability of the results to inner-city areas is therefore uncertain.

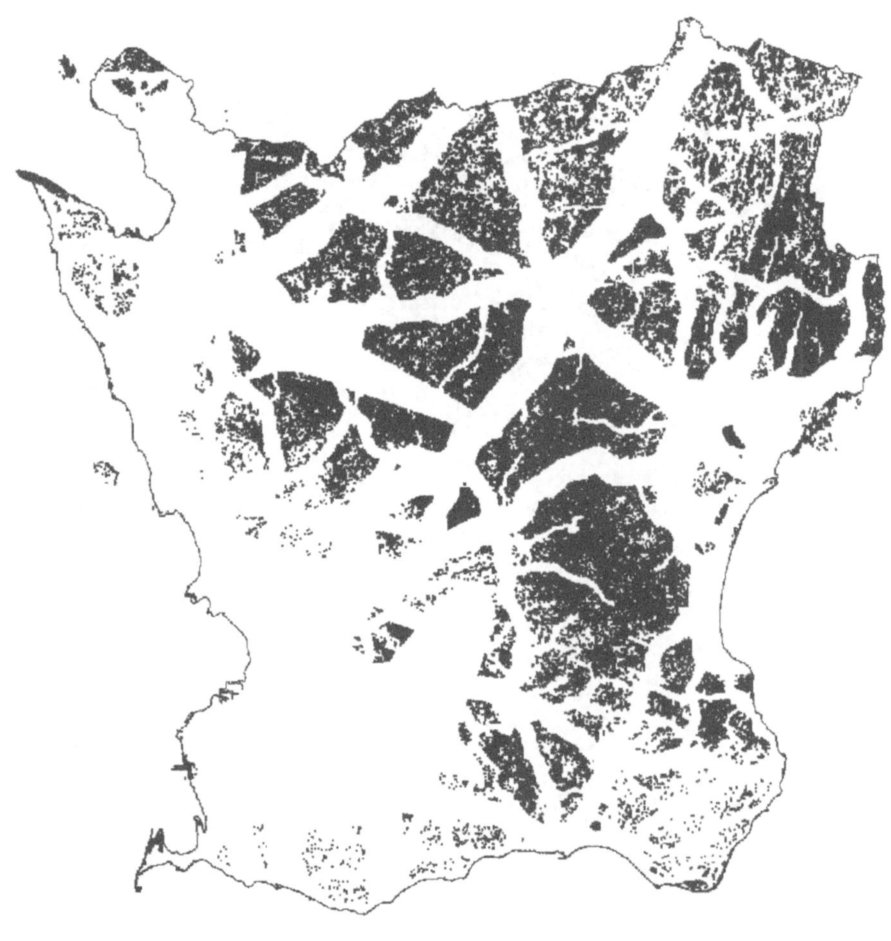

Map 1. Serene. The western part of Skåne Region consists mainly of arable land and cities. Silent (<30 dB(A) nature areas accessible for recreation are rare in that urbanized area. The eastern part has much more assessable nature land, however partly penetrated by road noise, which reduces the serenity.

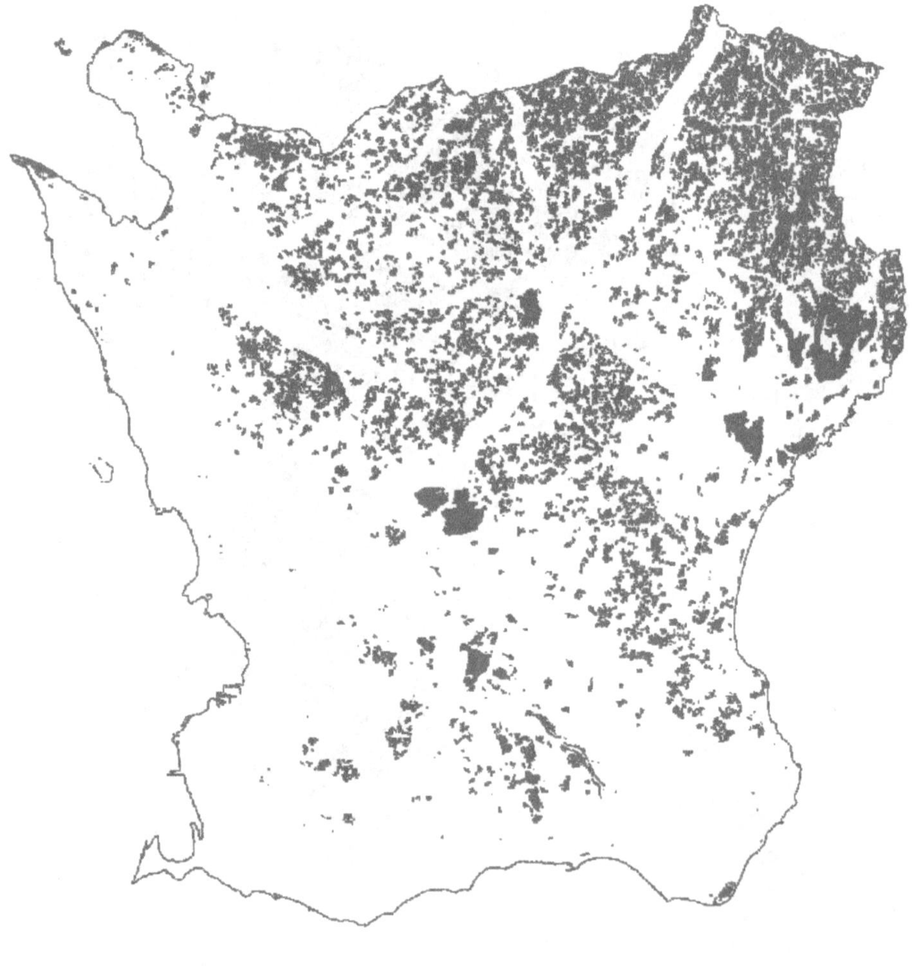

Map 2. Wild. The criteria for silence (<40 dB(A)) is not as heavy as for Serene. Some open land use categories from Serene are left out here.

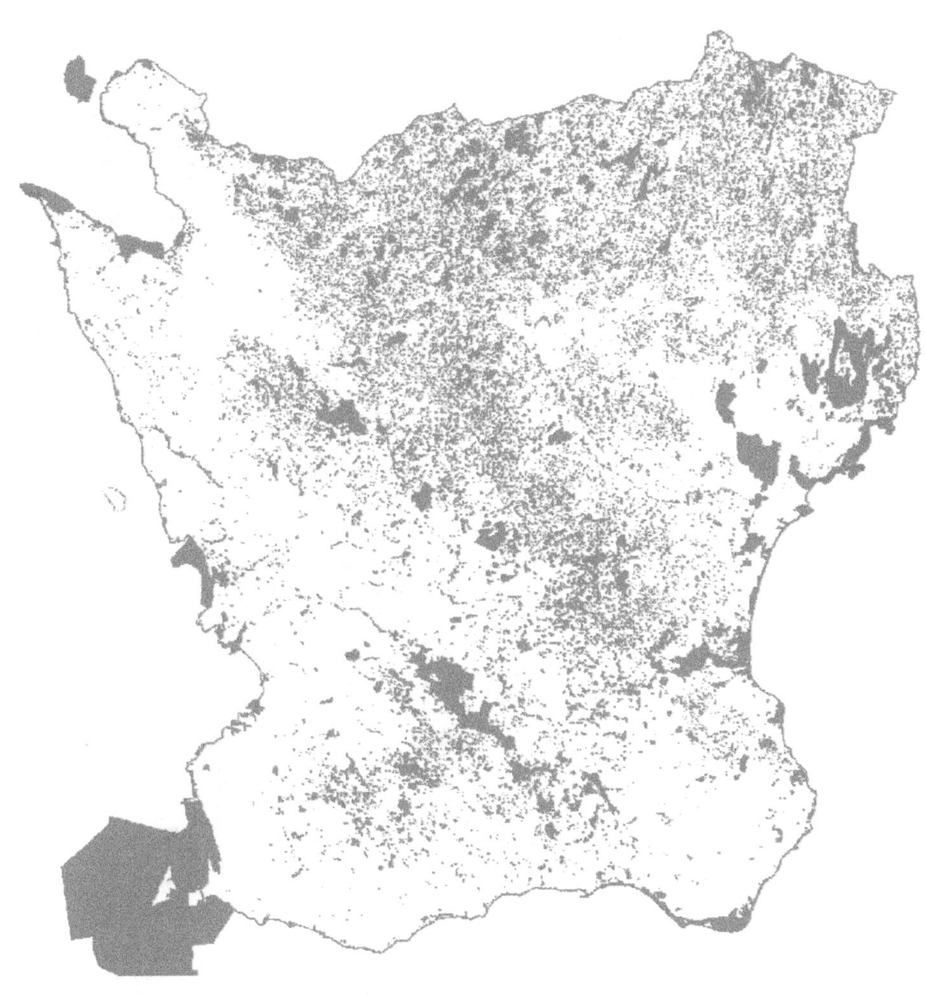

Map 3. Lush, rich in species. Small parts of nature. And some large nature reservation areas. No reduction of noise.

Map 4. Space. Mainly open and half open accessible land, e.g. classified in a national plan for preservation of farmland.

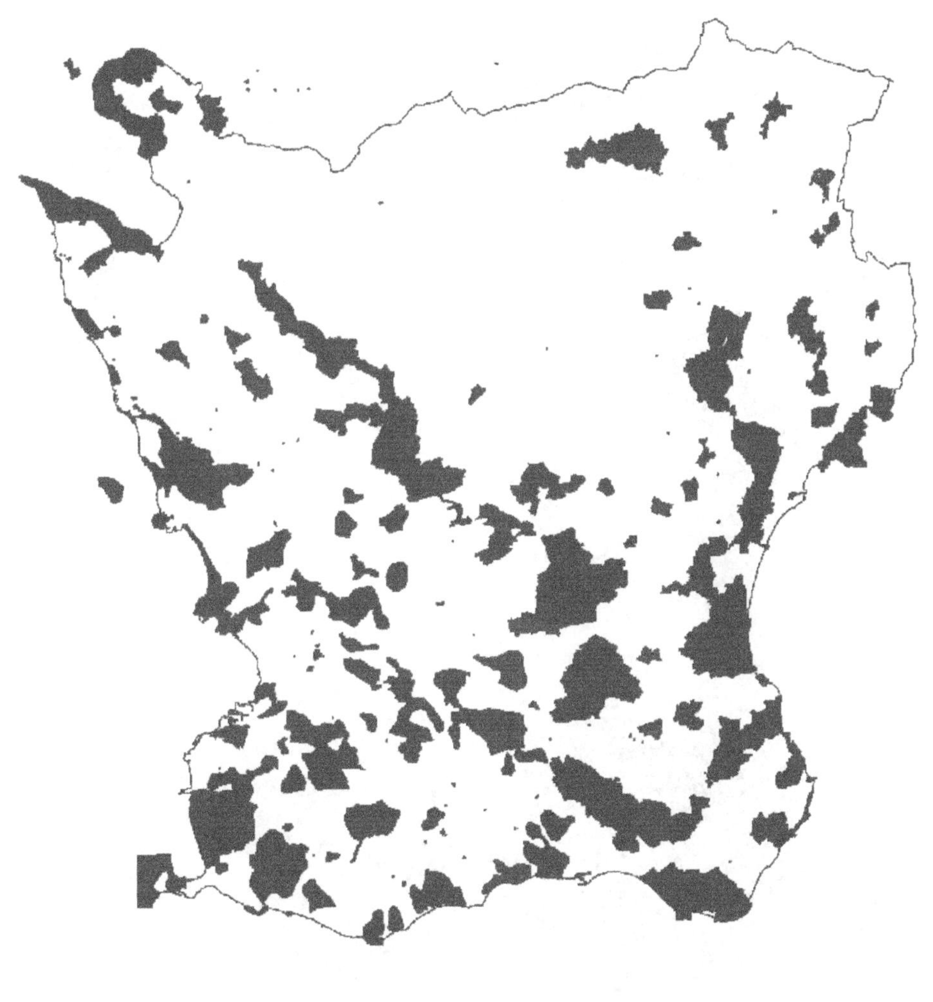

Map 5. Culture. Mainly identified from national and regional plans for preservation.

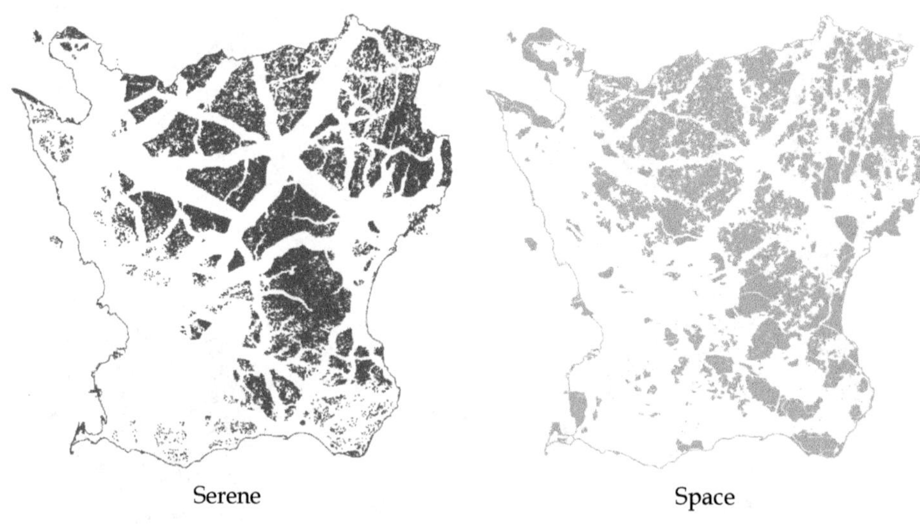

Serene Space

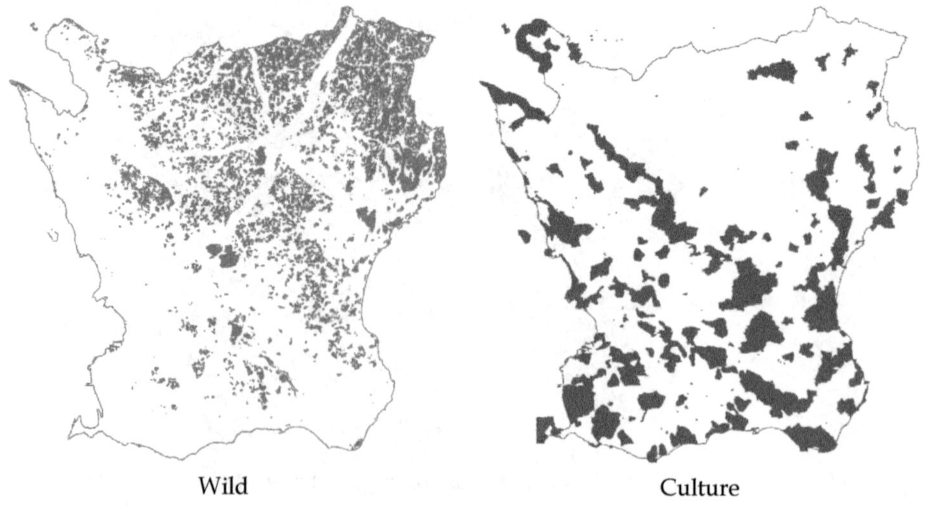

Wild Culture

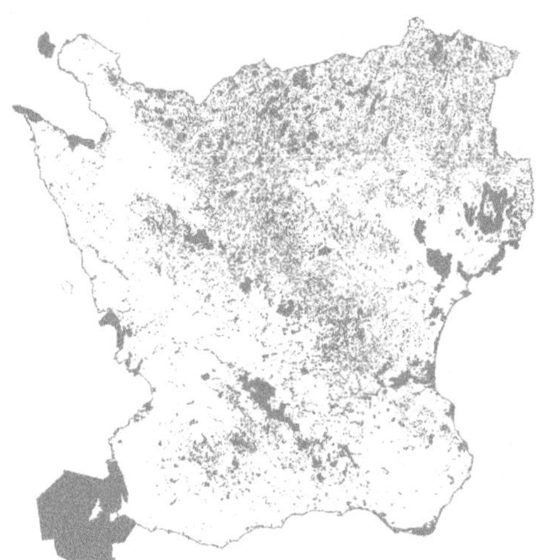

Lush, rich in species

6. References

Björk J, Albin M, Grahn P, Jacobsson H, Ardö J, Wadbro J, Östergren PO, Skärbäck E. april 2008, Recreational values of the natural environment in relation to neighbourhood satisfaction, physical activity, obesity and wellbeing. Journal of epidemiology and community health. 2008;2. (e-publ.) http://luur.lub.lu.se/luur?func=downloadFile&fileOId=1056501

CORINE. 1985. Coordination of Information on the Environment. European Commission. (http://reports.eea.europa.eu/COR0-landcover/en/land_cover.pdf).

Coss, R.G. 1991. Evolutionary Persistence of Memory-Like Processes. *Concepts in Neuroscience.* Vol 2, pp 129-168.

FHI. 2009. Grönområden för fler. Statens folkhälsoinstitut. R 2009:02

Grahn, P & Stigsdotter, U. 2003. Landscape Planning and Stress. Urban Forestry & Urban Greening Vol 2, pp 1-18 (2003). Urban & Fischer Verlag, Jena.

Grahn, P. Stigsdotter, U. & Berggren-Bärring, A-M. 2005. A planning tool for designing sustainable and healthy cities. The importance of experienced characteristics in urban green open spaces for people's health and well-being. In Conference proceedings *"Quality and Significance of Green Urban Areas"*, April 14-15, 2005, Van Hall Larenstein University of Professional Education, Velp, The Netherlands.

Hartig, T. 1993. Testing Restorative Environments Theory. Doctoral Dissertation. University of California, Irvine.

Hartig, T., Böök, A., Garvill, J., Olsson, T. & Gärling, T. 1996. Environmental influences on psychological restoration. Scandinavian Journal of Psychology 37:378-393.

Kaplan, R. & Kaplan, S. 1989. The Experience of Nature. Cambridge.

Lantmäteriverket. 2003. Svenska CORINE, Marktäckedata (SMD).

Ottosson, J. & Grahn P. 2005a. A Comparison of Leisure Time Spent in a Garden with Leisure Time Spent Indoors: On Measures of Restoration in Residents in Geriatric Care. *Landscape Research* vol 30 23 - 55.

Ottosson J. & Grahn P. 2005b. Measures of Restoration in Geriatric Care Residences. *Journal of Housing for the Elderly* vol 19 nr 3/4 229 - 258.

Ottosson, J. & Grahn, P. 1998. *Utemiljöns betydelse för äldre med stort vårdbehov.* Licentiatavhandling för Ottosson, J. Stad & Land nr 155. Alnarp.

Ottosson J., & Grahn, P. 2008. The Role of Natural Settings in Crisis Rehabilitation. How does the level of crisis influence the response to experiences of nature with regard to measures of rehabilitation? *Landscape research vol. 33.* pp. 51-70.

Rost T, och Ahlcrona E., 2005, Tematisk noggrannhet i svenska Marktäckedata. 2005-04-13. Lantmäteriet.

Skärbäck E. 2007. Planning for healthful landscape values. In Mander Ü., Wiggering H., and K. Helming (Eds.) Multifunctional Land Use - Meeting Future Demands for Landscape Goods and Services. Springer-Verlag Berlin-Heidelberg, pp 305-326.

Ulrich, R.S. 1993. Biophilia, Biophobia, and Natural Landscapes. *The Biophilia Hypothesis.* (Kellert, S.R. & Wilson, E.O. eds) pp 73-137.

Section 7

Animal Nutrition

12

Performance and Heat Index of West African Dwarf (WAD) Rams Fed with *Adansonia digitata* Bark (Baobab) as Supplement

Idayat Odunola Agboola
Federal University of Agriculture, Abeokuta, Ogun State, Nigeria

1. Introduction

Shortages of feed during the dry season and sometimes during the wet season put a constraint to livestock production in almost every production system in Nigeria. Where feeds are found in abundance, they may be low in nutritive value which may manifest themselves in form of nutritional deficiencies when fed to animals. Increasing livestock production depends to a large extent on the availability of suitable feed resources. Forages in various conservation methods play a significant role in the nutrition of ruminant animals in general.

In large measure, the current under production of animal protein in the developing world is caused by lack of forages. Trees and shrubs play a dual role in the forage supply serving both as shade for grasses and as forage themselves (Nas, 1979). In dry savannas in particular, shrubs and trees are very precious, without them, stock raising would probably be impossible for pasture grasses die when upper soil layers lose their moisture, but the tree roots exploit deep underground moisture and they continue to flourish.

During the dry season, trees and shrubs provide green fodder leaves, flowers, fruits often rich in protein, vitamins and minerals. In the absence of forages trees and shrubs, animal have only straw from nature grasses.

The trees and shrubs can be interplanted with grasses thus increasing the carrying capacity of pastoral areas and often supplying the grazing during droughts or periods of year when other food is normally scarce (Nas, 1979). Trees products have many and often competing uses. Foliage and young leaves is useful supplier of plants nutrients (Kang, et.al 1999). Trees such as *Gliricidia sepium* provide useful forage in the form of leaves and bark and is commonly used to supplement poor quantity and low protein roughage.

During the dry season, it may become a major source of feed for goats, sheep and cattle in the sub-humid zone (Kang, et.al 1999). The importance of browse plants as source of protein and energy to ruminants particularly during the dry season of the year has been extensively reviewed. (Wilson,1969). However, browse plant cannot constitute a complete feed when fed alone. They should be given adequate attention in the feeding management of sheep (Carew, 1983).

Small ruminant production is an important segment of agricultural sector that forms a significant component of most farming systems in the country whether pastoral or agricultural.

Nigeria has a population of 56.599 million small ruminants of which goats and sheep accounted for 34.4-85 million and 22.104 million respectively (FNPCPS,1980) of an estimated Africa small ruminant population of 349.4 million.

Information on the utilization of *Adansonia digitata* bark as feed for ruminants in general and WAD rams in particular has not been documented.

2. Literature review

2.1 *Adansonia digitata*

Some trees are valued as excellent sources of forage feed for ruminant animals while they are known as noxious weeds in some areas. A tree, such as Leucaena leucocephala has been shown to have great potential as a source of high quality feed for ruminants and also capable of improving intake of poor quality rouphage and live weight gain in large and small ruminants.

Adansonia digitata linn (English: baobab; Yoruba: Ose; Hausa: Kuka; Nupe: Machi; Kanuri: kuka; Bini: Usi.) is a tropical tree specie popularly called the Baobab tree. It has distinctive large flower and fruits hanging from long stalks. It is wide spread in the drier regions of Africa. It is much more widespread in the savanna of Nigeria where it is usually planted or presented. The tree is about 25m high with a very stout bole reaching 12m in girth. The bark is grey or purplish and thick. The bark produces a strong fibre resulting in its being stripped off. Hence, the trunk is often much more deformed. (Keay, et.al. 1965).

The pulp of baobab tree was found to be acidic and rich in ascorbic acid, iron, calcium and pectin. The pectin was mainly water soluble and had a low degree of esterification and a low intrinsic viscosity. (Nour, et.al. 1980)

According to purseglove (1968) the fruit pulp which contains tartaric acid is made into a drink and is also used as a fruit seasonal. In the Sudan, the pulp is commonly chewed, sucked or made into a drink. The kernels are edible and the seed contain 19% oil.

2.2 Importance and characteristics of sheep

Sheep together with goats, *Ilamas* and *alpacas* are small ruminants because they are ruminants; they eat low quality food, particularly fibrous vegetation which cannot be eaten by humans and non-ruminant animals such as pigs and poultry. People keep sheep because they produce meat, milk, wool or hair, skins and manure. Sheep are the only species of animal that produce wool, although goats, rabbits and alpacas sometimes produce similar high quality fibre. Therefore sheep are a way of converting poor quality food into desirable products. Breeds which have to survive along dry season often have a fat tail or rump which is a store of energy equivalent to the hump of camels or cattle (Gatenby, R.M, 1991).

2.3 Feeding

A variety of feeds are used throughout the tropics and sub-tropics. Sheep are known as herbivores; feeding readily on a wide range of plant except poisonous ones. Under certain condition they feed on every parts of the plants within their reach such as leaves, stem, flowers, seeds, barks and fruits. Some of the pasture that are common in Nigeria include *Cynodon emiensis, Andropogon gayanus, Panicum maximum.* Some other crop residues and agro-industrial by-products are also used in feeding sheep e.g. cassava peels, yam peels, maize offals, wheat offal, PKC, BDG, Bone meal and Cereal straws. There is no sufficient information on the nutrient requirement of livestock. (Akinyosinu, 1985).

3. Effect of heat stress on animal productivity

High ambient temperatures depress body activities which viewed homeostatically are biological mechanism for preventing overheating. The climatic condition also affect the amount of food and water intake, the availability of the potential energy in the ingested forage, the animals heat production system, the net energy available for productivity and the body composition of growing animal (Hafez, 1968).

Pulse rate which is expressed in beats per minutes is like the respiratory rate inversely proportional to the weight (W) of the animal, it can also alter rapidly due to external factor such as temperature or intense activity by the animal itself. Bianca and Findlay (1962) showed that exposure of sheep to severe heat for a short period increased pulse rate but exposure to relatively long period decreased pulse rate. Pulse rate is usually higher in small animals than in large animals due to the relatively high metabolic rate of small species (Bianca, 1968).

Respiratory rate can change rapidly and at the extreme, in a matter of minutes. It is indirectly influenced by the animal's activity and by environmental being inversely proportional to the volume of the animal.

Rectal temperature is taken to be equivalent to the body temperature. Body temperature is the best indicator of the good health of the animal and its variation above and below normal is a measure of the animal's aptitude to resist hardship factors of the environment.

4. Objectives of the study

The broad objective is to determine the performance of WAD Ram fed *Adansonia digitata* bark based concentrate supplement.

The specific objectives are

1. To determine the composition of experimental diets
2. To determine the proximate analysis of Adansonia digitata bark and Wheat Offal
3. To determine the performance of West African Dwarf (WAD) rams fed Baobab bark-based concentrate diet.
4. To estimate the effect on pulse rate, respiratory rate and rectal temperature.

4.1 Materials and methods

The study was carried out in the teaching and Research Farm, University of agriculture, Abeokuta, Nigeria. (Latitude 7° 5.5′N - 7° 8′N, Longitude 3° 11.2′ - 3°3.5′E and Latitude 76.

5. Animals and their management

Twenty healthy growing WAD rams of about 8-10 months of age, with average body weight of 10.2kg were used for the experiment. The WAD rams were separated from the rest of the flock in the small ruminant unit of the teaching and research farm for three weeks into previously disinfected, well ventilated and illuminated pens with wood shavings as litter materials. *Panicum maximum* was supplied in liberal amount. Five litres of water per day were supplied in each pen containing 5 rams. The animals were dewormed with Banminth II wormer (12.5g/kg body weight) and bathed with asuntol powder solution R (3g/litre of water) to eliminate possible ectoparasites. Clout R was applied at 4 weeks intervals along the spine of the animals to check against possible mange infection. At the end of 3 weeks pre-experimental management period, the animals were grouped into 4 with five animals per treatment, balanced for body weight.

6. Dietary treatment

The basal diet for the experiment was guinea grass (Panicum maximum). Four concentrate diets with different levels of *adansonia digitata* (0, 5, 10, and 15%) were prepared. (Table 1). The four groups of WAD rams were randomly assigned to the treatments using completely randomized designed (CRD) with the treatment as the only source of variability apart from the experimental error. The animals were supplied concentrate twice daily between the hours of 8 and 9a.m and 3 and 4p.m at the rate of 0.4kg/animal/day out which 0.2kg was supplied in the morning and 0.2 in the afternoon so as to control feed wastages. Each group was also supplied 5 litres of water daily. The dietary treatment lasted for 10 weeks excluding one week dietary adjustment period. The one week dietary adjustment period was to flush out the residues of the previous feed from the gut of the rams thereby eliminating carry-over effects of previous feed.

7. Method of data collection

After an adjustment period of one week, daily data collection followed. The animals were fasted by withdrawing feed only for 14- 16 hours and allowing them access to water. The body weight were taken after the fasting prior to the commencement of the experiment and taken again every fortnight. Feed residues and left-over water were recorded every morning before fresh feeds and water were supplied.

Pulse rate was taken for each animal by placing the finger on the femoral arteries on the medial aspect of the hind limb for one minute using a stop-watch, respiratory rate was taken for each animal by counting the number of flank movements per minute using stop-watch while the rectal temperature was taken using a clinical thermometer which was allowed to stay in the rectum of each animal for one minute before the reading was taken. (Fasoro, 1999). The physiological parameters (pulse rate, respiratory rate and rectal temperature) were taken for three consecutive days (Friday, Saturday and Sunday) before the commencement of the experiment and every fortnight on each animal between 7 and 9a.m on each day. The average air temperature and relative humidity recorded during this study were 35^{0}C and 54.5% respectively using wet and dry thermometer.

8. Results

Table 1 above showed the composition of experimental concentrate diet to be used for the experiment. Wheat offal was supplied at different percentages in the diets starting from 50, 45, 40 and 35 contained in the treatments 1, 2, 3 and 4 respectively, this was supplied in different levels because wheat offal had high percentage of dry matter content of 89% while adansonia digitata bark was low in dry matter content (Table 2). PKC, Bone meal, Salt and BDG was supplied in the diet with the same percentage of 20, 1, 1 and 28 in the treatments 1, 2, 3 and 4 respectively. Baobab bark was supplied in the diet with different levels 0, 5, 10 and 15 in the treatments 1, 2, 3 and 4 respectively.

Ingredients (%)	Treatments			
	1	2	3	4
Wheat offal	50	45	40	35
PKC	20	20	20	20
Bone meal	1	1	1	1
Salt	1	1	1	1
BDG	28	28	28	28
Baobarb Bark	0	5	10	15
Total	100	100	100	100

Note: PKC- Palm kernel cake, BDG- Brewers dried grains

Table 1. Experimental Diet Composition (%)

Table 2 showed that Adansonia digitata bark had very high moisture content (89.3%) while it is low in dry matter content (10.7%). Also it contained 10.7% of crude protein and is high in crude fibre (32.16%) and Ash content (7.02%) while wheat offal had high dry matter content and low moisture content.

	Adansonia digitata	Wheat offal
Moisture content	89.3	11.0
Dry matter	10.7	89.0
CP (Crude Protein)	10.7	15.0
CF (Crude Fibre)	32.16	23.1
EE (Ether Extract)	2.52	6.0
NFE (Nitrogen Free Extract)	47.6	20.7
ASH	7.02	8.2

Table 2. Proximate composition of adansonia digitata bark and Wheat Offal

Table 3 contains the results of the performance of West African Dwarf rams fed Baobab-based concentrate supplement. The grass intake was slightly higher than concentrate supplement intake in all the treatment groups including the control. (0% Baobab inclusion

Parameters	Diets			
	D1	D2	D3	D4
Final Av. Body wt (kg)	12.38±0.29	12.47±0.23	12.30+0.23	13.14±0.23
Av. Growth rate (g/day)	62.50±9.3^{b}	65.48±7.6^{b}	69.52±7.6^{b}	89.3±7.6^{a}
Average DM intake				
i.Concentrate diet(%BW)	1.40±0.02^{b}	2.0±0.02^{ab}	1.7±0.02^{b}	2.5±0.02^{a}
ii.panicum maximum(%BW)	1.7±0.004^{c}	2.3±0.004^{b}	2.0±0.004^{c}	2.7±0.004^{a}
Total DM intake (%BW)	3.1±0.002^{b}	4.3±0.024^{b}	3.7±0.024^{b}	5.2±0.024^{a}
Feed Efficiency (kg)	15.7±1.56	14.7±1.27	14.9±1.27	17.6±1.27

abc means in the same row with different superscripts are significantly different (P<0.05)

Table 3. Performance of West African Dwarf Rams on adansonia digitata based concentrate supplement

Table 4 showed that the differences among treatments for pulse rate, respiratory rate and rectal temperature at the end of the experiment were not significantly different from initial conditions. This shows uniformity in the environmental parameters, physiological state of the animals and lack of effect of dietary treatments on the animals.

Parameters	Diets			
	Initial/final 1	Initial/final 2	Initial/final 3	Initial/final 4
Average pulse rate (beats/min)	78/78	76/76	78/78	78/78
Average respiratory rate (beats/min)	34/36	36/36	34/34	36/36
Average rectal temperature (°C)	38.5/39.7	38.7/38.7	39/39	39.1/39.2

Table 4. Performance of the animals to Heat index

9. Discussion

The high level of moisture content (89.3%) in the baobab bark could be an advantage as sole feed for ruminants in the period of scarcity of water. The consumption of high water content forages reduces water intake by ruminant, but may however make for difficulties in obtaining a sufficiently high DMI (payne, 1990). The 10.70% crude protein in the Baobab bark appears to be adequate in the compounded ration of ruminant. A diet of 10% crude protein has been reported adequate in meeting the maintenance requirements of sheep and goats (NRC, 1980). Adansonia digitata bark would appear from the proximate component, to be adequate as sole feed supplement for sheep.

It appears that the grass was more preferred than the concentrate. The preference of the experimental rams for grass over concentrate supplement could be attributed to the more succulent nature of the grass than the dry and coarse concentrate supplement. The higher intake of grass in all the diets is an index of the better acceptability of the grass forage to the rams than concentrate diet containing baobab bark (Aina, 1998). However, the animals responded better in terms of ADG and DMI as the inclusion level increased in the concentrate diet than the control. The results also suggested clearly that the rams treated with D4 showed the highest total DMI, FCR and ADG compared with other treatment groups (Table 3). The increasing DMI of concentrate diet with increasing level of baobab bark inclusion up to the maximum (D4) is an indication that higher levels may still be accommodated by the animals. The increasing growth rate with increasing baobab inclusion in the diet suggested beneficial effects of the bark and an encouragement for better performance in the WAD rams.

The results of the environmental and physiological parameters (Table 4) show the uniformity in the environmental parameters and physiological status of the animals as well as lack of effect of dietary treatments on those physiological parameters of the animals. The pulse rate (beats/min) range of 76-78 agree with the records of Olusanya and Heath (1988) who stated that the heart rate of sheep falls into the range 60-120 beats/min. It can thus be inferred that the baobab bark inclusion in the diet of sheep up to 15% of the compounded diet is safe for consumption.

10. Conclusion

From this present study, it can be concluded that to get a better performance than control, the concentrate supplement must contain about 15% inclusion of Adansonia digitata in the diets of rams. However, higher level of inclusion of baobab bark beyond 15% in the concentrate supplement for sheep is recommended since the highest level in the trial (15%) induced the highest performance.

11. References

Nas, A.E (1979): Tropical legumes. Resources for the future. Washington- National Academy of science. Pg 383.

Kang, B.T., Attah-Krah, A.N and Reynolds, L. (1999). Alley farming. The Tropical Agriculturalist. CTA-Macmillan, IITA. Pg 29

Wilson, A.D (1969). A review of browse in the nutrition of grazing animals. Journal of Range management.

Carew, B.A.R. (1983). The potential of browse plants in the nutrition of small ruminants in the humid forest and derived savannah Zones in Nigeria. In: Browse in Africa. The current state of knowledge. Edited by H.N. Le Hoverou. Pg 307-311.

FNPCPS, (1980). National livestock survey, Federal department of livestock and pest control services, Abuja.

Keay, R.W.J., Onachie, C.F.A and Stanfield, D.P (1965). The Nigerian Trees. Federal Department of forest research, Ibadan, Nig Vol II. Pg 23 and 232.

Nour, A.A magboul, B.I and Khuri, N.H (1980). Chemical composition of baobab fruit- Trop. Sci. pg 383

Gatenby, R.M (1991). Sheep: The Tropical Agriculturalist. CTA-Macmillan, 1st edition

Akinyosinu, A.O. (1985). Nutrient requirements for sheep and goats in Nigeria. In: Proceedings of national Conference on small ruminant production. Zaria, Nigeria. October 6-10th, 1985 pg. 141-148.

Hafez, E.S.E. (1968). Environmental effect of animal productivity. Adaptation of Domestic Animals. Pg 74-93.

Bianca, W and Findlay, J.D. (1962). The effect of Thermally induced hypernoea on the acid-base status of the blood of calves. Res Vet. Sci 3: 38-49.

Fasoro, B.F, (1999). Determination of heat stress index in three breeds of goat. B.Agric project of the University of Agriculture, Abeokuta.

Payne, W.J.A. (1990). An introduction to animal husbandry in the tropics. Longman scientific and technical, pg 13.

NRC, (1980). Nutrient requirements of Dairy cattle (5th revised edition). National Academy of science, Washington D.C.

Aina, A.B.J (1998). Preliminary study on the use of Margaritaria discoidea leaf in the diet of West African Dwarf Goats. Nig. Journal of Animal Production 25(2): 169-172.

Olusanya and Heath, E (1998). Anatomy and physiology of Tropical livestock. ELBS/Longman, pg. 34.

Permissions

The contributors of this book come from diverse backgrounds, making this book a truly international effort. This book will bring forth new frontiers with its revolutionizing research information and detailed analysis of the nascent developments around the world.

We would like to thank Dr. Godwin Aflakpui, for lending his expertise to make the book truly unique. He has played a crucial role in the development of this book. Without his invaluable contribution this book wouldn't have been possible. He has made vital efforts to compile up to date information on the varied aspects of this subject to make this book a valuable addition to the collection of many professionals and students.

This book was conceptualized with the vision of imparting up-to-date information and advanced data in this field. To ensure the same, a matchless editorial board was set up. Every individual on the board went through rigorous rounds of assessment to prove their worth. After which they invested a large part of their time researching and compiling the most relevant data for our readers. Conferences and sessions were held from time to time between the editorial board and the contributing authors to present the data in the most comprehensible form. The editorial team has worked tirelessly to provide valuable and valid information to help people across the globe.

Every chapter published in this book has been scrutinized by our experts. Their significance has been extensively debated. The topics covered herein carry significant findings which will fuel the growth of the discipline. They may even be implemented as practical applications or may be referred to as a beginning point for another development.

The editorial board has been involved in producing this book since its inception. They have spent rigorous hours researching and exploring the diverse topics which have resulted in the successful publishing of this book. They have passed on their knowledge of decades through this book. To expedite this challenging task, the publisher supported the team at every step. A small team of assistant editors was also appointed to further simplify the editing procedure and attain best results for the readers.

Our editorial team has been hand-picked from every corner of the world. Their multi-ethnicity adds dynamic inputs to the discussions which result in innovative outcomes. These outcomes are then further discussed with the researchers and contributors who give their valuable feedback and opinion regarding the same. The feedback is then collaborated with the researches and they are edited in a comprehensive manner to aid the understanding of the subject.

Apart from the editorial board, the designing team has also invested a significant amount of their time in understanding the subject and creating the most relevant covers. They scrutinized every image to scout for the most suitable representation of the subject and create an appropriate cover for the book.

The publishing team has been involved in this book since its early stages. They were actively engaged in every process, be it collecting the data, connecting with the contributors or procuring relevant information. The team has been an ardent support to the editorial, designing and production team. Their endless efforts to recruit the best for this project, has resulted in the accomplishment of this book. They are a veteran in the field of academics and their pool of knowledge is as vast as their experience in printing. Their expertise and guidance has proved useful at every step. Their uncompromising quality standards have made this book an exceptional effort. Their encouragement from time to time has been an inspiration for everyone.

The publisher and the editorial board hope that this book will prove to be a valuable piece of knowledge for researchers, students, practitioners and scholars across the globe.

List of Contributors

Bnejdi Fethi and El Gazzeh Mohamed
Laboratoire de Génétique et Biométrie Faculté des Sciences de Tunis, Université Tunis, El Manar, Tunisia

Walma Nogueira Ramos Guimarães, Gabriela de Morais Guerra Ferraz and Vivian Loges
Department of Agronomy, Federal Rural University of Pernambuco, Recife, Pernambuco, Brazil

Luiza Suely Semen Martins
Department of Biology, Biochemical Genetics Laboratory/Genome, Federal Rural University of Pernambuco, Brazil

Luciane Vilela Resende
Department of Agronomy, Federal University of Lavras, Minas Gerais, Brazil

Helio Almeida Burity
Agronomic Research Institute of Pernambuco, Recife, Pernambuco, Brazil

Cointault Frédéric
AgroSup Dijon, France

H. Arnold Bruns
USDA-Agricultural Research Service, Crop Production Systems Research Unit, Stoneville, MS, USA

Margarita Nankova
Dobrudzha Agricultural Institute - General Toshevo, Bulgaria

Jerry L. Hatfield
National Laboratory for Agriculture and the Environment, USA

Xing-Quan Liu
School of Agriculture and Food Science, Zhejiang Agriculture and Forestry University, Hangzhou, P.R. China

Kyu-Seung Lee
Department of Bio-Environmental Chemistry, College of Agriculture and Life Sciences, Chungnam National Univeristy, Taejon, Korea

Chuan-Gen Lű
Jiangsu Academy of Agricultural Sciences, Nanjing, P.R. China

R. Swaminathan and Kan Singh
Department of Entomology, Rajasthan College of Agriculture, Maharana Pratap University of Agriculture and Technology, Udaipur, Rajasthan, India

V. Nepalia
Department of Agronomy, Rajasthan College of Agriculture Maharana Pratap University of Agriculture and Technology, Udaipur, Rajasthan, India

Cécile Levasseur-Garcia
Université de Toulouse, Institut National Polytechnique de Toulouse, Ecole d'Ingénieurs, de Purpan, Département Sciences Agronomiques et Agroalimentaires, Toulouse, France

Erik Skärbäck, John Wadbro and Patrik Grahn
SLU Alnarp, Sweden

Jonas Björk, Kim de Jong, Maria Albin and Jonas Ardö
LU Lund, Sweden

Idayat Odunola Agboola
Federal University of Agriculture, Abeokuta, Ogun State, Nigeria

Printed in the USA
CPSIA information can be obtained
at www.ICGtesting.com
JSHW011440221024
72173JS00004B/879

9 781632 390585